D1220802

The Immune System

1 · Past and Future

Festschrift in Honor of Niels Kaj Jerne
on the Occasion of his 70th Birthday

The Immune System

1 · Past and Future

Edited by Charles M. Steinberg and Ivan Lefkovits
with the assistance of Catherine di Lorenzo
Basel Institute for Immunology, Basel, Switzerland

41 figures and 17 tables, 1981

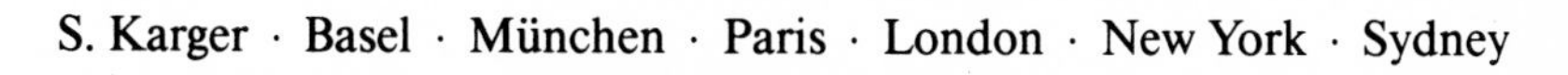

S. Karger · Basel · München · Paris · London · New York · Sydney

The cover of the dust jacket shows *Jean Tinguely's* 'birthday card' for Niels Kaj Jerne. The dust jacket was designed by *Hanspeter Stahlberger* and *Miroslav Pazdera* of the Basel Institute for Immunology. The latter has also designed the portrait of Niels Kaj Jerne.

National Library of Medicine, Cataloging in Publication
The Immune system
Festschrift in honor of Niels Kaj Jerne on the occasion of his 70th birthday
Editors, Charles M. Steinberg and Ivan Lefkovits. – Basel; New York: Karger, 1981
2 v.
Contents: v. 1. Past and future. – v. 2. The present
QW 504 I3205
ISBN 3–8055–3407–8

Printed in Switzerland by Thür AG Offsetdruck, Pratteln
ISBN 3–8055–3407–8

Contents

Vol. 1 · Past and Future

Prologue

Networks and Complex Systems

Evolution and Ontogeny

Antibodies and B Cell Function

Antibody Specificity and Diversity

T Cell Function

Major Histocompatibility Complex

Cell Membrane

Tolerance and Suppression

Allotypes and Idiotypes

Factors, Hormones, Interferon

Lymphoid Tissues

Growth of Cells in Culture

Cancer

Parasitology and Infectious Disease

Vol. 2 · The Present

Networks and Complex Systems

Evolution and Ontogeny

Antibodies and B Cell Function

Antibody Specificity and Diversity

T Cell Function

Major Histocompatibility Complex

Cell Membrane

Tolerance and Suppression

Allotypes and Idiotypes

Factors, Hormones, Interferon

Lymphoid Tissues

Growth of Cells in Culture

Cancer

Parasitology and Infectious Disease

Epilogue

This Festschrift was made possible by
the financial support of
F. Hoffmann-La Roche and Co.,
the founder and sponsor of
the Basel Institute for Immunology.

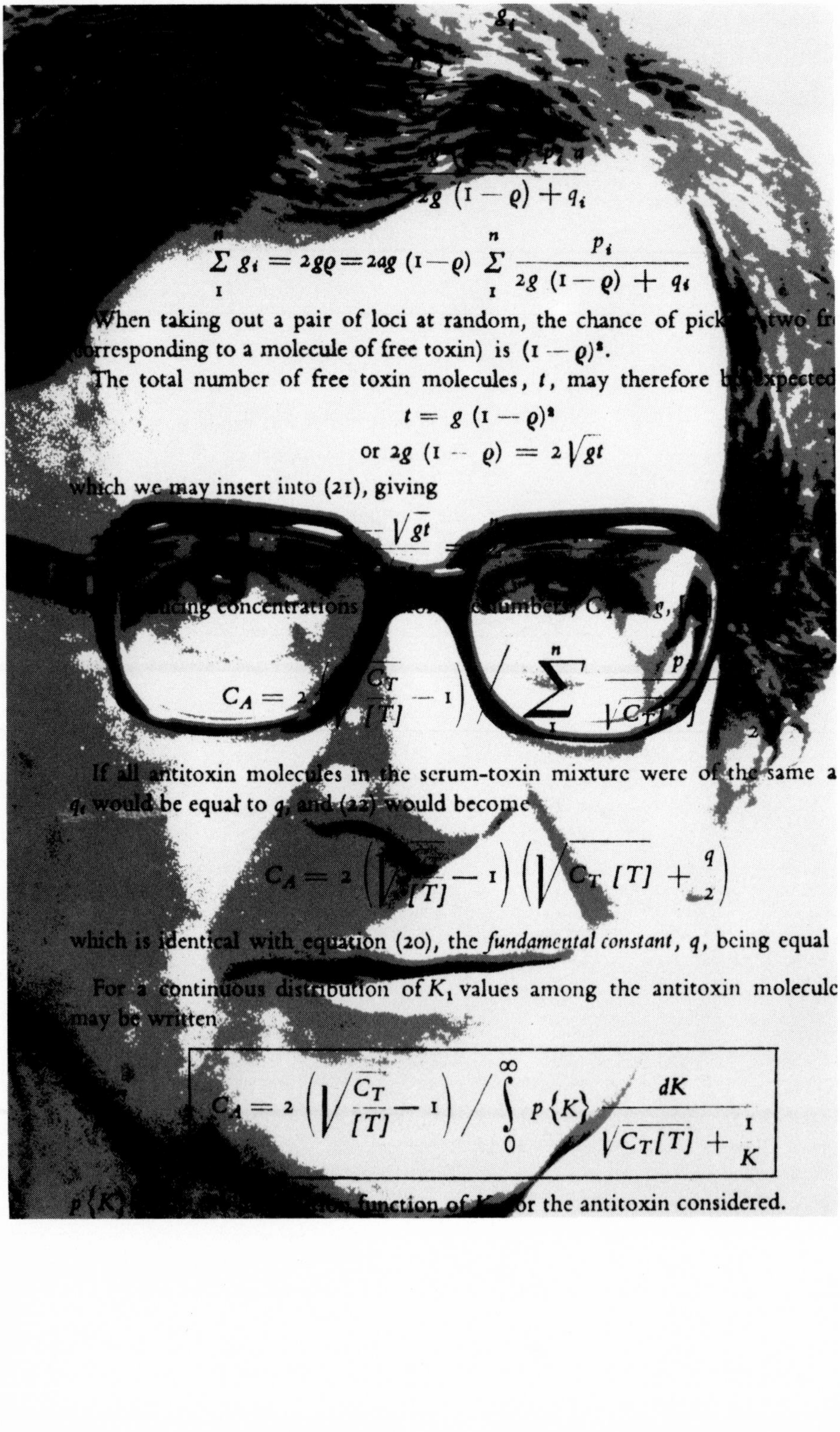

g_i

$$2g(1-\varrho)+q_i$$

$$\sum_{1}^{n} g_i = 2g\varrho = 2ag(1-\varrho)\sum_{1}^{n}\frac{p_i}{2g(1-\varrho)+q_i}$$

When taking out a pair of loci at random, the chance of pick... two fr... corresponding to a molecule of free toxin) is $(1-\varrho)^2$.

The total number of free toxin molecules, t, may therefore b... expected

$$t = g(1-\varrho)^2$$

$$\text{or } 2g(1-\varrho) = 2\sqrt{gt}$$

which we may insert into (21), giving

$$\sqrt{gt}$$

...ucing concentrations ... numbers, C_T ... g,

$$C_A = 2\left(\sqrt{\frac{C_T}{[T]}}-1\right)\Bigg/\sum_{1}^{n}\frac{p_i}{\sqrt{C_T[T]} \ldots}$$

If all antitoxin molecules in the serum-toxin mixture were of the same a... q_i would be equal to q, and (22) would become

$$C_A = 2\left(\sqrt{\frac{C_T}{[T]}}-1\right)\left(\sqrt{C_T[T]}+\frac{q}{2}\right)$$

which is identical with equation (20), the *fundamental constant*, q, being equal

For a continuous distribution of K_1 values among the antitoxin molecule... may be written

$$C_A = 2\left(\sqrt{\frac{C_T}{[T]}}-1\right)\Bigg/\int_0^\infty p\{K\}\frac{dK}{\sqrt{C_T[T]}+\frac{1}{K}}$$

$p\{K\}$ ... function of K for the antitoxin considered.

A Letter to Niels Kaj Jerne

Dear Niels,

These volumes are a birthday present to you from all of us:
- from all of your friends and colleagues who contributed articles
- from *Catherine di Lorenzo,* our editorial assistant, and many others at the Institute who helped us turn the raw material into the makings of a book
- from *Rolf Steinebrunner* and all of his helpers at Karger, who made the books
- from *Jean Tinguely,* whose 'birthday card' is on the dust jacket
- from *Miroslav Pazdera* and *Hanspeter Stahlberger,* who helped to design the rest of the dust jacket
- from Roche, whose generous subsidy made it all possible
- and, of course, from the undersigned.

We know that you will be interested in how these volumes came to be the way they are. The first idea for a Festschrift for your seventieth birthday was quite different in concept. The model would have been the Festschrift for *Max Delbrück* ('Phage and the Origins of Molecular Biology', eds. J. Cairns, G.S. Stent, and J.D. Watson, Cold Spring Harbor Laboratory, 1966) that you admired so much and to which you yourself contributed such a fine article. But somehow what was 'right' for Max, just didn't seem 'right' for you – for the founding Director of the Basel Institute for Immunology with its famous horizontal structure, for a man who not only studied the Immune System but who patterned his life after it, honoring Diversity above all else. No, a volume with articles by thirty carefully selected, distinguished scientists would not do. We would ask *everybody* to contribute. Then we remembered that you used to say that there were 10^4 immunologists in the world today. If any reasonable fraction (say one out of five) wanted to submit a manuscript ...

So to keep the dimensions of the Festschrift within some imaginable, if not reasonable, bounds we decided only to invite people who had actually worked in the same institute or department with you. We put together a list of almost three hundred names, *Sharon Buser* got out her machine that automatically types personal letters, and we sent them all an invitation to write something (i) that was 'in some way connected with "The Immune System"'; (ii) that they thought 'Niels would enjoy reading', and (iii) that was 'short'. For many, the first condition was too much; how many people have you deflated with 'what does that have to do with immunology?' The second condition was apparently the toughest; several would-be contributors are still trying to write something that you would really like to read. The third condition caused us, the editors, the most grief, and one contributor backed out because he was unable to write something short; conciseness is not the immunologist's vice.

And so we are left with 125 articles. We suspect that some of them don't have much to do with The Immune System, we are sure that there must be a few that you won't enjoy reading, and not all of them are short because their authors elected to pay page charges. But Diversity we wanted and Diversity we got! Within these two volumes are the past, the present, and the future of immunology, broad generalizations and tedious detail, rabbits and sheep, experiments and speculations, chickens and frogs, antibodies and complement, high philosophy and low comedy, lymphocytes and macrophages, mice and men, poetry and prose, women and work, genotypes and phenotypes, humor and the sort of grimness of which only dedicated scientists are capable.

How should we organize this plethora of Diversity? First we grouped the articles into sections according to the method, pioneered by *Ben Pernis,* that we use for the Annual Report of the Institute. Then we tried to distribute these sections between two volumes, but no matter how we did it, each volume seemed to have no coherence on its own. Next we tried a 'vertical structure'. That is, for each section, the articles were split into two groups. Into the first volume went articles of more general interest; into the second volume went articles for the cognoscenti. Articles in volume 2 were characterized by extensive bibliographies and lots of tables, while volume 1 articles tended to be free of these encumbrances. Volume 1 articles were often concerned with history and speculation, so we called it 'Past and Future'.

Volume 2 articles were often concerned with current research and the state of the immunological art, so we called it 'The Present'. Thus, we have two volumes, each with essentially the same subject headings, but with a different slant – two Immune Systems, distinguishable only idiotypically.

Comments from the contributors ranged from 'I congratulate you ... on the idea ... and feel honored to participate in this event' to 'I feel a bit sorry for Niels to have 125 articles to read for his birthday and if nothing else turns him off about immunology ... this certainly will!' (You would guess it, Niels, but the other readers need to be told – both quotations are from the same person.) But the most apt comment, a prescription for the use of this Festschrift, really, was made by *Jean Tinguely* and appears on the dust jacket. *Ingrid Schnurr* sent him the physical dimensions of the books, but somewhere along the way these got lost. So he wrote above his birthday card these words: 'Also wenn's zu gross oder zu klein ist, dann vergrössert oder verkleinert – mit der Schere – Viele Grüsse: Jean Tinguely.' (If it is too big or too small, then enlarge it or make it smaller – with your scissors – many greetings: Jean Tinguely.) If these volumes are too small for you, come out of your castle and add another chapter to the Immune System. If they are too big, don't panic, just get out your scissors.

With best regards and hopes of many more birthdays to come, we remain,

Your friends,
Charles Steinberg, Ivan Lefkovits

Prologue

The Immune System, vol. 1, pp. 1–5 (Karger, Basel 1981)

How It All Began

Ole Maaløe[1]

University Institute of Microbiology, Copenhagen, Denmark

The History of Science is a respectable subject of great antiquity. The stories of scientists and of how they themselves believe it all happened is a modern derivative from which a *Dick Cavitt* might some day produce a series of TV interviews. In recent years molecular biology has had its fair share of publicity, and the same may happen to immunology in the near future. If this comes true, *Niels Jerne* is an obvious candidate (or victim, if you like), and I want to encourage the writing of a fuller account by offering my own version of how he came to be the kind of an immunologist we all know.

This takes me back to the State Serum Institute in Copenhagen in the late 1940s, when *Niels* and I began working together there. The institute was already a big and multi-faceted organism and, like the Pasteur Institutes on which it was modeled, it produced sera and vaccines and performed all sorts of diagnostic tests. This involved a considerable amount of applied research and antibodies of course played a central role everywhere. Nevertheless, there was little activity of a kind that today we would call immunology, and even less genetics or biochemistry. From our present viewpoint this does not sound too interesting but there were definite redeeming features: the supplies of animals and of growth media were nearly inexhaustible and, even more important, we were expected to spend a good deal of time on research.

[1] *Ole Maaløe* was *Niels Jerne's* 'professor' in Copenhagen in 1949–1951, and they have remained friends and colleagues through the years. *Ole Maaløe* was a member of the Board of Advisors of the Basel Institute for Immunology from 1970 to 1974.

In Europe, it was in places like this that some of the groundwork was done from which molecular biology, including immunology, developed. It is also characteristic of the early period that basic research in microbiology and immunology was done largely by scientists with a medical degree; outstanding examples are *Burnet, Luria* and *Jacob.*

It was fate in the shape of *Jeppe Ørskov,* the benevolent, but absolute ruler of the Serum Institute, who brought *Niels* and me together. The director of the international laboratory for biological standardization had just left the institute, and I was put in his place, chiefly, I think, because *Ørskov* wanted to offer me plenty of opportunity for doing research; in his opinion biological standardization was a routine that did not require too much time or energy. My predecessor, with whom *Niels* had already been associated for some time, held a very different view and the two of them had worked intensively on the problems involved in assaying and in evaluating the potency of sera and vaccines.

The standardization department consisted of one medium-sized and three small attic rooms, and when *Niels* and I started working together there I was conscious of him being a couple of years my senior, and he was probably equally conscious of my seniority as an MD (for reasons I need not go into *Niels* began his medical studies rather late). Moreover, since *Niels* and our mutual friend *Jens Ole Rostock,* who for years had been responsible for much of the routine work, had to teach their new director his duties, we might have had a difficult time. In fact we did not, and as I see it in retrospect the reason was simple: *Niels* and I quickly discovered that we enjoyed talking to each other on many subjects but particularly about our work. We would spend hours in front of a small wooden blackboard and the discussion would drift from standardization proper to the possible reasons some standard assays gave ambiguous results. This was the beginning of *Niels'* preoccupation with 'avidity', the subject of his thesis. This fine piece of work clearly shows that an antiserum, far from being a well-defined reagent for potency determinations, represents a family of antibodies related through their specificity. Paradoxically, it was the routine of classical standardization that made this study possible: thus, innumerable rabbit-skin titrations of diphtheria antisera were recorded in the protocols of the standardization department, and each of them included dilutions of a standard toxin as a control of the sensitivity of the rabbit. Together this large number of measurements of the diameters of skin reactions produced a marvellously well-defined dose-response curve against which the effective concentration of free toxin in a toxin-antitoxin mixture could be read. I believe that the

Fig. 1. The Biological Standards research group at the State Serum Institute in Copenhagen in 1950. From left to right, *G. Stent, L. Larsen, O. Maaløe, J.O. Rostock, N.K. Jerne,* and *J.D. Watson.*

most useful result of our many and long discussions was the decision to apply this unique tool (the dose-response curve) to an analysis of the binding properties of different diphtheria antibodies. As I mentioned, this analysis became *Niels'* thesis [1], and already at that time he realized that he had identified one dimension of that much larger concept – antibody diversity – which was to occupy his thoughts from then on. *Niels'* first major theoretical contribution to immunology, his selection theory, probably grew out of his work on avidity. Essentially, the theory argues that an organism is capable of producing a large but finite number of different antibodies, and the drift *Niels* observed from low to high avidity during a course of immunization could readily be seen as a gradual selection of the 'best fits' from among a larger number of possibilities. As *Burnet* himself has acknowledged, the addition of the word 'clonal' to *Niels'* selection theory, although important, is in a sense secondary.

I already mentioned the continuous dialogue with *Niels* during the first years we worked together but, of course, we did not work in total isolation.

Two outside influences were important. First there were our frequent contacts with *Georg Rasch* and his younger colleague *Michael Weis Bentzon,* both mathematicians and experts on biological statistics. *Rasch* was a fine mathematician with the delicate touch of a calligrapher at the blackboard and with a roaring laughter. We enjoyed all three qualities through long hours of instruction during which we learned much about how to approach biological problems by making models and testing them quantitatively. *Rasch* died recently at 79.

The second important input came from our guests. After a visit to *Delbrück's* laboratory in 1949 I had begun doing phage work and for a year, beginning in late 1950, *Jim Watson* and *Günther Stent* came to work with us in the Serum Institute (fig. 1). It was an exciting and productive year during which *Niels* finished writing up his thesis, and at the end of which the group exploded: *Jim* went to Cambridge, *Günther* to Paris and I left for Pasadena. *Niels* stayed behind keeping watch over the standards.

I now return to the History of Science. In modern biology (and elsewhere) it is striking that a number of major achievements can be referred to by naming a *pair* of scientists; say, *Watson and Crick* or *Jacob and Monod.* This, I think, suggests that the intellectual interaction between poeple who communicate particularly well contributes very significantly to the growth of science. The DNA structure and the operon model speak for themselves, but there probably are many examples of interaction which has proved fruitful to two scientists, although no joint achievement resulted. I believe *Niels* and I provide such an example. In view of the fact that we interacted so well from the beginning it should not be a surprise that our research in later years developed along parallel lines. Thus, we have both built models purporting to show how the many parts of a complex biological system can be thought to function cooperatively. What this amounts to is to reconstitute a whole, such as the immune system, from its elements, and the tricky part of this game of course is that new elements keep coming up as the result of continued research in cell biology, molecular biology and genetics. An impressive model of the kind I'm thinking about is *Niels'* network theory.

The great sophistication of modern, experimental biology is something we accept with various degrees of enthusiasm. The philosophy behind model-building is less well established, and it may not be generally accepted. As I see it, the process of integration involved here still contains a strong element of intuition about the nature of biological systems. I don't know how far back this intuition can be traced, but a fine expression of what I have in mind can be found in *Kant's Kritik der Urteilskraft* [2]. In paragraph 66 he talks about the purposeful design of organisms, and he writes[2]: 'Ein organisiertes Produkt der Natur ist das, in welchem alles Zweck und wechselseitig auch Mittel ist.'

References

1 Jerne, N.K.: Acta path. microbiol. scand., suppl. 86 (1951).
2 Kant, I.: Kritik der Urteilskraft (1789).
3 Webster's Second New International Dictionary (Merriam, Springfield 1953).

[2] The 1953 edition of Webster [3] offers this translation: 'An organism is a being in which every part is at once a means and an end to every other.'

Prof. Ole Maaløe, University Institute of Microbiology, Øster Farimagsgade 2A, DK-1353 Copenhagen K (Denmark)

Networks and Complex Systems

The Immune System, vol. 1, pp. 6–13 (Karger, Basel 1981)

Programmatic Phenomena, Hermeneutics and Complex Networks

Gunther S. Stent[1]

Department of Molecular Biology, University of California, Berkeley, Calif., USA

> Wär nicht das Auge sonnenhaft,
> die Sonne könnt es nie erblicken,
> läg nicht in uns des Gottes eigene Kraft,
> wie könnt uns Göttliches entzücken?
> *Goethe*

Practicing scientists are wont to regard the philosophy of science as an activity of has-beens or parasitic ne'er-do-wells, a subject suitable at most for discussion over brandy after dinner. That view is not wholly unjustified since many lines of scientific work can be successfully pursued without any clear understanding of their epistemological basis. Moreover, in the two centuries since *Kant* during which philosophy has existed as a distinct professional calling, there have been very few known instances in which professional philosophers of science have actually made a productive contribution to scientific progress at the cutting edge of research. Nonetheless, there *have* been cases in which explicit philosophical considerations did play a crucial role in a fundamental scientific advance, such as in the development of relativity theory and of quantum mechanics. But here the philosophical ground work was done, not by professional philosophers, but by *Einstein* and by *Bohr* and *Heisenberg.*

Another such case is *Niels Jerne's* development of the natural selection theory of antibody formation. *Jerne* opened his 10-year retrospective essay

[1] *Gunther Stent* is a long-time friend and associate of *Niels Jerne* and is at present a member of the Advisory Board of the Basel Institute for Immunology.

on the theory in the *Delbrück* 60th Birthday Festschrift [5] as follows: 'Can the truth *(the capability to synthesize an antibody)* be learned? If so, it must be assumed not to pre-exist; to be learned, it must be acquired. We are thus confronted with the difficulty to which Socrates calls attention in *Meno,* namely that it makes as little sense to search for what one does not know as to search for what one knows; what one knows one cannot search for, since one knows it already, and what one does not know one cannot search for, since one does not even know what to search for. Socrates resolves this difficulty by postulating that learning is nothing but recollection. The truth *(the capability to synthesize an antibody)* cannot be brought in, but was already inherent.' By means of this quotation from *Kierkegaard's Philosophical Bits,* in which the word 'truth' had been replaced by the italicized words, *Jerne* presented the philosophical basis of the selective theory. That is to say, when translated into the parlance of Molecular Biology, the Socratic insight that truth inheres in every human being becomes the proposition that the potential to synthesize a particular antibody cannot be imposed upon nucleic acid, but must pre-exist. It was this a priori logical necessity of the selective theory which immediately convinced me of its truth when, as *Jerne* tells later in that same essay, he explained it first to me in 1954, while we waited for ^{32}P atoms to decay in T4 phage DNA in the experiments we were then doing together in Berkeley.

It is my belief that contemporary neurobiology happens to present yet another of those rare cases where some philosophical attention (or 'intellectual hygiene', as my teacher *André Lwoff* called it) would be of some benefit for further scientific progress. Two aspects of neurobiology in particular seem in want of conceptual clarification, namely the development of the nervous system and the functional analysis of its networks.

What Is a Programmatic Phenomenon?

In the mid-1960s, at the time of the vindication of *Jerne's* natural selection theory and the triumphant culmination of molecular biological research in the cracking of the genetic code, research projects began to be formulated that attempted to wed the disciplines of genetics and developmental neurobiology. Thus, the idea gained currency that the deep biological problem posed by the metazoan nervous system, namely how its cellular components and their precise interconnections arise during ontogeny, could, and even should, be approached by focusing on genes. In particular, the notion arose that the structure and function of the nervous system, and hence the behavior of an animal is 'specified' by its genes. Admittedly, it cannot be the case that the genes really embody enough information to permit explicit specification of a neuron-by-neuron circuit diagram of the nervous system [3]. But, all the same, the circuit might somehow be *implicit* in the genome, since purely quantitative information-theoretical arguments, such as those advanced by *Horidge* [4], against genetic specification of the nervous system are clearly invalid [8]. No, the problem with the notion of genetic specification is other than information-theoretical, for even if the genes did embody a neuron-by-neuron circuit diagram, the existence of an agency that reads the diagram in carrying out the assembly of the components of a neuronal Heathkit would still transcend our comprehension. So a seemingly more reasonable view of the nature of the genetic specification of the nervous system would be that the genes embody, not a circuit diagram, but a *program* for the development of the nervous system [1, 2]. But this view is rooted in a semantic confusion about the concept of 'program'. Once that confusion is cleared up, it becomes evident that development of the nervous system is unlikely to be a programmatic phenomenon.

Development belongs to that large class of regular phenomena which share the property that a particular set of antecedents generally leads, via a more or less invariant sequence of intermediate steps, to a particular set of consequences. However, of the large class of regular phenomena, programmatic phenomena form only a small subset, almost all the members of which are associated with human activity. For membership of a phenomenon in the subset of programmatic phenomena it is a necessary condition that, in addition to the phenomenon itself, there exist a second thing, the 'program', whose structure is isomorphic with, i.e. can be brought into one-to-one correspondence with, the phenomenon. For instance, the on-stage events associated with a performance of 'Hamlet', a regular phenomenon, are programmatic since there exists *Shakespeare's* text with which the actions of the performers are isomorphic. But the no less regular off-stage events, such as the actions of house staff and audience, are mainly non-programmatic, since their regularity is merely the automatic consequence of the contextual situation of the performance. Or, the operation of a digital computer has programmatic aspects, insofar as there exists a 'program', or set of instructions separate from the 'hardware', whose structure is isomorphic with the sequence of operations performed by the machine. It should be noted, however, that in the example of computer programs the demand for isomorphism has to allow for the possibility that the structure of the program is actually more elaborate than that of the phenomenon. Here the program often calls for one of two or more alternative operations at various stages of the process, depending on the result of earlier computations (and hence on the initial state). In such cases the phenomenon is evidently isomorphic with only part of the program.

One of the very few regular phenomena independent of human activity that can be said to have a programmatic component is the formation of proteins. Here the assembly of amino acids into a polypeptide chain of a particular primary structure is programmatic because there exists a stretch of DNA polynucleotide chain – the gene – whose nucleotide base sequence is isomorphic with the sequence of events that unfolds at the ribosomal assembly site. However, the subsequent folding of the completed polypeptide chain into its specific tertiary structure lacks programmatic character, since the three-dimensional conformation of the molecular is the automatic consequence of its contextual situation and has no isomorphic correspondent in the DNA. This example of the formation of proteins can serve also to clarify the distinction made earlier between the embodiment by the genes of a neuron-by-neuron circuit diagram of the nervous system on the one hand and a program for its development on the other. Evidently in the case of proteins, the genes do not embody an explicit atom-by-atom specification of spatial coordinates of the tertiary structure of proteins but merely a program for assembly of their primary structure from ready-made amino acid building blocks.

When we extend these considerations to the regular phenomenon of development we see that its programmatic aspect is confined mainly to the assembly of polypeptide chains (and of various species of RNA). But as for the overall phenomenon, it is most unlikely – and no credible hypothesis has as yet been advanced how this *could* be the case – that the sequence of its events is isomorphic with the structure of any second thing, especially not with the structure of the genome. The fact that mutation of a gene leads to an altered neurologic phenotype shows that genes are part of the causal antecedents of the adult organism, but does not in any way indicate that the mutant gene is part of a program for development of the nervous system.

The general notion of genetic specification of the nervous system is defective not only at the conceptual level but also represents a misinterpretation of the knowledge already available from developmental studies, including those that have resorted to the genetic approach. As *Székely* [9] has pointed out, we know enough about its mode of establishment already to make it most unlikely that neuronal circuitry is in fact pre-specified;

rather, all indications point to stochastic processes as underlying the apparent regularity of neural development. That is to say, development of the nervous system, from fertilized egg to mature brain, is not a programmatic but a historical phenomenon under which one thing simply leads to another. To illustrate the difference between programmatic specification and stochastic history as alternative accounts of regular phenomena, we may consider the immune response, or the establishment of ecological communities upon colonization of islands [7], or the growth of secondary forests [10]. All three of these examples are regular phenomena. In the case of the immune response the introduction of an antigen into the circulation of an animal regularly elicits production of antibody molecules having a high specific affinity for that antigen. As the articles in this volume make abundantly clear, in the immune response the presence of the antigen initiates a progressive and predictable change in the genetic character of the animal's lymphocyte population. And in the cases of island colonization and forest growth a more or less predictable ecological structure arises via a stereotyped pattern of intermediate steps, in which the relative abundances of various types of flora and fauna follow a well-defined sequence. But the regularity of these phenomena is obviously not the consequence of an immunological or of an ecological program encoded in the genome of the immunized animal or of the colonizing taxa. Rather in all three examples regularity arises via a historical cascade of complex stochastic interactions between various cell types or between various biota (in which genes play an important role, of course) and the world as it is.

Hermeneutics

To fathom the complex interactions that produce historical phenomena it is necessary to understand the context in which they are embedded. And upon recognizing the importance of contextual relations for the problem of development, we move into a domain of phenomenological analysis to which conventional scientific methodology is no longer fully applicable. That domain bears a strong epistemological affinity to the scholarly activity called 'hermeneutics'. This designation was originally given by theologians to the theory of interpretation of sacred texts, especially of the Bible. The name is derived from that of Hermes, the divine messenger. In his capacity as an information channel linking gods and men, Hermes must 'interpret', or make explicit in terms that ordinary mortals can understand, the implicit meaning that is hidden in the gods' messages. In recent years, scholars have applied the term hermeneutics also to the interpretation of secular texts, since there may be implicit meanings hidden even in the literary creations of ordinary men that need to be made explicit to their fellow mortals. But hidden meanings pose a procedural difficulty for textual interpretation, because one must understand the context in which implicit meaning is embedded before one can uncover hidden meanings in any of its parts. Here we face a logical dilemma, a vicious hermeneutic circle. On the one hand, the words and sentences of which a text is composed have no meaning until one knows the meaning of the text as a whole. On the other hand, one can come to know the meaning of the whole text only through understanding its parts. To break this vicious circle – which comes first, the chicken or the egg? – hermeneutics invokes the doctrine of *pre-understanding.* As set forth by *Rudolf Bultmann,* hermeneutic pre-understanding, or *Vorverständnis,* represents his life of experience and insights that the subject must bring to the task of interpreting a particular text.

It is apparent, of course, that *Bultmann's* recognition of the need for pre-understanding in textual interpretation is, like *Jerne's* recognition of the need for pre-existence of the genetic potential to synthesize a particular antibody, merely another special case of the Socratic lesson that truth must inhere in every human being, as set forth in *Jerne's* quotation from *Kierkegaard's Philosophical Bits.* In fact, it is the claim that prior knowledge is necessary for meaningful interaction with the world that distinguishes most clearly the philosophy of rationalism from its rival, empiricism, whose point of departure is the belief that the truth *can* be learned. Thus, as seen from the perspective of the history of ideas, *Jerne's* selective theory of antibody formation is part of the tradition of rationalism, as opposed to the earlier *Haurowitz-Pauling* inductive theory that belongs to the empiricist tradition.

In assessing the epistemological status of hermeneutic studies we may ask to what extent the concept of objective validity is applicable to their results. An objectively valid interpretation would presumably be one that has made explicit the 'true' meaning hidden in the text, i.e. the meaning intended by the author. But here we encounter the difficulty that the interpreter's pre-understanding is necessarily based on his own subjective historical, social and personal background. Hence, agreement regarding the validity of an interpretation could be reached only among persons who happen to bring the same pre-understanding to the text. Thus, because of the necessarily subjective nature of pre-understanding, there cannot be such a thing as an objectively valid interpretation. It is this evident unattainability of universal and eternal truth in interpretation that makes hermeneutics different from science, for which the belief in the attainability of objectively valid explanations of the world is metaphysical bedrock.

To what extent is this belief actually justified? According to such writers as *Kuhn, Lakatos* and *Feyerabend,* it is not really justified because resort is made also to subjective notions equivalent to hermeneutic pre-understanding in the search for scientific explanations of phenomena. Thus, in judging the objective validity of these explanations we must try to assess the degree to which preunderstanding enters into their development. Such an assessment can help us to understand why the belief in the attainability of objectively valid explanations does seem to be more appropriate in the 'hard' sciences than in the 'soft' human sciences, such as economics, sociology and psychology.

One of the main reasons for this epistemological difference between the 'hard' and 'soft' sciences is that the phenomena of which the 'soft' sciences seek to give an account are much more complex than those addressed by the 'hard' sciences. And the more complex the ensemble of events that the scientist isolates conceptually for his attention, the more hermeneutic pre-understanding must he bring to the phenomenon before he can break it down into meaningful atomic components that are to be governed by the causal connections of his eventual explanations. Accordingly, the less likely is it that his explanations will have the aura of objective truth. By way of comparing a pair of extreme examples – one very hard, the other very soft – we may consider mechanics and psychoanalysis. There is an aura of objective truth about the laws of classical mechanics because the phenomena which mechanics consider significant, such as steel balls rolling down inclines, are of low complexity. Because of that low complexity it is possible to adduce critical observations or experiments about rolling steel balls. By contrast, there is no comparable aura of truth about the propositions of analytical psychology, because the phenomena of the human psyche which it attends are very complex. Here there are no critical observations or experiments because the failure of any prediction based on psychoanalytic theory can almost always be explained away retrodictively, by considering additional factors or by modifying slightly one's pre-understanding of the phenomenon. Hence in psychoanalysis a counterfactual prediction rarely qualifies as negative evidence against the theory that generated it.

Neurobiology as well as immunology cover a broad range of this hardness-softness scale. At the hard end, neurobiology is represented by cellular electrophysiology and immunology by protein and nucleic acid chemistry, whose phenomena, although more complex than those associated with rolling steel balls, can still be accounted for in terms of explanations that are susceptible to seemingly objective proof. But at the soft end, both neurobiology and immunology are represented by the study of the function of large and complicated cellular networks. As has long been realized of course, the output of neural networks comprises phenomena whose complexity approaches that of the human psyche – in fact, that *include* the human psyche. But of more recent date is the realization [6] that also the immune response depends on a network of cellular interactions of enormous complexity. Hence at their soft end neurobiology and immunology take on some of the characteristics of hermeneutics: the student of a complex cellular network must bring considerable pre-understanding to the system as a whole before attempting to interpret the function of any of its parts. Accordingly, the explanations that are advanced about complex neural or immunological systems may remain beyond the reach of objective validation. Here, to paraphrase my employer, Governor *Edmund G. Brown, Jr.*, we may have to be satisfied with less.

References

1 Brenner, S.: Br. med. Bull. *29:* 269–271 (1973).
2 Brenner, S.: Genetics *77:* 71–94 (1974).
3 Brindley, G.S.: Proc. R. Soc. Lond. B *174:* 173–191 (1969).
4 Horidge, G.A.: Interneurons (Freeman, San Francisco 1968).
5 Jerne, N.K.: in Cairns, Stent, Watson, Phage and the origins of molecular biology, pp. 301–312 (Cold Spring Harbor Laboratory, New York 1966).
6 Jerne, N.K.: Harvey Lectures, ser. 70, pp. 93–110 (Academic Press, New York 1976).
7 Simberloff, D.S.: A. Rev. Ecology Systematics *5:* 161–182 (1974).
8 Stent, G.S.: Paradoxes of progress, pp. 169–189 (Freeman, San Francisco 1978).
9 Székely, G.: Trends Neurosci. *October:* 245–248 (1979).
10 Whittaker, R.H.: Communities and ecosystems (Macmillan, New York 1970).

Dr. Gunther S. Stent, Department of Molecular Biology, University of California, Berkeley, Wendell M. Stanley Hall, Berkeley, CA 94720 (USA)

The Immune System, vol. 1, pp. 14–20 (Karger, Basel 1981)

The Origin of Immunological Networks

H.M. Etlinger [1]

Basel Institute for Immunology, Basel, Switzerland

Universal Recognition

The bottommost layers of the immunological tel reveal a linkage of its origins to disease prevention. For example, the incipient awareness of the importance of antibody was provided by the demonstrations of *von Behring* and *Kitasato* that serum derived from immunized animals had the capacity to afford protection against an otherwise lethal challenge of bacterial toxin. It was in the same decade of the 1890s that *Bordet* and *Tchistovitch* exposed a principle construct of the immune system with the observation that the inducer of antibody production was not limited to toxic materials. Starting with the report by *Obermayer* and *Pick* around the turn of the 19th century that antibody could be produced which reacted with haptenic groups of artificially prepared protein conjugates and continuing on through to the present where it appears that antibodies can be elicited whose specificity is limited only by the imagination of the immunologist, the formal extension is made that the recognitive powers of the antibody-producing system are limitless.

The discernment which antibody displays in the recognition of small bits or antigenic determinants [epitopes; 11] of an antigen – i. e. the specificity of antibody – was emphasized by *Landsteiner* in a series of studies

[1] *Howard Etlinger* has been a member of the Basel Institute for Immunology since 1980.

which commenced in 1917. Around the middle of the 20th century, the structural basis for such antibody specificity was localized to special areas in the amino terminal end of the antibody molecule, the antibody-combining site [paratope; 11]. The paratypic areas of antibody are expected to be and, in fact, are quite diverse since, subject only to the limits of physical and chemical fit, they are complementary to a universe of epitopes.

Assault on the Universality of Paratypic Recognition?

Formal extensions notwithstanding, it seems appropriate to question the conclusion that the antibody-producing system displays essentially unlimited powers of recognition. In particular, since the system exists not merely to discern but also to destroy, there may be certain paratypic patterns which are not permissible. Perhaps the likeliest candidates to exclude from the repertoire are those paratopes which recognize epitopes displayed on self antigens. This consideration, if correct, could lead to a reduction in paratypic diversity minimally equivalent to the number of self epitopes. The heterogenous nature of self antigens, however, may preclude the conceptualization of a single rule which governs whether paratopes which recognize certain of these will, nevertheless, persist. The self epitopes which are of interest here are those found in the variable regions of antibody molecules.

The Idiotypic Determinant

It was already appreciated in the 1950s that, although myeloma proteins shared epitopes with normal serum immunoglobulins, there were certain antigenic determinants which seemed to be unique [17, 26]. In the early 1960s these observations were generalized when it was found that not only myeloma proteins but also populations of induced antibodies of restricted heterogeneity displayed an individuality [22]. These unique epitopes [idiotypic determinants or idiotypes; 22] which are carried by antibody molecules were localized, as the paratypic determinants are, to the variable region. It should be stressed, however, that such an equivalence of geography does not necessarily imply that the same constellation of atoms is under discussion. Subsequent to these initial analyses utilizing heterologous reagents, the crucial observations were made that idiotypic determinants are recognized and, therefore, that antibody specificity for this class of self antigen exists, within the repertoire of a single animal [16, 24]. The extension of these limited data calls for paratypic recognition of all self idiotypes. Other evidence and certain theoretical considerations led *Jerne* [12] to conclude that: 'formally, the immune system is an enormous and complex network of paratopes that recognize sets of idiotopes, and of idiotopes that are recognized by sets of paratopes'.

The Query

If it is true that the mature immune system is such a series of interconnected populations of cell-bound and soluble units of self-recognition, then we can inquire into the origins of such a structure – i.e. we would like to understand the ontogenetic basis for the connectivity of the network. At the extremes two possibilities could be imagined. On the one hand, there may be a developmental selection for fit such that newly arising lymphocytes which are not recognized by preexisting paratopes or which do not recognize preexisting idiotopes are discouraged from clonal persistence. On the other hand, early in ontogeny there may be no selection for idiotypic recognition in the sense of positive clonal induction; it is simply that the mature immune system has the capacity to detect all epitopes and, given the universality of recognition, what is required for the evolution of a closed, functional network is a failure to purge those clones demonstrating self anti-idiotypic specificity.

Induction of the Clonal Expansion of B Cells

Although the exact requirements for the expansion of B cell clones on the basis of idiotypic recognition are not clear, any theory which favors positive selection for idiotypic fit in the emerging lymphoid system requires a consideration of those determinants which may control such clonal expansion. Unfortunately, even in the mature lymphoid system there is limited data on this point; it is known, however, that heterologous anti-idiotypic antibody can initiate the expansion of idiotype positive B cell clones without inducing differentiation to antibody secretion [4]. Utilizing another approach to analyze potential network interactions, it was noted that the introduction of homologous IgM antibody into mice resulted in the production of antibody with equivalent specificity and that this phenomenon required T cells [8]. Furthermore, in certain epitopic systems there appears to be a requirement for idiotypic matching between T cells and B cells such that idiotype-positive B cells fail to differentiate into effector cells unless they are mixed with T cell populations from non-idiotypically suppressed mice [9, 28]. Finally, although anti-idiotypic antibody production, per se, may not equate with B cell clonal expansion, it is useful to consider the data on this point insofar as, minimally, they are reflective of B cell differentiation whose basis is idiotypic recognition. Considered together, the crucial element which emerges from these latter studies [10, 14, 25] is taken to be the requirement for T helper (Th) cells and it is concluded that the expansion of B cell clones and the subsequent differentiation to effector cells on the basis of idiotypic recognition is dependent on the participation of Th cells. These considerations lead to an examination of the ontogeny of Th cell function in the context of the clonal expansion of fetal B cells.

Ontogeny of Th Cell Function

That T lymphocytes may exist early in mouse ontogeny was suggested by the finding of thymocytes around day 11 of gestation [23] while by days 15–16 of gestation T (θ-positive) cells were detectable in fetal spleen [27]. Thus, the first requirement for Th cell function – namely the presence of T cells early in ontogeny – appears to be satisfied. However, studies which correlated Th cell function with age led to the conclusion that such function was absent in fetal and neonatal liver and spleen cell populations [2]. Moreover, regulatory cells and fluid constituents which were isolated from fetal and neonatal mice were found to suppress antibody responses to T-dependent as well as T-independent antigens [5, 20, 21]. The last point to be made here turns on the likelihood that Th cells must recognize epitope (idiotope?) and I region-encoded determinants prior to transmission of B cell induction signals. In situations where Th-B cell collaboration is H-2 restricted at the level of Th-B cell interaction, it follows that Ia-negative B cells may not be specifically inducible by Th cells; it is of interest, then, that as late as one day prior to parturition only 5% of splenic B cells and almost none in fetal liver were found to display Ia antigens [15]. Considered collectively, these data suggest that the substantial increase in B cell numbers which occurs from days 15 to 19 of gestation in the fetal mouse transpires in the absence of effective Th cell function.

Consequence of Paratypic/Idiotypic Recognition during the Fetal and Neonatal Period

T-dependent antigens have been utilized to analyze the functional consequences of epitope recognition by lymphocytes in the immature murine lymphoid system. The data demonstrate an antigen-specific, functional elimination of both B and Th cells [6, 19]. Similar conclusions are reached with regard to the introduction of anti-idiotypic antibody into neonatal mice; thus, both idiotype-positive B cells and Th cells are functionally inactivated [1, 13]. It appears, then, that the bias of the immature lymphoid system, when confronted with either T-dependent antigens or anti-idiotypic antibody, is against the expansion of Th and B cells which maintain either the same idiotypic or paratypic specificity, regardless of whether recognition occurs through paratypic or idiotypic means.

The Emerging Repertoire

Since it is to be expected that repertoire size is related to lymphocyte numbers, it is worthwhile to consider the forces which may drive the immune system so as to expand the numbers of B cells from 0% on day 11 to 1% of the fetal liver population by day 18 of gestation, while at the same time maintaining network connectivity. The postulate that B cell recognition of self-epitopes is the basis for an increase in both the diversity and numbers of B cells is not helpful for explaining positive selection of a network because the paratypic specificity of sufficient binding constant required to maintain connectivity may be lost. An alternative candidate for setting an initial idiotypic/paratypic pattern is maternal immunoglobulin. However, this possibility is excluded as describing a necessary condition by the absence of maternal immunoglobulin in fetal sheep, pigs and cows. A third mechanism which addresses itself to the developmental basis for the generation of a complete repertoire considers that certain of the earliest expressed B cell specificities recognize idiotypic determinants which are shared by antibodies and growth receptors on B cells and their precursors; it is through the interaction of these anti-idiotypic antibodies and growth receptors that the expansion of B cell numbers and diversity occurs [3]. The possibility that soluble molecules which possess anti-idiotypic specificity and the constant regions of immunoglobulin are functional in this model seems to be excluded by the observation in fetal sheep that at a point in time when there is no detectable serum immunoglobulin these animals are capable of producing antibody to a diverse series of foreign antigens [7]. Furthermore, in the mouse, similar frequencies of antigen-binding cells and range of avidities were found in splenocytes from 15-to 16-day-old fetal and adult animals [27], while it has been reported that even in 18-day-old fetuses immunoglobulin-secreting (plaque-forming) cells are almost absent from bone marrow and liver [18]. These results place quantitative constraints on the role of self-derived, soluble, anti-idiotypic mitogens, but do not address themselves to the possibility that repertoire generation is linked to the interaction of cell-bound, anti-idiotypic specificities and mitogen receptors on precursor B cells.

Concluding Remarks

The consequences of idiotypic recognition are seen as deriving not necessarily from recognition itself, but from the nature of the lymphocytes involved in the recognition process. It is considered that Th cell function is necessary for both the positive selection of lymphocyte clones which fit into a pre-existing paratypic/idiotypic pattern and the normal operation of a functioning network. Those systems which possess certain pieces of the immunological machine but which are incomplete with respect to Th cell function – for example, the mouse fetal lymphoid system or athymic mice – present immunological function without resorting to network interactions. The main point is that, in the mouse, early in ontogeny there is no positive selection for idiotypic fit and there may be a negative selection if self recognizing reactants exceed threshold concentrations. In this view, the ontogenetic diversity which is generated by the somatic mechanisms of: (1) random permutative pairing of germ-line V_H and V_L genes and germ-line V and their joining genes; (2) junction variations at the points of joining, and (3) variable region mutational events, taken together with the persistence of clones which recognize self idiotypes, can provide the basis for idiotypic connectivity of the lymphoid system.

References

1 Augustin, A.; Cosenza, H.: Eur. J. Immunol. *6:* 497 (1976).
2 Chiscon, M.O.; Golub, E.S.: J. Immun. *108:* 1379 (1972).
3 Coutinho, A.; Forni, L.; Bernabé, R.R.: Manuscript in preparation.

4 Eichmann, K.; Coutinho, A.; Melchers, F.: J. exp. Med. *146:* 1436 (1977).
5 Etlinger, H.M.; Chiller, J.M.: Scand. J. Immunol. *6:* 1241 (1977).
6 Etlinger, H.M.; Chiller, J.M J. Immun. *122:* 2564 (1979).
7 Fahey, K.J.; Morris, B.: Immunology *35:* 651 (1978).
8 Forni, L.; Coutinho, A.; Köhler, G.; Jerne, N.K.: Proc. natn. Acad. Sci. USA *77:* 1125 (1980).
9 Hetzelberger, D.; Eichmann, K.: Eur. J. Immunol. *8:* 846 (1978).
10 Iverson, G.M.: Nature, Lond. *227:* 273 (1970).
11 Jerne, N.K.: A. Rev. Microbiol. *14:* 341 (1960).
12 Jerne, N.K.: Annls. Immunol. *125C:* 373 (1974).
13 Julius, M.H.; Cosenza, H.; Augustin, A.A.: Nature, Lond. *267:* 439 (1977).
14 Klaus, G.G.B.: Nature, Lond. *278:* 354 (1979).
15 Kearney, J.F.; Cooper, M.D.; Klein, J.; Abney, E.R.; Parkhouse, R.M.E.; Lawton, A.R.: J. exp. Med. *146:* 297 (1977).
16 Kluskens, L.; Köhler, H.: Proc. natn. Acad. Sci. USA *71:* 5083 (1974).
17 Lohss, F.; Weiler, E.; Hillmann, G.: Z. Naturf. B *8:* 625 (1953).
18 Melchers, F.: Eur. J. Immunol. *7:* 476 (1977).
19 Metcalf, E.S.: Klinman, N.R.: J. exp. Med. *143:* 1327 (1976).
20 Mosier, D.E.; Johnson, B.M.: J. exp. Med. *141:* 216 (1975).
21 Murgita, R.A.; Tomasi, T.B.: J. exp. Med. *141:* 269 (1975).
22 Oudin, J.; Michel, M.: C.r. hebd. Séanc. Acad. Sci., Paris *257:* 805 (1963).
23 Owen, J.J.T.; Ritter, M.A.: J. exp. Med. *129:* 431 (1969).
24 Rodkey, L.S.: J. exp. Med. *139:* 712 (1974).
25 Schrater, A.F.; Goidl, E.A.; Thorbecke, G.J.; Siskind, G.W.: J. exp. Med. *150:* 808 (1979).
26 Slater, R.J.; Ward, S.M.; Kunkel, H.G.; J. exp. Med. *101:* 85 (1955).
27 Spear, P.G.; Wang, A.L.; Rutishauser, U.; Edelman, G.M.: J. exp. Med. *138:* 557 (1973).
28 Woodland, R.; Cantor, H.: Eur. J. Immunol. *8:* 600 (1978).

Dr. Howard Etlinger, Basel Institute for Immunology, Postfach, CH-4005 Basel (Switzerland)

The Immune System, vol. 1, pp. 21–27 (Karger, Basel 1981)

Individuality of Immune Systems: the Thousand Ways and One Way of Being Complete

Luciana Forni[1], *Antonio Coutinho*[1]

Basel Institute for Immunology, Basel, Switzerland; Department of Immunology, Umeå University, Umeå, Sweden

For Cartesians as we are, it should be feasible to present in a few pages the questions that we consider fundamental in immunology. Having learned it though from *Niels Jerne,* as we did, immunology acquires further perspectives which go much beyond the sterile exposition of facts, on which nonetheless it is based. To study immunology is to learn a fantastic story, how 'le hazard et la nécessité' have tinkered the immune system in so many different ways, because of so extremely varied reasons. We do not know the end of the story and we cannot describe our own immune systems – perhaps because there are many different solutions for each immune system or because lymphocytes are so special. Unlike a cell in the hepatic parenchyma, they have little or nothing to do, they go around in the whole body, may die next day or live for 20 years, and have individual markers of distinction. Lymphocytes are definitely too aristocratic and 'gâtés' to have their story, as we learned it from *Niels Jerne,* confined to the few pages of a summary.

Natural selection operates only on survivors. The survival value of the immune system rests on three properties: the ability to recognize all invasive pathogens, its promptness for an effective neutralizing response, and its ability to avoid autoaggression. We want here to examine these characteristics and to discuss how they are fulfilled by the different components of the immune system. Completeness is the determining characteristic of the antibody repertoire, rather than the promptness of the B cell compartment to respond, as documented by the cure of agammaglobulinemic patients by the

[1] *Luciana Forni* has been a Research Associate at the Basel Institute for Immunology since 1972, and *Antonio Coutinho* was a member of the Institute from 1975 to 1979.

passive transfer of normal natural immunoglobulin from other members of the species, and by the fact that normal individuals require T helper cells in order to develop antibodies to most antigens. The primordial importance of completeness means that antibodies cannot discriminate self from nonself, as in those patients who receive allogeneic natural immunoglobulin. In contrast, the T cell compartment does distinguish self from non-self, in particular from allogeneic nonself, and mature cells must learn what self is before the acquisition of competence. The T cell system, therefore, ends up incomplete. Promptness for an effective response appears to be the way of compensating that deficiency in the repertoires, and we make here the postulate that all helper cells in a normal individual, available in the periphery for reactivity to antigens, are in fact primed and expanded by the internal environment, in order to guarantee a sufficient frequency of clones and the necessary degree of differentiation to effector functions so that no limitation for the induction of antibody responses can result.

The ability of the immune system to produce specific protein molecules to any molecular pattern is still as striking now as it was 25 years ago when *Jerne* proposed the first selective theory of antibody formation. To be complete, a collection of antibody-combining sites need not be infinitely large, because antibody recognition is degenerated and the same paratope can recognize many epitopes with various degrees of precision. 15 years later, further considerations of *Jerne* on *potential* versus *available* antibody repertoires not only have excluded simplistic germ-line models for the origin of antibody diversity, but they also made it clear that each immune system has an individual way of being complete. The molecular biology of antibody genes tells us now that the vast majority of antibody specificities are somatically derived by recombination and mutation of germ-line DNA. It tells us, in fact, that there are no germ-line antibody genes, but pieces of coding sequences that can be mutated and recombined in many different ways. These possibilities make the available antibody repertoires limited only by the total number of cells in the individual at any given stage of development, and make it likely that each individual develops its own complete, available repertoire. The plasticity of the system, i.e. how different are the available antibody repertoires of the same individual at different points in its lifetime, is, on the other hand, provided by the enormous turnover of antibody-forming cells, characteristic of the normal immune system. Were it not important for the individual to adjust, at any moment, to new available repertoires, this large turnover of competent cells would appear purposeless. We conclude, therefore, that any somatic cell differentiating along the B lymphocyte lineage has an enormously large number of choices, as to the specificity of the antibody it is going to produce and, in addition, that a very large number of new individual choices are 'tested out' continuously (roughly 10^6/s in a human).

The genetic events involved in the expression of antibody genes are not all equally probable. It follows that some antibody specificities will be chosen very frequently and others extremely rarely. This imposes the requirement for selective mechanisms that operate on antibody-forming cell precursors and ensure the peripheral existence of the appropriate, complete repertoire. The tools used by the immune system to continuously select, from the innumerable possibilities, the components of the available repertoires are to be found in the elements of the immune system itself. A complete antibody repertoire cannot avoid recognition of its own elements, i.e. the normal antibody system can only be a network. Since every individual contains a complete combining site repertoire, it follows that the immune system can, in principle, always select for *any* available repertoire. On the other hand, from the individuality of available repertoires, it follows that any given available repertoire will select for new, defined sets of specificities, resulting in the characteristics defined by *Jerne* as the 'eigen-behavior' of the system, with maintenance of individuality.

The continuing selection of available repertoires, because it requires expansion of the desired specificities and extinction of the non-fitting ones, must necessarily be done on a functional basis. On the other hand, completeness of antibody repertoires confuses self and nonself in the same universe of recognizable molecular patterns, which implies that antibody recognition cannot provide discriminatory, functional signals for precursor cells, in terms of paralysis versus induction. It follows that the selection of available repertoires from the pool of precursor cells must have a molecular basis not in antibody specificities, but rather in the structures on antibody-forming cells responsible for triggering such cells into proliferation. Obviously, a system endowed with a complete set of 'recognizing' combining sites can select for any repertoire whatsoever, and on the basis of any given collection of molecules. Parenthetically, this postulate contains the explanation for the Promethean characteristics of the immune system. Selection of available repertoires is considered, therefore, to be mediated by what the system 'sees' on precursor cells, rather than by what these 'see' in their environment, ensuring the 1955 postulate that antibodies exist in the system (were selected to the competent cell pool) before exposure to antigen (in the absence of combining site recognition).

These considerations lead to an apparent paradox and a slight complication. The first has recently been solved experimentally, and it concerns the question of how the system can select functionally (on the basis of molecules other than cell-bound antibody receptors) for a given set of antibody specificities. The finding that such functionally relevent structures cross-react idiotypically with (some) antibodies solves that paradox inasmuch as the former are not equally represented in all precursor cells. The slight complication requires an ad hoc postulate, namely that selection on the basis of idiotypes can in fact be used to obtain a complete set of combining sites. This is made likely by the very 'eigen-behavior' of the system, as the selected will, in turn, be the selecting specificities, ensuring completeness of combining sites, but not necessarily of idiotypes, in the steady state.

These are properties of antibody repertoires, displayed by an isolated B cell system, as exemplified in the athymic mouse. We have no reason to suspect that nude mice do not develop a complete repertoire with the same diversity and plasticity as normal mice. The underlying mechanisms must, therefore, be essentially thymus-independent. In this thymus-independent precursor B cell induction and diversification, involving idiotypically cross-reactive immunoglobulin and non-immunoglobulin receptors, the origin of background plasma cells secreting natural IgM antibodies could be found. The effectiveness or promptness of antibody responses, in contrast, are severely affected in those mutants and, consequently, we ascribe to T cells these properties of the immune system. We are now considering helper cells exclusively, and we will disregard in the argument any other possible function of these cells except the induction of antibodies, i.e. we will look at T cells from the point of view of the B cell/antibody system. Here again, the selection of helper cell repertoires must be done in the absence of external antigens and, since all we want from helper activity is the induction of the available specific antibodies, the functionality and efficiency of the system is guaranteed by letting the repertoire of helper cells be specific for the antibodies that can be induced, i.e. for the available antibody repertoire. It follows that the specificity repertoire of helper cells is also continuously selected and adjusted to the ever-changing available antibody repertoire. Without arguing as far as to deny the existence of classical, 'carrier-specific' helper cells (which nonetheless can easily be done, since antibody-less mice do not develop helper activity and any antigen is internally represented), it appears logical to improve the promptness of antibody responses by allowing for selection and actual 'priming' of those helper cells, the activity of which is directly relevant for the response, namely those that are specific for available antibodies. An efficient selection and priming of anti-idiotypic T helper cells could be performed both by idiotypes on membrane-bound immunoglobulins of B cells and by natural antibodies. Since recognition of membrane-bound idiotypes in association with I antigens results in B cell induction, an extensive and continuous activation of the B cell compartment can be regulated by implying specificity of helper cells for idiotypes arising on IgM molecules by conformational changes upon reaction with the homologous ligand. Thus, only B cells recognizing an idiotype or being recognized by an anti-idiotype present in the natural antibody population will be targets of helper cell recognition. The incomplete repertoire of helper cells imposed in the thymus may then be further selected in the periphery for the most appropriate specificities. This might explain some of the current difficulties in ascribing predominant roles for either intra- or post-thymic selection.

This link between the completeness of antibody-combining sites and the incompleteness of T cell repertoires can be established, in principle, by clonal recognition in either direction. If antibody-combining sites would recognize T cell idiotypes in a functionally fruitful way, i.e. resulting in helper cell 'priming' and consequent activation of the same initiating antibody specificities in the steady state, the T cell system would be but a multiplying mirror, reproducing indiscriminately the complete combining site repertoire. This narcissistic solution, if ever tried, would have the same tragic consequences as its mythological counterpart, ending his days by the river at Thespies, since it would result in no self-nonself discrimination. Furthermore, this mechanism would result in complete helper cell functional repertoires, and this we know is not correct. Clonal recognition, therefore, must be effected (at least on functional terms) only in the opposite direction by the helper cell recognition of idiotypes. It follows that the participation of helper cells in the generation of available antibody repertoires cannot be fundamental, as an incomplete set of helper cell specificities interacts with another incomplete set of idiotypes. In contrast, this helper cell selection by a given set of idiotypes does have important consequences for the establishment of a mature T cell repertoire. Since idiotypes may in fact 'look like' any antigen, there is no a priori reason to exclude that some idiotypes mimic MHC antigens and, therefore, that apparently 'idiotype-restricted' helper cells may exist that in no way violate *Jerne's* postulates for the generation of T cell repertoires. As argued before, cross-reactivities between MHC antigens of the species and some antibody idiotypes might explain the predominance of such clones in the normal antibody repertoire and, simultaneously, include allo-reactive helper cells in the normal immune system and provide the phylogenetic link in the selective pressures driving germ-line antibody genes and those encoding T cell receptors.

In a closed situation, such as a germ-free, antigen-free mouse, the immune system produces conventional amounts of IgM but is 'blank' for all other Ig classes. This finding is fundamental for the support of all considerations above, as it demonstrates the intrinsic ability of the immune system to develop and maintain a steady state of activity that is self-perpetuating. The influence of maternal antibodies in this process can be excluded as to its essentials, because we now know that, while IgM antibodies induce the production of more antibodies with the same combining site and idiotype, IgG antibodies do not have this ability. On the other hand, the antigen-free mouse proves that IgG antibodies do not necessarily participate in the fundamental economy of the system and that they are always the result of exposure to nonself antigens. IgG-producing cells, as we know from other experimental approaches, are always the result of immunocompetent cell stimulation to extensive clonal expansion, and therefore they represent the individual cellular basis of memory. Regardless of how long they live, IgG-producing cells are the non-adjustable component of the antibody repertoire that are not internally generated, but rather the 'scars' from previous experiences in the continuing selection for ever-changing available IgM repertoire. The 'true' components of a closed immune system are therefore IgM. It is clear that the mechanisms driving IgM production in the closed immune system must be different from those leading to IgG secretion. Such differences in selective pressures could well result in a non-random representation of V regions in IgM and IgG, which would explain the differential effects exerted by the two immunoglobulin classes in both the regulation of immune responses and induction of antigen-independent antibody responses (both phenomena are in fact T-dependent and primarily mediated by V regions). The molecular basis for such different V region pools in IgM and IgG could reside either in restrictions of some V regions in switch rearrangements, or in further somatic diversification during B cell maturation. Or more simply, there may be no real difference between IgM and IgG V region pools – if conformational idiotopes are involved in T cell recognition, these could occur only on IgM, due to properties of the molecule other than those of V regions. It is tempting to involve helper cells in the solution for these questions, but this might not be correct; after all, we expect that antigen-free, athymic mice would produce normal levels of IgM. It is clear that class diversification in the immune system introduces higher levels of complexity that we only now start to define. With or without the postulate of non-random distribution of variable regions in the various antibody classes, the major problems can perhaps be reduced to the influence of non-IgM antibodies in the selection of available IgM repertoires and to the origin of natural IgM in the normal immune system. This is, after all, to bring the questions to the same point where they were after *Niels Jerne* solved immunology in 1955.

The choice of the subjects for these considerations and the way we developed them is simply the continuation of the many hours we spent with *Niels Jerne* discussing and arguing these issues, sharing excitement and doubts. It was inevitable, on the other hand, that a discussion on the essence of immune systems would turn out to be a wandering from one to another of the milestones marking *Niels Jerne's* scientific journey [1–7].

References

1 Jerne, N.K.: Proc. natn. Acad. Sci. USA *41:* 849–857 (1955).
2 Henry, C.; Jerne, N.K.: J. exp. Med. *128:* 133–152 (1968).
3 Jerne, N.K.: Eur. J. Immunol. *1:* 1–9 (1971).
4 Jerne, N.K.: A Ciba Found. Symp., pp. 1–15 (ASP, Amsterdam 1972).
5 Jerne, N.K.: Ann. Immunol., Paris *125C:* 375–389 (1973).
6 von Boehmer, H.; Haas, W.; Jerne, N.K.: Proc. natn. Acad. Sci. USA *75:* 2439–2442 (1978).
7 Forni, L.; Coutinho, A.; Köhler, G.; Jerne, N.K.: Proc. natn. Acad. Sci. USA *77:* 1125–1128 (1980).

L. Forni, Basel Institute for Immunology, Postfach, CH–4005 Basel (Switzerland)

The Immune System, vol. 1, pp. 28–34 (Karger, Basel 1981)

On Network Theory, Ly Phenotypes and Connectivity

Geoffrey W. Hoffmann[1]

Departments of Physics and Microbiology, University of British Columbia, Vancouver, B.C., Canada

Ly Phenotypes and Connectivity

Recent work on the Ly antigenic markers indicates that helper T cells characteristically have the Ly $1^{+}2^{-}$ phenotype, while those that suppress are mainly Ly 2^{+} [1]. Because of the observed differences in function, a natural and widely drawn conclusion has been that the Ly markers are differentiation antigens, and that an Ly $1^{+}2^{-}$ cell therefore has intrinsic properties that an Ly 2^{+} cell lacks, and vice versa.

The symmetrical or 'plus-minus' network theory of regulation of the immune system [6, 8] is a theory that is based on *Jerne's* [9] seminal network hypothesis. In the symmetrical network theory, only one basic type of regulatory T cell needs to be invoked, rather than two or more regulatory T cells that are fundamentally different from each other.

In the context of the symmetrical network theory, the observed differences between T cells with different Ly phenotypes are most simply ascribed to a quantitative difference in a parameter called connectivity [8]. Connectivity is a measure of the extent to which a cell (or clone) is connected with the rest of the network via idiotype-anti-idiotype interactions. As such, it is a property of the clone's network environment, rather than a property of the clone itself. I originally used the word 'connectance' in this context [8], but have since opted for the term connectivity, because connectance has been used in a slightly different sense by mathematical ecologists [3, 12]. They have used connectance to denote the probability that two randomly chosen

[1] *Geoff Hoffmann* was a member of the Basel Institute for Immunology from 1974 to 1979.

species of the ecological network interact. The strength of the couplings is then denoted by a second parameter, s. This ecological connectance is a single parameter for the entire network, and does not refer to individual species. It does not take population levels into account. In the treatment of network aspects of the immune system, we shall find it useful to have a variable, to be called connectivity, which reflects firstly the concentration levels of clones with anti-idiotypic specificity in the environment of the clone of interest, and secondly the strength (affinity) of the idiotype-anti-idiotype interactions. A reasonable mathematical definition of connectivity in this sense will be given below. But first we will use the above qualitative description of connectivity to describe some of the main features of the symmetric model in terms of that parameter, and to indicate how some recent experimental results can be interpreted in terms of the theory.

The core of the symmetrical network theory is a simple set of plausible postulates that leads to the existence of four stable states for any antigenic or idiotypic specificity [6]. The stable states include a suppressed state with high connectivity for both antigen-specific ('plus') and anti-idiotypic ('minus') cells, and an immune state characterized by relatively low connectivity for antigen-specific clones. The model predicts that T cells with high connectivity would tend to suppress immune responses, while those with low connectivity would preferentially help [8; due to an undetected printer's error (half a sentence was omitted), the paper referred to contains the erroneous statement (p. 207) 'T cells [with] low connectance [connectivity] would preferentially suppress', which should read 'T cells with low connectance would preferentially help, and those with high connectance would preferentially suppress']. In the suppressed stable state there are elevated levels of both T_+ (antigen-specific) and T_- (anti-idiotypic) cells, and the specific T cell factors they produce inhibit any stimulation that could otherwise cause the secretion of corresponding plus and minus antibodies. This distinctive feature of the theory (elevated levels of both T_+ and T_- in the suppressed state) received some important support from recent experiments by *Hirai and Nisonoff* [5]. They showed that the T cells of idiotypically suppressed mice secrete factors capable of specifically suppressing the corresponding idiotypic component of an immune response, and that the factors consist of a mixture of antigen-specific and anti-idiotypic factors.

The limiting dilution method [10] is a powerful method for differentiating between those properties a cell exhibits because of its network environment and intrinsic properties of the cell itself. Let us consider a situation in which we have a mixed bag of T cells, consisting of some antigen-specific idiotypes with their corresponding anti-idiotypes in the suppressed state, and some other idiotypes (specific for the same antigen) in the virgin or the immune stable state. We then do a limiting dilution experiment, adding small numbers of the T cells to cell cultures, and look for help. If, as we have postulated, the T cells in the suppressed state are not fundamentally different from those in the virgin state, the limiting dilution experiment should reveal a set of helper T cells occurring at high frequency, which are then suppressed when the cultures have sufficient cells to ensure the presence also of the corresponding anti-idiotypic cells. The cells in the virgin or immune state might then be revealed at a higher cell dose, corresponding to their being present at a lower concentration. Precisely this phenomenology has been reported recently in elegant experiments by *Eichmann* et al. [2]. They expanded limiting numbers of T cells using T cell growth factor, and assayed the cells for help in the TNP-strep system. The population that occurred at high frequency and was suppressed at the higher T cells doses had the Ly-123 phenotype, and the helper that became evident only at higher doses was an Ly-1 cell. That system clearly has great potential for clarifying these matters further. So far it does not prove that suppressor cells can also provide help, but the data can certainly be interpreted most simply that way.

Towards Measuring Connectivity

Is connectivity just a useful buzzword for hand-waving theorizing, or is it something that can be measured, and hence take its place in the realm of real science? We will now define connectivity mathematically, and suggest a model-dependent means for measuring the connectivity of cells in the virgin state. We will formulate our equations in terms of the cellular concentrations of a representative pair of interacting clones x_i and x_j, and then sum over i and j to derive relationships that involve the entire repertoire. We are thus progressing to a more detailed description of the system than the plus-minus approximation used before, in which a single pair of variables (x_+ and x_-) was used to characterize antigen-specific and anti-idiotypic clones, respectively [4, 7].

A reasonable definition for the equilibrium connectivity C_i of a clone i would be a weighted sum of the population levels of the clones j that have a finite affinity for i:

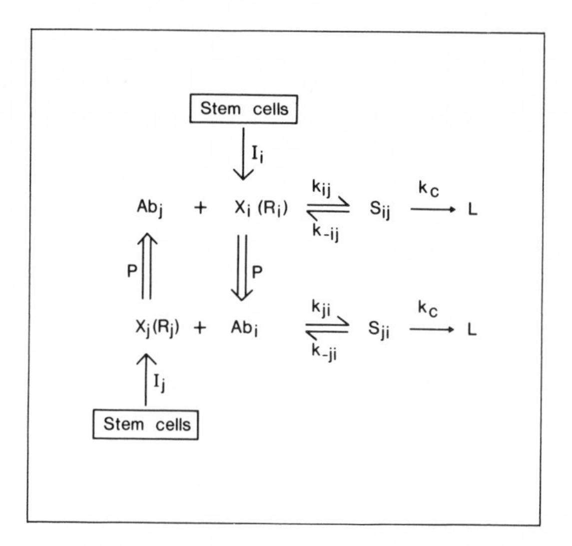

Fig. 1. Kinetic scheme of the interactions between a representative pair of clones that are in the virgin stable state. IgM antibodies of clone j kill cells of clone i and vice versa.

$$C_i = \sum_{j=1}^{M} K_{ij}x_j \quad (1)$$

where the x_js are the concentrations of each of the j clones, K_{ij} is the equilibrium binding constant that characterizes the interaction between idiotypes i and j, and the summation is applied over all the clones (M different ones) present in the particular immune system. Since most clones do not interact with each other, K_{ij} is a sparse matrix.

The virgin state of the symmetrical network theory is a steady state in which there is essentially a balance between non-specific influx and idiotype-specific killing of cells by a linear cytotoxic mechanism. (By a linear cytotoxic mechanism we mean one with a rate that depends linearly on the concentration of the anti-idiotypic cells, or the antibodies they secrete.) Killer T cells and IgM plus complement-mediated killing are possible candidates for that role. For the sake of being definite, we will consider the case that just IgM plus complement-mediated killing is important; the other possibility can be treated in an analogous way.

An appropriate kinetic model for the interactions that are then important in the virgin state is shown in figure 1. I_i denotes the rate of influx of cells of specificity i into the system. X_i and X_j represent the cells of specificity i and j, respectively, R_i and R_j are the associated receptors, S_{ij} denotes a site for idiotype-specific complement-mediated attack on an i cell (that is, a complex between an i receptor and a j antibody), and L denotes lysed cells. The constants k_{ij} and k_{-ij} are the forward and reverse kinetic rate constants for the binding of antibody j to receptor i, and k_c is a lumped catalytic rate constant for the elimination of cells due to killing by IgM antibody plus complement. P is the average rate of secretion of antibodies by individual virgin cells. The form of this.scheme is essentially that of Michaelis-Menten enzyme kinetics [11]. The antibody plays the role of the enzyme, and the cell is the substrate. Complement can be regarded as a milieu, in which the 'enzyme-substrate' complex is unstable, and decays to L with rate constant k_c.

We will use the small letters x_j, a_i and r_i to denote the concentrations of cells of clone j, antibodies of clone i, and receptors on cells of clone i, respectively. The variable s_{ij} denotes the concentration of complexes of Ab_j and R_i (sites for complement-mediated attack). The kinetic rate equation for a_j is assumed to have the form:

$$\dot{a}_j \simeq Px_j - Da_j. \qquad (2)$$

Here D is the rate constant for the decay of antibodies (IgM antibodies have a short half-life), and we are assuming that only a very small fraction of the antibody is bound to anti-idiotypic receptors. The kinetic equations for s_{ij} and x_i are:

$$\dot{s}_{ij} = k_{ij}r_ia_j - k_{-ij}s_{ij} - k_cs_{ij}, \qquad (3)$$

$$\dot{x}_i = I_i - \sum_{j=1}^{M} k_cs_{ij}. \qquad (4)$$

The vast majority of receptors can be assumed not to have anti-idiotypic antibodies attached to them, so we have

$$r_i \simeq Nx_i, \qquad (5)$$

where N denotes the number of receptors per cell. In the steady state we have $\dot{a}_j = \dot{x}_i = \dot{s}_{ij} = 0$. We can then deduce from equations 2–5 that

$$I_i = \frac{Nk_cx_iP}{D} \sum_{j=1}^{M} \frac{k_{ij}x_j}{k_{-ij} + k_c}.$$

The equilibrium constant K_{ij} is equal to the ratio k_{ij}/k_{-ij}. The constant that is relevant for determining the level of interaction between clones i and j is however the reduced ratio $k_{ij}/(k_{-ij} + k_c)$, which may be thought of as a 'steady-state constant' as opposed to an equilibrium constant. We accordingly define a 'steady-state connectivity' C_i' of clone i by

$$C_i' = \sum_{j=1}^{M} K_{ij}' x_j,$$

where $K_{ij}' = k_{ij}/(k_{-ij} + k_c)$. Then the total rate of influx I_T of virgin cells into the system is given by

$$I_T = Nk_c \frac{P}{D} \sum_{i=1}^{M} C_i' x_i.$$

We can now readily derive a value for the average steady-state connectivity

$$\overline{C}' \equiv \sum_{i=1}^{M} C_i' x_i / \sum_{i=1}^{M} x_i$$

in terms of the steady-state total concentration of IgM secreted by virgin B cells, to be denoted by $[Ab]_T$. The result is

$$\overline{C}' = \frac{I_T}{k_c [Ab]_T N}. \tag{6}$$

We have thus derived a model-dependent expression for the average steady-state connectivity of cells in the virgin state. The parameters I_T, k_c, N and $[Ab]_T$ are all readily measurable, and my laboratory is presently engaged in determining $\overline{C}'$ in this way. The steady-state connectivity constitutes a lower limit for the equilibrium connectivity. If it turns out that k_{-ij} is typically much greater than k_c, the equilibrium connectivity and the steady-state connectivity will have essentially the same values. The measurement of connectivity by the brute force, model-independent method of measuring binding between the idiotypes of a sufficiently large number of different clones, and determining the size of each of the clones is much more difficult. In principle it could however be done, and would then provide a quantitative test of the model. Equation 6 can be useful for systematic measurements of connectivity. We will be able to determine whether, for instance, connectivity changes with age, and whether the strain dependence of connectivity correlates with other known regulatory phenomena.

A Word of Thanks

The formulation of the network hypothesis by *Niels Jerne* [9] was a unique event in the history of science. Never before has a scientist, at a point in time so close to his retirement, instigated a revolution in the theoretical basis of an entire field. I feel privileged indeed to have witnessed early stages in that revolution at first hand, and am grateful for having had the opportunity to play a part in it. Furthermore, the other great creation of *Jerne* (the Basel Institute for Immunology) provided an environment that is dominated by stimulatory rather than inhibitory interactions. Thank you *Niels.*

References

1 Cantor, H.; Boyse, E.A.: Cold Spring Harb. Symp. quant. Biol. *41:* 23–32 (1976).
2 Eichmann, K.; Falk, I.; Melchers, I.; Simon, M.M.: J. exp. Med. *152:* 477–492 (1980).
3 Gardner, M.R.; Ashby, W.R.: Nature, Lond. *228:* 784 (1970).
4 Gunther, N.; Hoffmann, G.W.: Submitted for publication.
5 Hirai, Y.; Nisonoff, A.: J. exp. Med. *151:* 1213–1231 (1980).
6 Hoffmann, G.W.: Eur. J. Immunol. *5:* 638–647 (1975).
7 Hoffmann, G.W.: in Bruni et al., Lecture notes in biomathematics, vol. 32, pp. 239–257 (1979).
8 Hoffmann, G.W.: Contemp. Top. Immunobiol. *11:* 185–226 (1980).
9 Jerne, N.K.: Annls Immunol. *125C:* 373–389 (1974).
10 Lefkovits, I.; Waldmann, H.: Limiting dilution analysis of cells in the immune system (Cambridge University Press, Cambridge 1979).
11 Lehninger, A.L.: Biochemistry, pp. 153–157 (Worth, New York 1970).
12 May, R.M.: Stability and complexity in model ecosystems; 2nd ed., pp. 62–68 (Princeton University Press, Princeton 1974).

Geoffrey W. Hoffmann, PhD, Departments of Physics and Microbiology,
University of British Columbia, 2075 Wesbrook Mall, Vancouver, B.C., V6T 1W5
(Canada)

The Immune System, vol. 1, pp. 35–42 (Karger, Basel 1981)

Some Thoughts on Information Processing by the Nervous System

W. Reichardt[1]

Max-Planck-Institut für biologische Kybernetik, Tübingen, FRG

Introductory Remarks

The nervous systems of animals and humans respond adequately, like the immune system, to a great variety of signals. These signals are received from the environment by the receptors of various sensory modalities and by so-called proprioreceptors that sense the state of the organism. The effector organs usually respond to the receptor inputs after the input signals have been 'intelligently' processed and possibly compared with information already stored in the nervous system.

Our present knowledge of the principles underlying uptake, transduction and processing of sensory information is still limited and mainly confined to the mechanisms of so-called preprocessing of information that is taken up by the receptors of the various sense modalities.

For the last 20 years, the analysis of these principles has been strongly based on concepts and research strategies which have found application in the analysis of complex systems, consisting of very many elements that interact with each other in an intricate form of non-linear information exchange.

[1] *Werner Reichardt* was a colleague of *Niels Jerne* at Caltech in Pasadena in 1955.

Concepts and Strategies

Generally speaking, complex systems like the nervous system or even parts of it can be studied and understood at several different levels. Only three of them are considered here. The highest level is characterized by the nature of the overall computations, expressed in the behaviour of an organism. At the second level we deal with the algorithms that implement these computations. At the third level we look at the neural circuitry by which these algorithms are realized.

Generally speaking, the nature of the overall computation that controls the behaviour of an organism is determined and restricted by the problem to be solved by the organism, whereas the particular algorithms involved in the computation depend on the problem and the available neural mechanism.

The scheme presented here suggests a top-down approach, in spite of the fact that most investigators in the neurosciences believe that such an analysis, if applicable, does not lead to an understanding of the functional properties of the nervous system. They argue that we underestimate our ignorance of cell biology and cell-cell interaction and therefore cannot even proceed from the bottom towards the top: from molecules to cells, from cells to cell assemblies, up to an understanding of the processes of information handling performed by the nervous system.

I do not share the scepticism about a top-down analysis of complex systems. My arguments are borrowed from computer science under the assumption that the computational systems under investigation are a priori unknown.

The conclusion derived at each of the different levels of understanding, mentioned before, will be perfectly true and yet very different. That is, there are several distinct yet equally valid ways of understanding a system, depending on one's domain of discourse.

One may ask how well relations can be seen among these various conclusions. It is not too easy to see how understanding at one level may help that at another level. For example, if one wishes to understand the operation of the system at the programming level it is immaterial whether the system under investigation is based on this or that hardware. Those who are trying to understand at the algorithmic level need to know essentially nothing about programming languages. And those who try to understand how a piece of hardware functions are not helped either by, for instance, a specific programming language. Indeed, there are logical and causal relationships among these levels, but there seems to be very little useful extrapolation from one level to another. In the neurosciences, it seems to me equally hopeless to accept the argument that one must start at the lowest level before going on the greater system complexity. If possible, extrapolation in the other direction, namely from top to bottom, seems to be easier. Knowledge at a higher level may help to develop strategies for the analysis at a lower level. The difference is primarily one of hierarchical levels of code and language, i.e. the descriptions of information flow and process are entirely different at these various levels; the simpler signal structures at lower levels do not yield a prediction of those at higher levels.

Some Principles of Information Processing

Range Adjustment of Sensory Channels

I begin my consideration with the problem of dynamic range adjustment which *Delbrück and Reichardt* [1] tackled many years ago in Pasadena when I met *Niels Jerne* for the first time.

Sense organs, in general, can adjust their sensitivity to a wide range of signal levels, a property also called adaptation. The phenomenon is well known and by no means confined to sense organs of higher organisms. Even at the level of a single cell, such as, for instance, the sporangiophores of the mould *Phycomyces,* the growth response of these cells shows the property of an adjustment to wide ranges of light intensities. The importance of that adjustment can be easily seen: reflected light intensities from objects in the environment differ by a factor of about 20. This does not depend on the brightness of the illuminating light if the spectral composition of the light does not change. When one measures the average reflected light intensities during the night and during bright sunshine one finds that they differ by about a factor of at least 10^6. A light sense organ that collects and sends signals into information channels with a dynamic range of about 1–20 has to match a range of $1–10^6$ with a range of 1–20. Generally it is this process of range adjustment or adaptation that is so typical for the first step of information uptake from the environment. How is it accomplished? The processes involved are by no means completely understood, at least not in molecular terms. But what is known is the structure of the signal flow during adaptation. Two different channels are involved. The primary channel represents the receptor signals I(t), whereas the secondary channel represents a running time average of the primary channel which is called the level of adaptation A(t). I(t) is 'measured' in units of A(t), so that i(t) = I(t)/A(t) is the input variable to the sensory channel. This variable fulfils the so-called Weber and Fechner laws, typical for the process of automatic range adjustment or adaptation. It is also typical for processes responsible for information handling by biological systems that are mostly non-linear. Since A(t) and I(t) are related to each other, i(t) is a non-linear 'image' of I(t).

Transformation of Sensory Data and the Process of Lateral Inhibition

Sense organs like the vertebrate or invertebrate eyes consist of very many photoreceptors whose numbers per unit solid angle define the primary optical resolution of the organ. A point source of light mounted in the optical axis of one receptor is usually 'seen' not only by the on axis receptor but also by at least its neighbours if not by a field of receptors surrounding the on axis receptor. The overlap of the receptive fields of neighbouring light receptors leads to a transformation of the optical environment on the receptor plane of the eye. In spite of the fact that the receptive fields of the receptors overlap, the transformation or imaging process has the property of a one-to-one correspondence. That is to say, one intensity distribution at the receptor plane is generated by one and only one brightness distribution in the optical environment; consequently the transformation or imaging process from the optical environment on the receptor plane is not accompanied by a loss of sensory information.

Once the optical information is represented at the level of the light receptors it undergoes a sequence of further transforming steps before it reaches the more central parts of the visual system. For example I mention here the transforming principle of lateral inhibition which has been studied in detail in the lateral compound eyes of the horseshoe crab *Limulus* [2, 5]. The anatomy of the eye is such that collaterals of the sensory cell axons make recurrent synaptic contacts with the other sensory cells. Their physiological influence is expressed by a linear inhibition of the generator potentials of the receptor cells in proportion to the nervous activity of the sensory axons that form the optic nerve. In principle the transformation by lateral inhibition can regenerate the light distribution at the receptor plane to a neural equivalent of the brightness distribution present in the optical environment. Usually, however, as also in the case of *Limulus,* the transformation by lateral inhibition overcompensates the transformation based on the overlap of the visual fields of the light receptors. The overcompensation leads to a strong exaggeration of those parts of the optical environment where contrast changes occur. In this sense lateral inhibition yields a one to one correspondence of the brightness distribution in the environment with the excitation distribution across the optic nerve in such a way that important optical information is emphasized. This is a powerful principle at least for representation and extraction of optical data.

In this connection it is worthwhile mentioning another point of general interest: neurons usually transport information that is encoded in instantaneous nerve spike frequencies. When a neuron receives signals from only one synaptic input it has to 'wait' for at least two spikes until it can measure the instantaneous frequency. When however, as in the case of lateral inhibition, a neuron receives signals from many synaptic inputs, it can determine the incoming information flux in a much shorter time. The advantage is due to the ensemble average generated by the signals from the many synaptic inputs. In this sense, lateral inhibition is a principle that compensates for the disadvantages of encoding sensory information into spike frequencies.

Position and Movement Detection

Behavioural as well as electrophysiological experiments at the cellular level clearly show that organisms are able to detect, to measure and to represent position as well as movement information from moving objects in the optical environment [4, 7]. The computations that are responsible for the evaluation of position and movement are local, i.e. they receive their information from each of one or each of two photoreceptors of the eye.

In the fly's visual system, which has been most intensively studied with respect to position and movement detection, the position computation seems to be bound to any single photoreceptor which sends signals via information channels to the central parts of the visual system. The channels are parametrized by the coordinates of the receptors or by their locations in the eye.

Due to the fact that movement is a vector, the direction-selective extraction of movement information must be and is bound to at least each of two receptors. The analysis has led to the conclusion that the signals from the environment received by each receptor are transformed by a kind of running time average which in turn is correlated with the signals from each of the other receptors of the double receptor systems. The process underlying the evaluation of movement is therefore non-linear, since the data correlated are at least statistically dependent. Interestingly, the correlation of the optical data selectively destroys 50% of the information carried by the signals which are received from the optical environment. It seems to me that this is a special case of a general principle: namely the formation of an abstraction, like the evaluation of movement, is necessarily accompanied by a selective destruction of sensory information received from the environment. If the process of selective destruction were known in detail it would be possible to specify the rules (invariance class) for changing sensory information without changing the result of the computation responsible for the formation of an abstraction.

The Computation of Relative Movement

An object in front of a textured background can be detected if the two (object and ground) move relative to each other. A behavioural analysis of the reactions of a fly to object and ground movements has already led to a characterization of the algorithm used to perform this computation, which amounts to the detection of discontinuities in the optical flow field [8]. The analysis of this behaviour established that the algorithm relies on an inhibitory correlation-like operation between the elementary movement detectors stimulated by the ground texture and the object. The main point here is that the visual system of the fly, like possibly the visual systems of other animals, can detect and compute relative movement and by the same process discriminate figure from ground if they are moved relative to each other.

Another point of interest is the following: the fly's visual system is equipped with neuronal devices (modules) that measure the coherence of the movement field, computed by local movement detectors. It has been shown more recently that this computational process is in a sense analogous to the interference with light waves that leads to the formation of a hologram. The analogy, however, has to be considered with some caution. The photoreceptors of the fly receive light from small solid angles that overlap only with the receptive fields of neighbouring photoreceptors. A typical point on a photographic plate, where the hologram is stored, may in principle receive radiation from a solid angle of 2π. This property is a necessary requirement for the holistic feature of information storage whereas in the fly the collection of optical data is confined to rather local regions of neighbouring receptive fields of photoreceptors. The conclusion is that the evaluation of relative movement makes use of computations analogous to those that take place during the process of hologram formation, without leading, however, to a complete holographic pattern of the optical environment. The important point, however, is the experimentally established fact that the fly is equipped with neuronal components whose functions could be implemented for the processing of information if the nervous system were to process and possibly store data in holographic fashion [6].

A Differentiation between Self and Non-Self

Jerne has, at various times, explained to me in great detail what is known about the immunological mechanisms that are responsible for a differentiation between self and non-self. For that reason it might be appropriate to say a few words about a similar problem in the field of the neurosciences. As already discussed by *Helmholtz,* the image of a moving object on the retina produces the same receptor stimulation as an equivalent movement of the eyes or the entire body if the object is at rest. How then can we visually differentiate between object movement or self movement? A solution of the problem has been given by *von Holst and Mittelstaedt* [3]; they called it the reafference principle. Its application to the case mentioned here led to the following conclusions. When the eye is actively moved a copy of the efference signal is stored in the nervous system and compared with the signal (the reafference) which is generated by the moving image of a resting object on the retina. When reafference and efference copy compensate each other, we have the impression that the object did not move. When, however, the object is moved and the eyes are at rest, the efference copy is zero and the signals from the retina lead to the impression that an object is moving. The reafference theory is based on various different and independent observations which led to the assumption that it could be called a general principle for the differentiation between self and non-self. That this in general is not so has been shown more recently by new experiments with flies. We have demonstrated that only the retinal displacement of the image of an object is responsible for the generation of a behavioural response by the fly. How then can a fly differentiate between the two possibilities: the object moves in space and the fly is at rest, or the object is at rest and the fly moves in space. The solution of the problem has been given in connection with the computation of relative movement between object and ground. When the object moves against the ground and the fly is at rest, it receives signals from those receptors of the retina of the two eyes which are stimulated by the image of the moving object. When, however, the fly moves actively, the image of the contrast elements of ground and object move across the retinae. In this case the flow field of the relative moving ground serves as a reference for the eigenmotion of the fly and the flow field of the moving object is related to the ground's flow field by an evaluation of the coherence of the velocity signals of object and ground.

The reafference theory and the theory for the detection of relative movement allow differentiation of self and non-self motion. The reafference theory is based on the observation that self is established by an internal signal, the efference copy, whereas the theory for the computation of relative movement takes the information on self from the flow field of the moving ground image on the retinae. In summary then: the nervous systems have developed different mechanisms for differentiating between self and non-self to enable and maintain the orientation characteristic of the organism.

References

1 Delbrück, M.; Reichardt, W.: in Cellular mechanisms in differentiation and growth, pp. 3–44 (Princeton University Press, Princeton 1956).
2 Hartline, H.K.; Ratliff, F.; Miller, W.H.: Nervous inhibition. Proc. Int. Symp., pp. 241–284 (Pergamon Press, New York 1961).
3 von Holst, E.; Mittelstaedt, H.: Naturwissenschaften *37:* 464–476 (1950).
4 Poggio, T.; Reichardt, W.: Q. Rev. Biophys. *9:* 377–438 (1976).
5 Reichardt, W.: Kybernetik *1:* 57–69 (1961).
6 Reichardt, W.: Naturwissenschaften *67:* 411 (1980).
7 Reichardt, W.; Poggio, T.: Rev. Biophys. *9:* 311–375 (1976).
8 Reichardt, W.; Poggio, T.: Biol. Cybernet. *35:* 81–100 (1979).

Prof. Werner Reichardt, Max-Planck-Institut für biologische Kybernetik,
Spemannstrasse 38, D–7400 Tübingen 1 (FRG)

The Immune System, vol. 1, pp. 43–50 (Karger, Basel 1981)

Some Similarities in the Way the Immune System and Nervous System Work

A.J. Cunningham[1]

Ontario Cancer Institute, Toronto, Ontario, Canada

Niels Jerne has made enormous contributions to immunology, among them: championing 'selection' against 'instruction'; realizing the need for, and providing a convenient assay which detects single antibody-forming cells; proposing the idiotype network. A most attractive feature of his writing has been his willingness to ignore trivia and encompass global issues in the subject. Not surprisingly, therefore, he has from time to time [6, 7] touched on the intriguing similarities between immune and nervous systems. The present paper is a brief attempt to press this comparison a bit further.

Main Functions of the Two Systems

Since *Darwin,* it has been respectable to rephrase teleological questions in evolutionary terms. Thus we may ask: for what purposes have immune and nervous systems evolved? Both seem to be primarily concerned with preserving the individual as an integral unit against external forces which seek to tear it down. The immune system guards the body from invasion and decomposition by parasites, and perhaps also from tumours; the nervous system coordinates the immense complexity of functions going on within an individual, and allows effective interaction with the outside world.

[1] *A.J. Cunningham* was a member of the Basel Institute for Immunology in 1974–1975.

For both immune and nervous systems the ability to distinguish self from non-self is of prime importance. Many lymphocytes react to self antigens: they are deleted or otherwise controlled. Perhaps the genes for certain particularly troublesome anti-self specificities are removed from the germline, but the capacity to learn self-tolerance must be available, since no infant can 'know' what antigens it will inherit from its parents. Similarly the nervous system can inherit many response patterns, from simple reflexes to complex instinctual behaviour, but it too has to learn by experience to distinguish what is 'out there' and so potentially threatening or useful, from what is 'in here'. This process of defining the self as opposed to the rest of the world is particularly interesting in the ego development of the human infant [9].

The external world consists of an immense variety of potential stimuli, both perceptual and antigenic: no organism can hope to inherit receptors specific for all of them. This leads us to consider the central functional property shared by immune and nervous systems (and distinguishing them from other specific, adaptive recognition systems like enzymes): their ability to react to 'unexpected' stimuli, i.e. to things which neither the responding individual nor its ancestors may have encountered before. This was pointed out by *Cohn* [3]. It is obvious for the nervous system – we can all learn a new face or language – but has often been neglected in immunology. The individual inherits many (not all) potential response patterns in the DNA, but must learn many more from experience, an acquired wisdom which is not passed on genetically (apart from the occasional inheritance of genes with outstanding and constant survival value).

Structurally, immune and nervous tissues are rather different, of course. The human nervous system comprises about 10^{11} cells, the great majority of which make no direct sensory or motor connection with the rest of the body and the outside world, but are fixed in the brain where they have computing and controlling functions. They receive patterns of stimulation (information), coordinate these, and send instructions to the body. The nervous system is particularly concerned with the positional relations between things, self and non-self and parts of self: it therefore has to be relatively rigid in structure. By contrast the immune apparatus is much more responsive to the chemistry of the external and internal environments. It consists essentially of about 10^{12} free floating lymphocytes which guard the individual against not the sabre-toothed tiger but the insidious micro-organism. Yet the two systems have a great logical similarity: they are both primarily processors of information rather than of energy, unlike most other functions

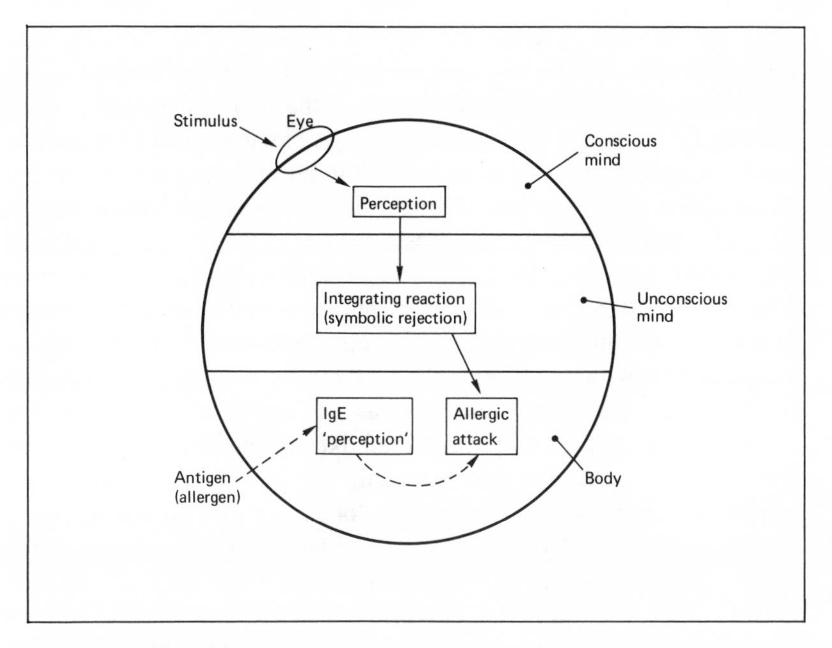

Fig. 1. 'Information flow' involved in an asthmatic attack produced by a visual stimulus (solid lines) or an antigenic stimulus (dashed line).

in the body (for example digestion or muscular contraction) which are predominantly energetic. Both respond specifically and adaptively to information from the environment, as discussed below. Chemical and positional information from the environment can induce remarkably similar effects in the body at times: figure 1 shows how the same kind of asthmatic attack can be provoked in certain individuals by either an allergenic molecule or the mere sight of a conditioned stimulus such as a vase of plastic flowers [5].

Specificity

Perhaps the most striking property of both immune and nervous systems is their ability to recognize specifically a great variety of stimuli. The information allowing such discrimination is of course stored in different ways in the brain and in the lymphocyte population. The nervous system relies mainly on the position and interrelations of the neurones activated, while the immune system depends more on intrinsic chemical variety among the cells which make contact with the antigenic stimulus.

It appears, however, that perception of the environment is a very complex process in both systems. Activation of a small number of sensory neurones may cause general arousal through input into the reticular formation in hind- and mid-brain and through multiple synapses with many thousands of neurones in the cortex. Most neurones in the brain are connected to most others via a very small number of intermediate cells. The immune system was, until recently, thought to work quite differently. One antigenic determinant stimulated one or a few lymphocytes only, and these expanded independently, without affecting the great majority. Perhaps the most exciting change in immunology during the last few years has been the realization that immune responses are probably not like this, but much more like brain events. A single antigen may stimulate, directly or indirectly, a large proportion of the lymphocyte pool (as shown, for example by the sharp, large rise in Ig-secreting cells). A relatively specific final product emerges as the result of a dynamic interaction between the foreign antigen, antibody molecules and receptors, their idiotypes, and probably MHC and other self antigens. According to the idiotype network idea [6], lymphocytes, like neurones, are extensively interconnected.

How much specific information (i.e. specific response capacity) is inherited and how much acquired? For the brain it has always been obvious that much must be acquired but earlier notions that much of the connectivity of the brain was functionally selected from a randomly generated set of connections are now giving way to a more 'germ line' view: most connections seem to be precisely established at an early stage of development, sometimes down to the level of the individual neurones [4]. Neurones in the optic pathway, for example, are specialized as to whether they will detect light or dark spots, rays or lines at different angles, and are wired to a specified part of the retinal field. Nevertheless, the anatomical patterns of development of some neurones and synaptic connections do depend on environmental stimulation. For example, a recognizable, discrete aggregate of neurones in the cortex of the mouse develops to receive input from each whisker on its snout, and if one line of whiskers is destroyed shortly after birth, the corresponding neural aggregates do not appear [4]. Immunologists, likewise, have for many years debated the relative importance of the genome and of environmental experience in establishing a repertoire of specific immune responsiveness. As for the brain the answer seems to be 'a bit of each'. There are many more germ-line V genes than some of us originally thought, but there is also a somatic production of many new specificities by recombination within V gene components, and by mutation (as shown in mouse λ chain).

The initial recognition of a stimulus also displays some interesting analogies between nervous and immune mechanisms. How a visual or other sensory stimulus is perceived depends very much on its surroundings; for example, a new fact is much more readily comprehended and remembered if it can be associated with something already known. In a perhaps logically similar way recognition of an antigen depends on past experience, and at least two lymphocytes must recognize the antigen molecule at the same time ['associative recognition'; 1].

It might pay us to look more closely at neural mechanisms in sensory perception to get ideas which could apply to the immune system. For example, neurones detecting retinal stimulation are connected by excitatory and inhibitory circuits whose purpose seems to be to sharpen the contrast between light and dark areas in the visual field. A dark spot or line can be enhanced because detector neurones stimulated by adjacent brighter areas inhibit the firing of the neurones connected with less strongly stimulated areas of the retina. This lateral inhibition has been known for a long time; it occurs in all sensory modalities. Immunologists might look for analogous 'focussing' or sharpening mechanisms in the generation of specific antibody responses. Perhaps a 'dominant clone' results from inhibition of related clones.

When we descend to the biochemical level we might expect analogies between specific nervous and immune mechanisms to break down. Yet even here the logical similarities are remarkable: both systems transduce external stimuli from one form to another (antigen to idiotype; electromagnetic or pressure stimuli to an electrical impulse to chemical reactions at the synapse); in both cases stimulation of a responding cell depends on the power of a specific agent, antigen or neurotransmitter, to induce changes in the shape of a stereocomplementary receptor in the cell membrane. We might even draw an analogy between Ig classes and lymphokines in the immune system, and the approximately 30 kinds of neurotransmitter in the nervous system. This cruder level of diversity may exist in both cases to allow specific groups of target cells in the body to be selectively activated, e.g. sympathetic and parasympathetic tracts in the nervous system, suppressor or helper T cells in the immune system. The recent definition of neuromodulatory substance (e.g. endogenous opioids) in nervous tissue suggests that we might expect to find relatively non-specific molecules influencing the progress of each immune response. In any case, immunologists are being weaned from their simple single-hit ideas and stimulated by a vision of the immune system, like the nervous system, as a hierarchy of 'shells' of regulation with extensive connections within and between levels.

Adaptation

Memory could be defined as a persisting structural change based on experience. A 'trace' or change in a magnetic tape, on an area of skin (a callus), in the brain or in a lymphocyte population can all result in a specifically altered reaction to the second encounter with an environmental stimulus. The neurophysiological basis of short-term and long-term memory in the brain is still poorly understood and seems to offer immunologists little help with their own problems, but within the field of cognitive psychology [8] there are some interesting concepts that might be relevant to immunology.

Psychologists distinguish non-associative and associative forms of memory. Non-associative memory refers to the storage of events in particular locations in a defined temporal sequence, as on a recording tape. In associative memory events are recorded and stored only once in a network of units whose connections are strengthened by repeated stimulation. Long-term memory in the brain is of this second kind. This is an economical way to store information since it allows direct access to the data, but at a sacrifice of any accurate record of temporal relationships.

A few years ago one would have said that immunological memory was entirely non-associative, and based simply on increased numbers of specific competent cells. Recall would have been seen as initiated by a simple 'scanning' of the recirculating lymphocyte pool by antigens trapped in a lymph node. Now, because of idiotype network ideas, it seems that immune memory may in some respects resemble the associative memory of the brain, with its 10^{11} neurones, each making roughly 10^4 connections with other neurones. As many authors, from *Jerne* onwards, have pointed out, antigenic stimulation may cause a readjustment of the relative strengths of idiotype-related events in the 'net', e.g. may increase or decrease anti-idiotype-bearing cells. Subsequent exposure to the same antigen may elicit an increased response not primarily because of increased numbers of cells specific for the antigen, but because of events within the net. This seems to be analogous to the associative trace of a neurological network. Comparison with psychological phenomena provide some interesting predictions. For example a strong immunization with antigen X early in ontogeny may so change the form of the net that later stimulation with many different antigens might tend to produce recall production of some anti-X antibody.

Turning to tolerance, most immunologists identify two levels of mechanism: (a) direct action of antigen on lymphocytes, as in receptor blockade and deletion of cells; (b) regulatory (suppressor) tolerance. These may have neurological analogues, (a) being 'habituation' or temporary exhaustion, by repeated stimulation, of receptor cells, or of neurones near the periphery in a sensory pathway, and (b) representing more complex events that may weaken earlier-established associations. An example of the latter would be the interference phenomenon in psychology: having learned to associate A and B, the subsequent learning of AC makes recall of either B or C more difficult. Parallels in the immune system may be antigen competition and 'exhaustion' [2]. It would be worth studying the rate of decay of immunological memory under conditions of relatively intense or slight stimulation with irrelevant antigens.

Pathology; Levels of Study

If immune and nervous systems are so similar in the logic of their operation we might expect to see analogous pathological states associated with each. Failure to distinguish self from non-self adequately is a prime source of dysfunction in both systems: autoimmune disease may be compared with some kinds of schizophrena and with the kinds of damaging distortion of self that lead to psychosomatic disease (e.g. hysterical conversion, peptic ulcer). Inadequate reactions against elements of the external world lead to chronic infection (e.g. by metazoan parasites) when the immune system is 'at fault' and to symptoms of deficit like depression in psychiatry. Under the heading of inappropriate, exaggerated reactions to the world we might compare immune allergies with emotional allergies, i.e. phobias! Both systems have their genetic defects, e.g. immune deficiency states and Huntington's chorea (a dominant gene effect causing atrophy of the corpus striatum).

Finally it is intriguing to note the comparable levels at which these two complex regulatory systems have been studied. The recent molecular analysis of antibody V genes in the DNA seems to have no equivalent in research on the nervous system but both disciplines have made impressive advances in biochemistry, for example antibody and receptor structure and definition of Ig classes in immunology, synaptic events and neurotransmitters in neurochemistry. At the cell and tissue levels, we have in immunology work on antibody-forming cells and cell interaction which finds a neurophysiological equivalent in the recording of electrical events in single neurones and the definition of functional nervous pathways. Work on the overall anatomy of the brain might be compared with studies on lymphoid architecture and lymphocyte recirculation. Psychology, the study of behaviour in a 'black-box' fashion, has similarities with early studies on antibody formation where what was varied was the antigenic input, the measured result being antibody response. However, it would be a mistake to assume the obsolescence of either psychology or of studying the immune system as a whole instead of in fragments. Just as one would not predict consciousness by experiments on individual neurones, so we must expect the lymphocyte net to have properties that will only emerge by examining the immunocyte population as a integral unit. The challenging problem of immune regulation demands a holistic study, and might be illuminated by cross-comparisons with cognitive psychology and other complex events to which a systems approach has also been applied.

Acknowledgements. Drs. *Bob Adamak* (neurophysiologist), *Michel Silberfeld* (psychiatrist) and *Constantine Poulos* (psychologist) made many helpful suggestions. The author's research is supported by MRC and NCI of Canada.

References

1 Bretscher, P.; Cohn, M.: Science *169:* 1042 (1970).
2 Byers, V.S.; Sercarz, E.E.: J. exp. Med. *127:* 307 (1968).
3 Cohn, M.: in Plescia, Brown, Nucleic acids in immunology, p. 671 (Springer, New York 1968).
4 Cowan, W.M.: Sci. Am. *241:* 113 (1979).
5 Dekker, E.; Pelser, H.E.; Groen, J.J.: J. psychosom. Res. *2:* 97 (1957).
6 Jerne, N.K.: Annls Immunol. *125C:* 373 (1974).
7 Jerne, N.K.: in Miller, Stability and origin of biological information, p. 201 (Wiley & Sons, New York 1975).
8 Wickelgren, W.A.: Cognitive psychology (Prentice-Hall, Englewood Cliffs 1979).
9 Winnicott, D.W.: Int. J. Child Psychother. *1:* 7 (1972).

Dr. A.J. Cunningham, Ontario Cancer Institute, 500 Sherbourne Street,
Toronto M4X 1K9 (Canada)

The Immune System, vol. 1, pp. 51–56 (Karger, Basel 1981)

Antibodies and Learning: a New Dimension

Nicholas Cohen[1], *Robert Ader*

Departments of Microbiology and Psychiatry, University of Rochester School of Medicine and Dentistry, Rochester, N.Y., USA

In 1967, *Jerne* [13] published a paper entitled *Antibodies and Learning: Selection versus Instruction.* Although the emphasis of this manuscript was decidedly on those words to the right of the title's colon, *Jerne* was well aware that heuristically valuable analogies between components of each physiological system could and would be made [9, 12, 14).

We have purposefully purloined the format and some of the words of the aforementioned title for our own. The emphasis of our contribution, however, will be on the words to the left of the colon – antibodies and learning. We will present one example [reviewed in 6, 7] that supports the most intriguing and increasingly well-documented concept that the immune system and the central nervous system are functionally and dynamically interwoven in the fabric of homeostasis [3, 7]. This intimate interrelationship can be considered independently of any structural and functional similarities of the two systems.

The example we have selected deals with the application of the principles and techniques of behavioral conditioning (learning) to animals that are challenged with antigen. Although this is but one of several converging avenues that point to the involvement of CNS processes in modulating immune responses, conditioning may be the oldest experimental approach to studying this interaction since it was initiated more than 50 years ago by Soviet scientists [17, 18]. Even though these early studies were poorly designed, the phenomena they revealed could be independently replicated under a variety of circumstances, with a variety of antigenic stimuli, and in

[1] *Nicholas Cohen* was a member of the Basel Institute for Immunology in 1975–1976.

a variety of species [16]. A recent review of this early literature [4] provides a fascinating historical perspective for the studies that we will now review.

Our own investigation of conditioned immunosuppression stemmed from observations made in the course of experiments on taste aversion learning [1]. In the jargon of the behavioral psychologist, taste aversion learning is a one-trial passive avoidance conditioning paradigm in which an animal consumes a novel drinking solution and, at the same time, is injected with a noxious agent that elicits, for example, gastrointestinal disturbances. When the animal is subsequently presented with just the novel drinking solution, it avoids it. In the particular protocol that was being investigated, the consumption of varying amounts of a novel sodium saccharin drinking solution was paired with a constant dose of intraperitoneally injected cyclophosphamide (CY). As expected, animals under this condition developed an aversion to the saccharin solution. During the course of extinction trials (which involved repeated reexposure to saccharin in the absence of CY), some of the rats in this experiment died. It was subsequently noted that the rate of mortality was related to the volume of saccharin consumed on the single conditioning trial. In an attempt to explain this unexpected observation, it was hypothesized that the pairing of a novel, distinctively flavored drinking solution with an immunosuppressive drug could result in a conditioned immunopharmacologic effect, i.e. a conditioning of immunosuppression.

In our initial examination of this hypothesis [5], rats were conditioned by pairing saccharin consumption with an injection of CY. The animals were subsequently injected with sheep red blood cells (SRBC). Conditioned animals that were reexposed to saccharin at the time of antigen presentation and/or 3 days later were found to have titers of hemagglutinating antibody 6 days after immunization that were lower than controls. Controls included: conditioned animals that were not reexposed to the saccharin; nonconditioned animals (treated with CY after drinking plain water) who were subsequently provided with saccharin, and a placebo group. An attenuated antibody response in conditioned animals that are reexposed to the conditioned stimulus has been independently confirmed in other laboratories [15, 19, 20]. This phenomenon, then, is consistent with our hypothesis of a conditioned immunosuppressive response.

The attenuation of antibody titer is relatively small and it is transient. When we use the same dose of CY (50 mg/kg) and the same antigen, hemagglutinating antibody titers are low 4 days after stimulation and there are no group differences. At 6 days, we again observe an attenuated antibody response, but the asymptotic levels attained by 8 or 10 days postimmunization do not differentiate between experimental and control groups. When we increase the dose of CY to 75 mg/kg, we change the kinetics of the antibody response; titers are still low 6 days after antigen administration, but there is an attenuated response in experimental animals 8 days after antigen. It would appear that the transient delay that occurs in the complex chain of events that culminate in the production of antibody is a function of some interaction among the parameters of conditioning, the immunopharmacologic agents, and the diversity of immune responses to the antigen in question.

Although *Wayner* et al. [20] were able to repeat our findings with SRBC, they were unable to document a conditioned suppression of the antibody response to *Brucella abortus*. In two studies in mice, however, we were able to observe an attenuated antibody response to another T cell-independent antigen, TNP-LPS [11]. The generalizability of the phenomenon is further extended by our observation of conditioned suppression of a cell-mediated immune response [10]. In this instance, the experimental protocol was modified to take advantage of the observation that multiple, low-dose injections of CY could suppress a graft-versus-host (GvH) response [21].

Rats were conditioned by pairing saccharin consumption with an injection of 50 mg/kg CY *48 days* before immunogenic stimulation. Lewis × Brown Norwegian hybrid recipients were injected subdurally into the plantar surface of a hind footpad with a suspension of splenic leukocytes obtained from Lewis donors. On the same day, conditioned recipients were reexposed to saccharin and injected with saline. On the following day, they were again reexposed to saccharin and injected with 10 mg/kg CY, and on the next day they were reexposed to saccharin for the third time and injected with saline. Three low-dose injections of CY markedly reduced the weight of popliteal lymph nodes harvested 5 days after injection of the cellular graft. A single, low dose of CY, however, resulted in only a modest attenuation of the GvH response in nonconditioned animals (exposed to saccharin) and in conditioned animals that were not reexposed to the conditioned stimulus. In contrast, conditioned rats that received a single, low-dose injection of CY *and* reexposure to the conditioned stimulus displayed a significant suppression of the local GvH response relative to the control groups – *and they did not differ from animals that were injected with CY on 3 successive days.*

We had speculated [5] that the attenuated antibody response seen in conditioned animals might be mediated by an elevation in adrenocortical steroid level. It was subsequently demonstrated that the conditioning of a taste aversion could result in the concomitant conditioning of an adrenocortical response [2], but it does not now appear that this transient increase in the level of circulating steroid can account for the observed attenuation in immune responses. LiCl, like CY, for example, is an effective stimulus for inducing a taste aversion and for conditioning an adrenocortical response, but it is not an immunosuppressive drug and does not result in an impaired immune response when it is substituted for CY in our paradigm [5].

We then reasoned that an elevation in circulating steroids, superimposed upon the residual immunosuppressive effects of CY, might be responsible for the phenomenon of conditioned immunosuppression. To test this possibility, rats were conditioned with CY and an independent group of conditioned animals was injected with LiCl or with corticosterone instead of being reexposed to the conditioned stimulus. In neither instance, however, did the exogenous elevation of steroid levels reduce antibody titer relative to controls [8]. Finally, in several of our studies, taste aversion behavior was assessed using a preference testing procedure in which animals are presented with plain water in addition to the distinctively flavored solution previously paired with CY. In contrast to providing the flavored solution alone, this procedure (which also equates the total fluid consumption of experimental and control animals) does not result in differences in steroid level. One can, nevertheless, observe conditioned suppression of T cell-dependent and T cell-independent antibody responses and a GvH response given the parameters of behavioral and immunologic stimulation used in our studies. These data, then, provide no support for the hypothesis that adrenocortical responses mediate the observations of conditioned immunosuppression.

To date, conditioned immunosuppression has been a reliable and independently reproducible phenomenon. Using rats or mice, injecting T cell-dependent or T cell-independent antigens, and evaluating cellular as well as humoral immunity, an attenuation of immunologic reactivity has been invariably observed in conditioned animals reexposed to a conditioned stimulus that had previously been paired with an immunosuppressive drug. In one or another of our studies [6], we have varied the taste stimulus and immunosuppressive drug, varied the dose of drug, the number of conditioning trials, and the number of reexposures to the conditioned stimulus, eliminated cues that might be associated with drug treatment in control groups, reduced the residual suppressive effects of CY by lengthening the interval between conditioning and antigenic stimulation, equated fluid consumption in experimental and control groups, and titrated serum antibody at different intervals after antigenic stimulation. Although the basic phenomenon occurs under a variety of experimental circumstances, these refinements in methodology have not increased the magnitude of the differences between conditioned animals reexposed to the conditioned stimulus and the several control groups; they remain consistent, but small.

Considering the multiple effects of immunopharmacologic agents and the complexity of the immune system, there are several testable hypotheses as to why the effects of a variety of environmental or host factors might be relatively small. The addition of a behavioral or experiental dimension to the analysis of immunologic reactivity further multiplies the complexity of the adaptive processes involved. Within the framework of conditioning, for example, antigenic stimuli are unconditioned stimuli for the elicitation of immune responses. In our experimental paradigm, however, we are eventually pairing a conditioned stimulus for suppression of immunologic reactivity with an unconditioned stimulus for the elicitation of the same response. Our data indicate that the coincidental presentation of these conflicting signals results in a transient delay in the defensive response, reflecting, we presume, the effects of the less prepotent conditioned stimulus. This descriptive analysis suggests, further, that a *potentiation* of immunologic reactivity might more dramatically demonstrate the impact of conditioning in modulating an immune response.

The use of conditioning techniques to study immunologic reactivity can be viewed as a major extension of one of the more significant frontiers in the application of principles derived from the behavioral sciences to issues of health and disease, namely the conditioning of automatic, visceral, and pharmacologic responses. Methodologically, the application of conditioning procedures provides a means for studying the relationship between CNS and immune processes in the intact organism. Clinically, an elaboration of this phenomenon could lead to new regimens of immunopharmacotherapy. Conceptually, the capacity of experiential events, including conditioning, to suppress or enhance immunologic reactivity raises new issues with respect to the normal functioning and modifiability of the immune system and the mediation of susceptibility and responses to disease.

Although psychoimmunologic research is just beginning, the results obtained thus far reinforce the notion that the immune system is integrated with other physiologic systems to form an integrated network of defensive mechanisms that is sensitive to psychosocial events and subject to regulation and modulation by the central nervous system.

Acknowledgement. This work was supported by USPHS Research Grant NS 15071 and USPHS Research Scientist Award to RA, K05 MH-06318.

References

1 Ader, R.: Psychosom. Med. *36:* 183–184 (1974).
2 Ader, R.: J. comp. physiol. Psychol. *90:* 1156–1163 (1976).
3 Ader, R. (ed.): Psychoneuroimmunology (Academic Press, New York, in press, 1981).
4 Ader, R.: in Ader, Psychoneuroimmunology (Academic Press, New York, in press, 1981).
5 Ader, R.; Cohen, N.: Psychosom. Med. *37:* 333–340 (1975).
6 Ader, R.; Cohen, N.: in Ader, Psychoneuroimmunology (Academic Press, New York, in press, 1981).
7 Ader, R.; Cohen, N.: in Gentry, Handbook of behavioral medicine (Guillford Press, New York, in press, 1981).
8 Ader, R.; Cohen, N.; Grota, L.J.: Int. J. Immunopharmacol. *1:* 141–145 (1979).
9 Adinolfi, M.: Devl Med. Child Neur. *20:* 509–515 (1978).
10 Bovberg, D.H.; Cohen, N.; Ader, R.: Psychosom. Med. *42:* 73 (1980).
11 Cohen, N.; Ader, R.; Green, N.; Bovbjerg, D.H.: Psychosom. Med. *41:* 487–491 (1979).
12 Cunningham, A.J.: in Ader, Psychoneuroimmunology (Academic Press, New York, in press, 1981).
13 Jerne, N.K.: in Quarton, Melnechuk, Schmitt, The neurosciences, pp. 200–205 (Rockefeller University Press, New York 1967).
14 Jerne, N.K.: in Miller, Stability and origin of biological information, pp. 201–204 (Halsted Press/Wiley, New York 1975).
15 King, M.: Personal communication (1979).
16 Luk'Ianenko, V.L.: Esp. Sovrm. Biol. *51:* 170–187 (1961).
17 Metal'Nikov, S.; Chorine, V.: Annls Inst. Pasteur, Paris *40:* 893–900 (1926).
18 Metal'Nikov, S.; Chorine, V.: C.r. Séanc. Soc. Biol. *99:* 142–145 (1928).
19 Rogers, M.P.; Reich, P.; Strom, T.B.; Carpenter, C.B.: Psychosom. Med. *38:* 447–454 (1976).
20 Wayner, E.A.; Flannery, G.R.; Singer, G.: Physiol. Behav. *21:* 995–1000 (1978).
21 Whitehouse, M.W.; Levy, L.; Beck, F.J.: Agents Actions *3:* 53–60 (1973).

N. Cohen, PhD, Department of Microbiology, University of Rochester School of Medicine and Dentistry, Rochester, NY 14642 (USA)

The Immune System, vol. 1, pp. 57–61 (Karger, Basel 1981)

Network Concepts in Science and the Arts

P.G.H. Gell[1]

Department of Pathology, University of Cambridge, Cambridge, England

When *Niels Jerne* started talking about immunological networks in 1973 – it is amazing to think that it was only 7 years ago – he was characteristically showing a precocious awareness of a motif which has been cropping up all over the place since, both in the arts and the sciences. There seems to be a vast difference, a far greater sophistication, if we think about causation in terms of a web and not of a chain. All I hope to do in this brief, necessarily dogmatic and perhaps rather off-beat contribution to his Festschrift, is to draw attention to some of these wider implications, both inside and outside the domain of science.

In all of those life sciences concerned with whole organisms, from genetics to economics, there is a widespread though undefined feeling of dissatisfaction with the conventional analytical method normal to them, the method by which all variables are eliminated from a situation to leave only two, which can be subjected to simple mathematics and displayed on a graph. This gives one a sensation of increased knowledge but no one denies that it distorts the 'real' situation in a living organism. The question is, is the distortion so great that our knowledge is as unreal as the situation, a mere artefact? It was thought that one could make a construct of a number of these oversimplified, unreal situations to develop a model which could still represent in a useful way the nature of a whole organism. May it be however that owing to the untruth of our initial oversimplification we can never in fact generate a really useful model of a real situation?

In branches of biological sicence which are intrinsically abstract such as pure biochemistry this need not trouble us: but in the life sciences proper,

[1] *Philip Gell* was a member of the Basel Institute for Immunology in 1972.

physiology, pathology, ethology, sociology and so on, it may be of major importance. In genetics in particular, there seems to be a small battle on between the naïf Mendelians, who talk about gene effects as if they were normally as simple to unravel in the phenotype as the difference between smooth and wrinkled peas, and the biometrists who consider most phenotypic properties to result from the interaction of many genes, and ultimately from the whole genotype – a network in fact. Non-geneticists working in cognate life sciences such as anthropology and ethology tend to stick to the simple Mendelian concepts they learnt at school, with I believe dire consequences. Equally the optimistic assumption that drugs cure diseases, just like that, leaves us wide open to the criticisms of *Ivan Illich* et hoc genus omne; while in economics, surely crude monetarism is a prime example of the failure to make use intelligently of interactionist thinking. At the other pole of knowledge, it is no news to physicists that a full understanding of the interior of the atomic nucleus ought to take account of both the present state and past history of the whole universe.

We need then to think in terms of network rather than of chain causation, and in discussing this it might be useful to exemplify a few useful though not unfamiliar words. If we are describing the evolution of a species, or of a specific protein, this looks to be dendritic, a process of divergence like the branches of a tree, with the abortion of many sprigs. But if we analyse the elements in a process of natural selection the system is reticulate: it entails the feedback effects, within a species, of mutation on the rest of the individual genotypes, of species on species, and of species on environment, as well as the straightforward effect of environment in procuring the evolution of a given species by means of the preservation of random mutations. Broadly, dendritic processes are diachronic, along the time stream, while reticulate processes are also synchronic, across the time stream; one is conditioned by the relevant past, the other also by the relevant present.

In a network system the word relevant is a key one. In principle network systems have no edges, i.e. it is not strictly possible to draw a firm line round the influences upon a phenomenon, saying, inside this line they are relevant, outside they are irrelevant. Yet this act of isolating phenomena, of drawing lines round them, is the accepted process of analytical science. In a network no element can be precisely defined because its connectivity is indefinite; if you separate it off it bleeds to death. Of course we can define precisely a molecule, and perhaps even a cell; but as soon as a dynamic process enters the picture we have to think in terms of unbounded interactions, if we are to remain in the real world. This means abandoning at some stage the analytical method, apart from its use for mere description: when our hypotheses and discussions entail matters of causation (i.e. are meant to be of predictive value) then we may have to accept a certain imprecision. But at least this imprecision will be conscious and recognised, rather than unrealised or ignored as in the use of a simplistic analytical system.

A network system, though this is not perhaps essential to it, is much enriched by postulating inhibitory as well as promotional effects, feedback inhibition in fact. This generates a system of very low probabilities, that is the probability that an effect will reach any particular node in the network is small. For particular specific phenotypic effects these probabilities may summate, e.g. the probability in a given situation that antibody production will continue, a result which may be reached by many routes of stimulation or suppression of cells, some operating synchronically, may be quite high. Civil engineers are no doubt quite familiar with this problem, which they cope with by way of the still largely empirical science of stress analysis; but the best practical engineer is the garden spider.

Two current influences make for interactionist thinking, the computer and the study of the nervous system. A computer program will allow us I suppose to work in many more than two dimensions, that is to handle simultaneously many variables; but these variables have to be specified in the program, that is to say a line has to be drawn round the phenomena somewhere, and although we can no doubt make allowance for a factor of indeterminacy outside this line we are only a little better off from the point of view of arbitrariness than we were with the old Cartesian graph. Especially is this so in systems such as the gene network, where the effects of remoter nodes on certain sorts of resultant – say phenotypic behaviour – is still strong, i.e. the effects do not 'dilute out' with distance, as the effects of remoter nodes on a simple biochemical parameter such as the in vivo activity of cholinesterase would do.

The nervous system is of course the prototype network: our understanding of even the apparently simple phenomenon of visual perception now appears to be controlled by a dispersed network that goes all the way back to infantile memories and needs. We can never get within reach of an understanding of this by sticking to simple cause-and-effect analysis; indeed it is here I believe that the ultimate inadequacy of computer simulation, in spite of the many fascinating emergent properties of computer programs, tends to be thrown into relief. A computer is a useful partial model of, not a substitute for, interactionist thinking.

A loosening up of the thought processes of scientific thinking, in abandoning to some extent rigidly analytical methods, has brought it closer to the thought process of artistic creation. A network system also appears in making the choices when constructing a picture or a poem. A work of art has to be coherent, that is to say it has to satisfy some sort of probability function. This probability is not in the sense of something rational, but of something determined by its *relata,* a probable consequence of its environment. Thus, the pink patch in a painting by *Gauguin* is the resultant of a choice conditioned by an unbounded network of influences – the painter's reaction to the model, to colour in general, to his own other paintings, his personal history and socio-historical position, the history of art in Europe including perhaps a half-forgotten picture by *Rubens,* and so on. Every single choice in the picture is the outcome of a process as complicated as this. The more the artist tries to evade such choices, denying any commitment to the work, the more the picture becomes boring, as in some of the later works of *Jackson Pollock* for example.

In a poem it is the merest banality to say that it is the connotations not the denotations of words which are important. It is perhaps not quite so obvious to suggest that the connotative network used by the writer in constructing the poem (whether consciously or not) must coincide at a number of points with the connotative network aroused in the reader on reading it; it is in this that the whole process of communication in the arts rests. There may, or must, be a certain impoverishment, since some nodes in the network are inevitably 'private' to the writer: on the other hand the reader may reach many of these nodes by a route different from that of the writer but still perfectly valid. Too simple a coincidence – too banal and obvious a poem – may fail to stir the emotions by inability to activate these distant nodes. It has often seemed to me a puzzle that something so private as a poem can ever be understood by anybody except its creator. Perhaps not just the poet's recorded experience counts, our own is as, or more, important. These points, with similar applications to the other arts, are obvious enough; they are stressed only to show that one can talk about art as an interactionist process in just the same terms as we can talk about science.

Do not get me wrong: I certainly do not wish to imply that we should, or can, altogether abandon analytical thinking, which is part of the essential process of understanding the world in intellectual ontogeny. Putting it very crudely, and following more or less in the footsteps of *Piaget,* one can say that the infant's first step in visual perception is the holistic reception of a meaningless visual field. Experience of movement, distance and so on defines objects (nodes) in this field, a face, a woolly bear toy. This is to some extent an analytical process, but the field, the total connectivity of the newly defined objects, remains intact. This is the stage, encapsulated, of a work of art; there are objects but they are embedded and unique. When the object is detached from its context, when the woolly bear of one field is recognised as the same woolly bear as that of another, that is truly analytical, scientific, an anti-art discovery.

If however analytical perception is not mandatory for science, and holistic perception not unique to art, if network processes can be incorporated in science, just as analytical processes are incorporated in say an engineer's drawings, though usually excluded from art, if we can accept that science is not excluded from the connotative nexus in spite of referring to the real world rather than to the world of the imagination, then the chasm in our culture caused by the division and opposition between science and art may be in process of filling in; not by art drawing closer to science as the futurists used to hope, but by science drawing closer to art.

Prof. P.G.H. Gell, Department of Pathology, University of Cambridge,
Tennis Court Road, Cambridge CB2 1QP (England)

The Immune System, vol. 1, pp. 62–68 (Karger, Basel 1981)

Le Têtard et l'Anticorps

Louis Du Pasquier [1]

Basel Institute for Immunology, Basel, Switzerland

Amphibians have long been considered among the best models for experiments in embryology. It was logical that with the emergence of immunology they became an interesting system for the study of the ontogeny and phylogeny of immunity [17]. This paper describes how the amphibian system was developed and used at the Basel Institute for Immunology during the past 10 years. The text is divided in two sections under headlines borrowed from *Niels K. Jerne's* questions or comments that, at the appropriate time, gave impetus to research done by my colleagues and me.

'How many lymphocytes are there in a tadpole?'
[N.K. Jerne, *Nov. 1967*]

On the basis of the clonal selection theory [4], which is predicated on the commitment of the lymphocyte to make a single species of antibody, the following prediction is possible: at any moment in time, the number of antibody specificities that an individual can make is not greater than the number of lymphocytes that it has. This hypothesis can be tested by studying antibody responses in animals having a small, possibly limiting number of lymphocytes. Would the antibody repertoire of a small animal containing, for instance, 1×10^6 lymphocytes be smaller than that of a large individual with 1×10^{12} lymphocytes? At various stages of their develop-

[1] *Louis Du Pasquier* has been a member of the Basel Institute for Immunology since its founding.

ment tadpoles of the midwife toad *(Alytes obstetricans)* were injected with sheep red cells, and the immune response was measured by the hemolytic plaque assay, by immunocytoadherence, and by serum antibody titers. It turned out that 2- to 3-week-old tadpoles, having as few as one million lymphocytes (only about 10^4 in the spleen), were capable of mounting what seemed to be a specific immune response. The specificity was demonstrated by the apparent lack of cross-reactivity between sheep and human red cells [6]. These experiments suggested that amphibian tadpoles did not suffer a major gap in their antibody repertoire.

Such findings were interpreted by some people (including myself) as suggesting that lymphocytes were pluripotential. Either they could express several antibodies at the same time, or they could rapidly change potentiality from one generation to the other. The same findings were interpreted by others (including *Jerne*) as a suggestion that tadpole lymphocytes were also unipotential, but that the antibodies made by the lower vertebrates were of such a low affinity that they would cross-react with many antigens. Therefore, *Haimovich* and I compared the affinities and specificities of mammalian and tadpole antibodies. By inhibiting the antibody-mediated inactivation of dinitrophenylated (DNP) bacteriophage with various monovalent cross-reacting ligands, we measured the affinity of antibodies from small bullfrog tadpoles weighing 0.5–1 g and containing about 1×10^6 lymphocytes. Such tadpoles made specific anti-DNP antibodies and anti-trinitrophenyl (TNP) antibodies when immunized with DNP and TNP conjugates, respectively. Moreover, the affinities of their IgM anti-DNP antibodies for the immunogen were similar to those of isolated goat anti-DNP antibodies, ruling out the possibility that tadpole antibodies were generally of low affinity [10, 15].

At that time nothing was known about the heterogeneity of the antibody populations in tadpoles and frogs. Was it possible that the repertoire of the frog would be proportional to the small number of lymphocytes of its immune system without invoking pluripotentiality? The heterogeneity of larval and adult individual *Xenopus* and *Rana* was studied in collaboration with *Wabl*, who adapted to amphibians the technique of isoelectrofocusing (IEF) [24]. Both larvae and adults had a very restricted anti-DNP antibody repertoire when compared to rabbits or mice. Not more than 10–15 anti-DNP-antibody IEF spectrotypes corresponding to probably fewer clonal products could be detected in *Xenopus laevis*. This homogeneity of the response was also found within the species, using the isogenic animals made available by *Hans Rudolph Kobel* [18]. Unlike inbred mice or rabbits that can make up to 500 spectrotypes of antibodies toward DNP (with hardly any common spectrotype from animal to animal), isogenic *Xenopus* showed a high degree of spectrotypes and idiotype sharing – many animals gave identical antibody patterns [14]. Thus, somatic mutations seem to be less frequent (and perhaps unimportant) in lower species. So the frog system can be used to estimate the number of germ line variable region genes.

The repertoire being smaller, it was no longer necessary to postulate pluripotentiality of lymphocytes. The estimate of 5×10^4 antibody variable regions made from tadpole experiments [15] in the hypothesis of the uni-commitment of the lymphocytes was thus likely to be true. It was somehow a surprise that such a small repertoire actually covers all the antigen specificities tested with antibodies of high affinity.

The IEF analyses of tadpole serum also revealed an interesting difference between larvae and adults. Individuals of a clone of *Xenopus* differing only in age were immunized with DNP, and their IEF antibody patterns were compared. Adults, as a group, gave very similar responses, as did larvae as a group, but the larval antibody pattern differed markedly from that of adults [12]. However, if the animals were immunized as tadpoles and boosted as young adults, several larval patterns could be expressed. This observation is interesting in the context of the interaction of adult antigens with the larval immune system and their possible effect on the expression of the B cell repertoire. Sometimes an immune response against adult specific antigens could be elicited in larvae [20, 22]. We confirmed such observations by demonstrating a larval anti-adult mixed leukocyte reaction (MLR) within clones of isogenic animals [12, 19]. How does the tadpole metamorphose harmoniously without dangerous autoimmune phenomena? We knew from earlier experiments [3, 5, 23] that metamorphosis might be a period when immune capacities are modified. In collaboration with *Chardonnens* we investigated the capacity of *Xenopus* to reject skin grafts at various stages of its ontogeny. Grafts exchanged between animals differing at minor histocompatibility loci are rejected slowly by adults, but when performed in metamorphosing animals they were not rejected at all and survived indefinitely. Such a stage of immune deficiency, paralleled by a partial suppression of MLR during metamorphosis [8], led *Du Pasquier and Bernard* [7] to test the hypothesis of the possible existence of active suppression of the immune system during metamorphosis. Lymphocytes from the thymus of metamorphosing isogenic animals were transferred (2×10^6 per injection, twice a week) into 1-year-old adults bearing a graft differing from the host by minor histocompatibility antigens only. As long as the injections

Fig. 1. Lake Mutanda (southwest Uganda). This lake is situated in a region where a new tetraploid species of *Xenopus (X. wittei)* was discovered.

were maintained, the grafts were not rejected or were rejected very slowly, demonstrating an active suppressor role of the injected cells. As soon as the injection of cells from metamorphosing animals was stopped, the grafts started to be rejected.

'Was Darwin right?' [N.K. Jerne, *summer 1974*]

'You know, Louis, you should try to answer more basic questions such as "was Darwin right?"'. This critical comment of *Jerne* came just at a time when I was debating whether or not to orient the laboratory towards genetics and evolution. This was shortly after successful attempts [9] to find in *Xenopus* the homolog of a major histocompatibility complex (MHC), which has been further characterized in recent years [1, 2]. A year later, with *Jerne's* benediction (and Institute funds), I joined an expedition to Central Africa (fig. 1) to collect specimens of *Xenopus*. From talks and collaboration with my colleagues *Kobel, Fischberg,* and *Tinsley,* I became more familiar with *Xenopus* systematics and with the possibilities of this system.

Several species of *Xenopus* are apparently related by natural hybridization and polyploidization. The diploid number of chromosomes varies: 20 *(X. tropicalis),* 36 *(X. laevis),* 72 *(X. vestitus),* 108 *(X. ruwenzoriensis).* This can be exploited to study the effect of gene duplication in evolution [21]. Furthermore, some *Xenopus* interspecies hybrids lay fertile diploid eggs which, when fertilized by normal sperm, develop into laboratory-made polyploids (in this case a triploid) [13]. Thus, it is possible to compare the expression of various genes in natural polyploids with laboratory-made polyploids that natural selection has not had time to modify. In this system we have studied the expression of MHC and of antibody genes. The species with 20, 36, and 72 chromosomes all express a single MLR locus, a marker of the *Xenopus* MHC. In the species with 108 chromosomes *(X. ruwenzoriensis)* polysomic inheritance of MLR loci is maintained. More than one locus (probably 3) segregated in families of this species. However, one of the loci gave rise to stronger stimulation indices than the other, suggesting that the loci were not all equivalent and that the trend toward the expression of only one MLR locus can still be detected in this hexaploid species. In contrast to the natural polyploids, the laboratory-made polyploids always expressed all the MLR and other histocompatibility loci of their constituting haplotypes. There seems to be no mechanism automatically leading to the expression of only one MHC locus as in the case of the X chromosome in mammals.

Is this trend toward functional diploidization a generalized phenomenon? *Blomberg* and I investigated this question further by studying the expression of antibody diversity. The species with the lowest number of chromosomes *(X. tropicalis)* also had the lowest number (about 6–10) of low molecular weight anti-DNP antibody IEF spectrotypes. The species with 36 and 72 chromosomes both expressed more antibody spectrotypes than did *X. tropicalis* (i.e. about 20 and 30, respectively), but the species with 72 chromosomes did not express many more IEF spectrotypes than the 36 chromosome species. *X. ruwenzoriensis* (108 chromosomes), expressed up to 40 anti-DNP spectrotypes. In summary, for antibody genes there seems to be maintenance of polysomic inheritance in the polyploid species, as is the case for albumins. This differs markedly from the observations on the MHC. Using laboratory-made polyploids we showed that mere autopolyploidy by itself does not increase antibody diversity. Laboratory-made hybrids between *X. ruwenzoriensis* and *X. laevis* expressed an anti-DNP pattern intermediate between the ones observed in each species. These animals can increase their antibody diversity through hybridization and retain the genes that may be useful while at the same time functionally diploidizing other genes such as those of the MHC.

We have also exploited the relationships among *Xenopus* species to obtain aneuploid individuals that can be used for linkage analysis and chromosome mapping [11]. As stated previously [18], female hybrids between *X. laevis* and *X. gilli* (LG) lay diploid eggs. Such eggs can be fertilized with L sperm and give rise to triploid LLG females. Their large eggs are diploid, and their small eggs are aneuploid and contain a euploid haploid set of 18 L chromosomes plus 0–18 G chromosomes. After fertilization by L sperm these eggs give rise to aneuploid hyperdiploid *Xenopus* that are diploid for L and contain a random number of G chromosomes. With a panel of different aneuploid individuals exhibiting different karyotypes, it should be possible to assign genes to chromosomes.

Some preliminary experiments have been done using this system with the following results: MHC and immunoglobulin heavy chain genes are not linked. The Ig heavy chain genes are on one of the 6 acrocentric chromosomes. Chromosome 9 is involved in the expression of the anti-DNP response. This method of gene mapping is complementary to somatic cell hybridization, since in the latter case the chromosomes that are analyzed come from differentiated cells. The recent production of functional mouse-*Xenopus* hybrids [16] makes a comparison possible.

Conclusion

The *Xenopus* system has been successfully developed in Basel during these 10 years thanks to a happy concatenation of circumstances, among which is my great fortune in having the assistance of *Chantal Guiet.* This development was mainly due to the constant and discrete support of *Niels K. Jerne.* Actually, the most important factor has been his trust, and when I contemplate what has happened, I have mixed feelings. On the one hand, I am happy but feel a little bit guilty to have enjoyed such unbelievable freedom in *Niels Jerne's* Institute and to have had so much fun with my colleagues over the past 10 years. On the other hand, I am proud that the trust was deserved because the *Xenopus* system has given information of general interest in several areas of modern biology such as immunology, genetics, embryology, and zoology. I would like to stick to the last point, but I am not sure I can do it with one hundred percent sincerity!

References

1 Bernard, C.C.A.; Bordmann, G.; Blomberg, B.; Du Pasquier, L.: Immunogenetics *9:* 443–454 (1979).
2 Bernard, C.C.A.; Bordmann, G.; Blomberg, B.; Du Pasquier, L.: Eur. J. Immunol. (in press, 1981).
3 Bernardini, N.; Chardonnens, X.; Simon, D.: C.r. hebd. Séanc. Acad. Sci., Paris *269:* 1011–1014 (1969).

4 Burnet, F.M.: The clonal selection theory of acquired immunity, p. 209 (Vanderbilt University Press, Nashville 1959).
5 Chardonnens, X.; Du Pasquier, L.: Eur. J. Immunol. *3:* 569–573 (1973).
6 Du Pasquier, L.: Immunology *19:* 353–362 (1970).
7 Du Pasquier, L.; Bernard, C.C.A.: Differentiation *16:* 1–7 (1980).
8 Du Pasquier, L.; Chardonnens, X.: Immunogenetics *2:* 431–440 (1975).
9 Du Pasquier, L.; Chardonnens, X.; Miggiano, V.C.: Immunogenetics *1:* 482–494 (1975).
10 Du Pasquier, L.; Haimovich, J.: Immunogenetics *3:* 381–391 (1976).
11 Du Pasquier, L.; Kobel, H.R.: Immunogenetics *8:* 299–310 (1979).
12 Du Pasquier, L.; Blomberg, B.; Bernard, C.C.A.: Eur. J. Immunol. *9:* 900–906 (1979).
13 Du Pasquier, L.; Miggiano, V.C.; Kobel, H.R.; Fischberg, M.: Immunogenetics *5:* 120–141 (1977).
14 Du Pasquier, L.; Wabl, M.R.: Eur. J. Immunol. *8:* 428–433 (1978).
15 Haimovich, J.; Du Pasquier, L.: Proc. natn. Acad. Sci. USA *70:* 1898–1902 (1973).
16 Hengartner, H.; Du Pasquier, L.: Science (in press, 1981).
17 Hildemann, W.H.; Haas, R.: J. Immun. *83:* 478–485 (1959).
18 Kobel, H.R.; Du Pasquier, L.: Immunogenetics *2:* 87–91 (1975).
19 Kobel, H.R.; Du Pasquier, L.: in Solomon, Horton, Developmental immunobiology, pp. 299–306 (Elsevier/North-Holland, Amsterdam 1977).
20 Maniatis, G.M.; Steiner, L.A.; Ingram, V.M.: Science *165:* 67–69 (1969).
21 Ohno, S.: Evolution by gene duplication (Springer, New York 1970).
22 Triplett, E.L.: J. Immun. *89:* 505–510 (1962).
23 Volpe, E.P.: Am. Zool. *11:* 207–218 (1971).
24 Wabl, M.R.; Du Pasquier, L.: Nature, Lond. *264:* 642–644 (1976).

Dr. L. Du Pasquier, Basel Institute for Immunology, Postfach, CH–4005 Basel (Switzerland)

The Immune System, vol. 1, pp. 69–75 (Karger, Basel 1981)

Immuno-Ornithological Conversation

J.R.L. Pink[1], *C. Koch*[1], *A. Ziegler*[1]

Department of Pathology, University of Geneva, Geneva, Switzerland; Institute for Experimental Immunology, University of Copenhagen, Copenhagen, Denmark; Immunology Laboratory, Medical University Clinic, Tübingen, FRG

All gourmets know that it is useless to crack open chicken bones in the hope of finding much marrow. The advantage to a flying chicken of having hollow bones is accompanied by the advantage, to immunologists, of having the bird's B cell-producing tissues concentrated in the bursa. The idea that the bursa is an evolutionary adaptation to the needs of flying bodies led to the following conversation between two immunologically minded ornithologists (*Imm* and *Orn*).

Imm: Having a bursa may be an aerodynamic advantage, but are chickens as good antibody producers as mammals? Consider *Jerne's* [13] notion of antibody repertoire. The repertoire of frogs to a hapten such as DNP is very limited [5], whereas that of inbred mice to the same hapten is enormous [21]. Maybe other vertebrate phyla don't have as big an antibody repertoire as mammals.

Orn: One can estimate antibody repertoire by measuring antigen-binding B lymphocyte frequencies, which are not greatly different in birds and mammals [8, 10, 15, 18]; or by using idiotypic analysis or isoelectric focusing (IEF) to count different antibodies. For example, if one immunizes chickens with low doses of streptococcal group A vaccine, they respond by making antibodies which can be distinguished by IEF. The repertoire of

[1] *Richard Pink* was a member of the Basel Institute for Immunology from 1971 to 1977, *Claus Koch* was a member of the Institute in 1978 and 1979, and *Andreas Ziegler* was a student at the Institute from 1973 to 1976.

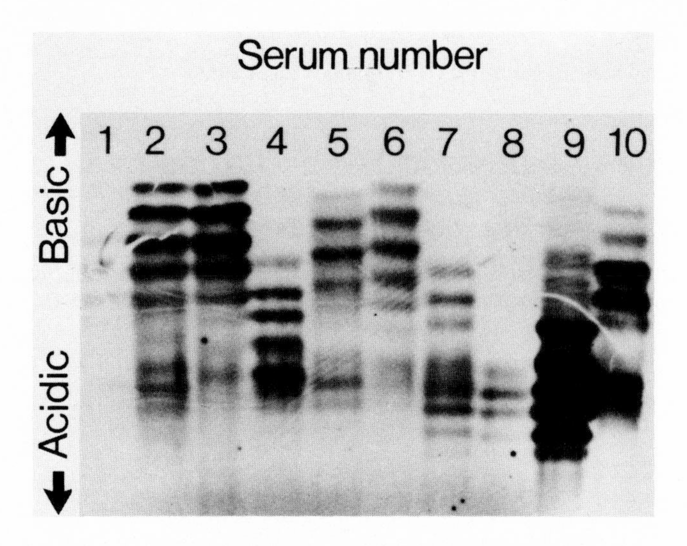

Fig. 1. Isoelectric focusing patterns of chicken anti-streptococcal group A carbohydrate antibodies. White Leghorn chickens were injected intramuscularly with 0.25 ml of J17A4 group A vaccine (5 μg rhamnose/ml) [3]. After 4 weeks, the chickens were bled and the antibodies in 20 μl serum focused in pH 5–9 gels [21] containing 1*M* urea. Antibodies were detected by overlay of the gels with ^{131}I-group A carbohydrate [3]. On the basis that the 10 sera analysed contained 14 well-defined clone products, of which at most two pairs (basic bands in sera 2 and 3, and perhaps also acidic bands in sera 7 and 9) had indistinguishable focusing patterns, the chickens' anti-group A repertoire was estimated [21] as about 50. This is similar in size to the anti-group A repertoire of BALB/c mice [3] and much larger than that of A/J mice [6].

these birds to the group A antigen is at least as large as the anti-streptococcal repertoire of mice (fig. 1).

Imm: Is anything known about the repertoire of chickens to other antigens?

Orn: Hardly anything. However, *Hála and Hašek* [7] tested patterns of reactivity of chicken anti-histocompatibility antigen (anti-B) sera, and were surprised to find that no two sera from 24 individual inbred birds (10 producing anti-B2 and 14 anti-B9) gave identical reactivity patterns when tested on a panel of cells from outbred birds. So the repertoire of anti-B antibodies is probably also large (even if one takes the multiplicity of the B antigens themselves [22] into account).

Imm: I would't expect that, if antibody germ line V genes are small in number and code for the V regions of anti-B antibodies [12].

Orn: Monoclonal antibodies against rat histocompatibility antigens also show a very wide repertoire of reactivity patterns [24], so I doubt that *Jerne's* original idea of a small germ line anti-B V gene repertoire is correct.[2]

Imm: If, as *Jerne* [12] and *Cohn* et al. [2] suggested, V gene mutation rates were normal (10^{-5}/V gene/cell division), and many mutants were being produced from a few genes, there would have to be a strong selection against non-mutated products. Is there any evidence of extensive cell death in growing bursae?

Orn: This is difficult to exclude on morphological grounds, and no detailed study of cellular kinetics in the bursa has ever been made. However, one can see from figure 2 that bursal growth is very rapid before hatching. At this time, bursal cells are probably dividing about once every 10 h, which corresponds closely to the cell population doubling time [18], leaving little room for strong selection. On the other hand, selection could take place earlier in the maturation of IgM-bearing cells, when not all IgM-negative cells may differentiate into permanent IgM-positive clones; or after hatching, when the bursal cell number increases only slowly (many cells being exported to peripheral lymphoid organs) although cell division remains very rapid.

But there exists other evidence against the existence of strong selective pressures on the growth of mutant antigen-binding cells which might arise after hatching. *Tufveson and Alm* [27], *Lydyard* et al. [18], and *Moriya and Ichikawa* [20] determined the numbers of antigen-binding cells in bursae and spleens of embryonic and newly hatched chickens: for example, the proportions of cells from these organs forming rosettes with sheep erythro-

[2] One must still explain the observations of *Longenecker and Mosmann* [16] that natural antibodies to chicken B antigens are present, in amounts of perhaps 1 μg/ml, in sera from mice, rats, humans, alligators and allogeneic chickens. *Longenecker and Mosmann* suggest three possible explanations: that the hypothesis of *Jerne* [12] is correct; that bacteria or viruses either can carry histocompatibility antigens from one host to another, or themselves can cross-react with these antigens (unlikely, as the natural antibodies were also present in germfree rats); or that the antibodies are autoimmune in origin, and directed against altered (for example, denatured) self histocompatibility antigens. We prefer this last explanation, since *Ivanyi* et al. [9] have shown that one can induce anti-H-2 antibodies in mice by syngeneic immunization. In addition, the generality of the Longenecker-Mosmann phenemenon is in doubt, as chicken natural anti-B antibodies are essentially all directed against the unusual B-G antigen [*Koch and Simonsen,* unpublished], which differs both structurally and in its tissue distribution from classical major histocompatibility antigens [22].

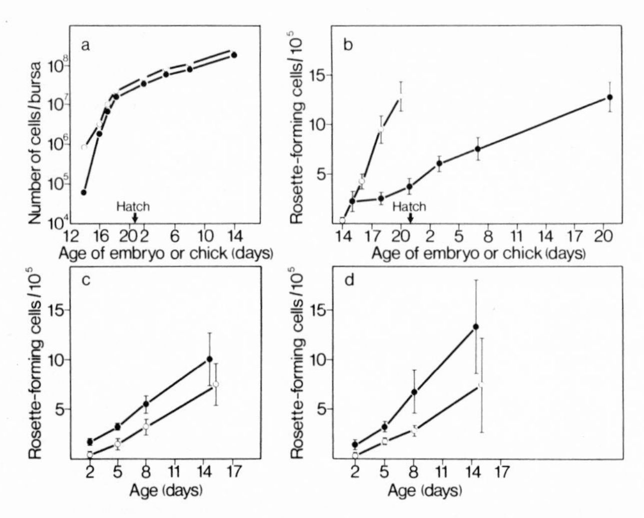

Fig. 2. Ontogeny of IgM-positive cells, and of cells forming rosettes with sheep erythrocytes, in chicken bursa. *a* Numbers of surface IgM-bearing cells (●), and total numbers of dissociable cells (○), in bursae at various ages [after ref. 18]. *b–d* Frequencies of bursal or spleen cells forming rosettes with sheep erythrocytes at various ages, given as numbers of rosette-forming cells (RFC) per 10^5 mononuclear, non-erythroid cells. *b* RFC in bursae, according to *Tufveson and Alm* [27] (●) and *Moriya and Ichikawa* [20] (○). Bursal cell suspensions were prepared by different methods by the different authors, *Moriya and Ichikawa* [20] using a Ficoll-Hypaque centrifugation step (which might increase the sensitivity of detection of RFC). *c* RFC in bursae (●) and spleens (○) of conventionally reared chickens [18]. *d* RFC in bursal (●) and spleens (○) of 'germ-free' chickens. The percentages of IgM-bearing cells in spleens of either conventional or germ-free chicks increase approximately linearly from 10 to 36% between days 2 and 15 [18]. The formation of RFC is inhibited by anti-Ig antiserum, and RFC are absent from spleens of bursectomized birds. Data for RFC are given as mean ± standard error.

cytes are plotted against age in figure 2. In most experiments[3], this proportion rises approximately linearly with time, exactly like the proportion of selectively neutral mutants in a growing bacterial culture. In fact, assuming that bursal cells continue to divide once every 10 h, the slope of the appropriate line gives the rate at which bursal rosette-forming cells (RFC) are being produced: about 1 in every 2×10^6 cell divisions. If the cells are being

[3] But not in all: *Lydyard* et al. [18] report, in one experimental series, that the frequencies of rosette-forming cells in bursal and spleen increase exponentially with time – in contrast to their two experimental series shown in figure 2, which are better explained by the assumption of a linear increase in frequency with age.

generated by mutation, this seems a very high mutation rate, since it concerns only mutations to a particular specificity: the overall mutation rate must be several orders of magnitude higher.

In addition, the frequencies of RFC are very similar in the bursae of germ-free, conventional, or deliberately immunized chickens [although RFC in the blood of immunised birds increase at least 300-fold in frequency; 18]. So I cannot believe that exogenous antigen is a strong selective force in somatic generation of antibody diversity [2].

Imm: All experimentalists working on the ontogeny of the mammalian immune system would agree on this[4] [4, 23].

Orn: Finally, *Lydyard* et al. [18] estimated the clone sizes of antigen-binding cells by overlaying frozen bursal sections with radioactive antigens (^{125}I-KLH or ^{125}I-TG-AL). They observed either single cells or clusters of 2–4 cells binding antigens, distributed in 0–5 follicles in sections containing 30–50 of the bursa's approximately 10^4 follicles. Their results are again consistent with the idea that the antigen-binding cells are not strongly selected for rapid growth [although unfortunately the data are not comprehensive enough to permit an estimate of mutation rate from a quantitative analysis of clone size variation; 17].

Imm: The variants might arise either from mutation, or because different V genes are expressed in variant cells. Although each follicle is probably populated by one or a very few stem cells [14, 25], it seems reasonable that the progeny of these cells can express different V gene combinations.

Orn: This is true in the embryo; however, after hatching, the great majority ($\geq$85%) of bursal cells express IgM, so the later variants most probably arise from cells in which V gene translocation has already occurred. It seems unlikely that new V genes are expressed after the original translocation has taken place. Also, the evidence for somatic mutation from mouse λ chains is clear – that about one third of these chains have undergone somatic point mutation [2].

[4] However, it has been very difficult to demonstrate a rise in the frequency of antigen-binding B lymphocytes in the spleens of growing mice, the only evidence for such a rise being that of *Sigal and Klinman* [23] for fluorescein- and phosphocholine-reactive cells. In all experiments on mammals, the evidence that the antigen-binding cells are immunoglobulin-bearing antigen-specific small lymphocytes is less strong (or rather, more weak) than the evidence for chicken RFC. Also, if taken at face value, the results of *D'Eustachio and Edelman* [4] would suggest that the 10^4 B lymphocytes they detect in the spleen of an 18-day mouse fetus have the same range of antigen-binding specificities as the 5×10^7 lymphocytes in an adult spleen, which seems unlikely.

Imm: The nucleotide sequences of mouse λ V genes show interesting homologies between the three hypervariable regions of the chain [26]. Also, when κ and λ V gene sequences are compared, the DNA sequences immediately adjacent to the hypervariable regions have been strongly conserved during the separate evolution of κ and λ chains [28]. These homologies may exist because the hypervariable regions and their flanking sequences are recognized by a mutation enzyme system, which might increase mutation rates of specific DNA regions by several orders of magnitude above the average rate [1, 19]. For example, if hypervariable region mutants are produced at a rate of 10^{-2}/gene/cell division, the frequency of such variants would rise in a few weeks to values approaching 0.3 (as found in mouse λ chains).

Thus, the two participants in this conversation convinced themselves that the antibody repertoire of chickens is comparable to that of mammals ($\geq 10^6$); that this repertoire is produced in the bursa, without strong selection of new variants, at a rate of at least 10^{-3} new specificities per cell division; and that an important generator of diversity is somatic mutation, acting specifically on hypervariable region sequences at a particular period in early B cell development. But doubts remain. Have the cellular kinetics of the bursa been adequately studied at different time points? Do cells binding other antigens appear at rates comparable to those obtained for RFC? Is the evidence that all these cells are B lymphocytes strong enough? Why have increases in frequency of antigen-binding cells been so difficult to demonstrate in spleens of fetal and newborn mice? Is the long-awaited End of the Antibody Variability Saga in sight? We, following *Jerne* [11] and *W.H. Auden,* are looking forward to its coming.

Anthropos apteros, perplexed,
To know which turning to take next,
Looked up and wished he were a bird
To whom such doubts must seem absurd.
(From *W.H. Auden: The Maze*)

References

1 Ben-Sasson, S.A.: Proc. natn. Acad. Sci. USA *76:* 4598–4602 (1979).
2 Cohn, M.; Blomberg, B.; Geckeler, W.; Raschke, W.; Riblet, R.; Weigert, M.: in Sercarz, Williamson, Fox, The immune system. 3rd ICN-UCLA Symp. on Molecular Biology, pp. 89–117 (Academic Press, New York 1974).
3 Cramer, M.; Braun, D.G.: J. exp. Med. *139:* 1513–1528 (1974).

4 D'Eustachio, P.; Edelman, G.M.: J. exp. Med. *142:* 1078–1091 (1975).
5 Du Pasquier, L.; Wabl, M.R.: Eur. J. Immunol. *8:* 428–433 (1978).
6 Eichmann, K.: Eur. J. Immunol. *2:* 302–307 (1972).
7 Hála, K.; Hašek, M.: in Lengerová, Vojtíšková, Immunogenetics of the H-2 System, pp. 334–340 (Karger, Basel 1970).
8 Hämmerling, G.J.; McDevitt, H.O.: J. Immun. *112:* 1726–1740 (1974).
9 Ivanyi, P.; Van Mourik, P.; Breuning, M.; Melief C.J.M.: Immunogenetics *10:* 319–322 (1980).
10 Jensenius, J.C.; Crone, M.; Koch, C.: Scand. J. Immunol. *4:* 151–160 (1975).
11 Jerne, N.K.: Cold Spring Harb. Symp. quant. Biol. *32:* 591–603 (1967).
12 Jerne, N.K.: Eur. J. Immunol. *1:* 1–9 (1971).
13 Jerne, N.K.: in Edelman, Cellular selection and regulation in the immune response, pp. 39–48 (Raven Press, New York 1974).
14 Le Douarin, N.M.; Houssaint, E.; Jotereau, F.V.; Belo, M.: Proc. natn. Acad. Sci. USA *72:* 2701–2706 (1975).
15 Lefkovits, I.: Curr. Top. Microbiol. Immunol. *65:* 21–58 (1974).
16 Longenecker, B.M.; Mosmann, T.R.: Immunogenetics *11:* 293–302 (1980).
17 Luria, S.E.; Delbrück, M.: Genetics *28:* 491–511 (1943).
18 Lydyard, P.M.; Grossi, C.E.; Cooper, M.D.: J. exp. Med. *144:* 79–97 (1976).
19 Mäkelä, O.; Cross, A.M.: Prog. Allergy, vol. 14, pp. 145–207 (Karger, Basel 1970).
20 Moriya, O.; Ichikawa, Y.: Immunology *37:* 857–861 (1979).
21 Pink, J.R.L.; Askonas, B.A.: Eur. J. Immunol. *4:* 426–430 (1974).
22 Pink, J.R.L.; Droege, W.; Hála, K.; Miggiano, V.C.; Ziegler, A.: Immunogenetics *5:* 203–216 (1977).
23 Sigal, N.H.; Klinman, N.R.: Adv. Immunol. *26:* 255–337 (1978).
24 Smilek, D.E.; Boyd, H.C.; Wilson, D.B.; Zmijewski, C.M.; Fitch, F.W.; McKearn, T.J.: J. exp. Med. *151:* 1139–1150 (1980).
25 Sorvari, T.; Toivanen, A.; Toivanen, P.: Transplantation *17:* 584–592 (1974).
26 Tonegawa, S.; Maxam, A.M.; Tizard, R.; Bernard, O.; Gilbert, W.: Proc. natn. Acad. Sci. USA *75:* 1485–1489 (1978).
27 Tufveson, G.; Alm, G.V.: Int. Archs. Allergy appl. Immun. *48:* 537–546 (1975).
28 Wu, T.T.; Kabat, E.A.; Bilofsky, H.: Proc. natn. Acad. Sci. USA *76:* 4617–4621 (1979).

Dr. Richard Pink, Basel Institute for Immunology, Postfach, CH-4005 Basel

The Immune System, vol. 1, pp. 76–80 (Karger, Basel 1981)

Was an Idiotype Network Working in the Evolutionary Past?

Takeshi Matsunaga [1], *Susumu Ohno* [1]

Division of Biology, City of Hope Research Institute, Duarte, Calif., USA

We would like to state an argument that might place *Niels Jerne's* seminal idea of an idiotype network of the immune response [4] in a more appropriate evolutional context.

Most immunologists accept the idea that the immune response must be adequately regulated. If not, animals would suffer unfavourable consequences such as autoimmune diseases caused by anti-self antibodies or uncontrolled lymphocyte proliferation leading to malignancy. If it is correct that idiotype and anti-idiotype network interactions are essential in regulating the immune response, a failure in forming a functional network would result in the breakdown of the normal performance of the immune system.

For a functional idiotype network to be formed within the immune system of an animal, a substantial degree of antibody or T cell diversity (immune repertoire) is required. With a typical laboratory mammalian species such as the mouse, which can generate at least several millions of different antibody molecules, one might intuitively feel that the idiotype repertoire of the animal could be completely recognized by the anti-idiotype repertoire of the same animal, and a network would function. Indeed, within the several years of investigation after the network theory was first proposed, a considerable amount of evidence has been obtained in support of its basic validity. Thus, for example, a tiny amount of anti-idiotypic antibody molecules injected into animals can drastically perturb the immune response pattern [3]. Even V_H-linked idiotype inheritance is the out-

[1] *Takeshi Matsunaga* was a student at the Basel Institute of Immunology from 1971 to 1973, and *Susumu Ohno* was a member of the Institute in 1976.

come of the network interaction demonstrated at least in the T helper cell population [1]. In most of these idiotype experiments, however, the crucial information came from the mouse or rabbit.

What would happen if the size of the immune repertoire was far smaller than that of mice? If an animal is capable of making only hundreds or thousands of antibodies as its total repertoire, is a functional network formed? Maybe not. Such an animal would have great trouble keeping the immune response under control.

The situation of a small immune repertoire could occur either in evolution or in ontogeny. Because the time period of the former may be from millions to hundreds of millions of orders of magnitudes longer than the latter, we only consider the possibilities arising in evolution. In all probability, the immune system would be able to cope with the temporal, short-term problem in ontogeny by utilizing various ad hoc mechanisms (e.g. nonspecific suppression, noninducibility).

Although modern mammalian species like the mouse have evolved many V genes which are maintained in the germ line, animals must have begun with a single or very few primitive V genes at the dawn of creation of adaptive immune system. Phagocytosis by macrophages and nonspecific humoral substances were employed as major lines of defense at this stage of evolution. However, with increasing pressure from invasion by pathogenic viruses and bacteria, acquisition of an adaptive immune mechanism based on V gene specified recognition, however clumsy in its handling of pathogens in the beginning, should be advantageous. Once started, such primitive V genes duplicate and mutate, contributing to the enrichment of repertoire. On the other hand, due to the small size of repertoire, the idiotype network must have functioned very poorly if it was present at all. Apparently, in order to maintain regulation, particularly in the face of the problem of autoimmunity, some mechansim other than that of the idiotype network were needed. The simplest one, we suggest, would be that by which V genes with anti-self activities are forbidden, i.e. individuals who happened to have such V genes were eliminated as a consequence of bearing the heavy burden of autoimmune diseases. On the contrary, V genes that specify activities against common pathogens were positively selected for and maintained in the population. Unfortunate individuals with no useful V genes against prevailing epidemics were also eliminated. In other words, each V gene was under the censorship of natural selection.

However, this kind of adaptive V gene immune system was not adequate in coping with invading microorganisms. The reason is that viruses and bacteria can escape immune recognition by generating new antigenic mutants. Note that even today human influenza viruses can nullify the effect of a vaccination by producing antigenic variants. If the highly evolved, powerful, modern immune system still has trouble, the more primitive, old one we are concerned with must have had more. Microorganisms challenge the immune system by generating antigenic mutants, thus taking advantage of their much shorter generation time than that of vertebrate species. What was the next strategy the immune system devised to meet this difficulty?

The combining sites of antibody molecules are built up using both heavy and light chain V domains. At a genetic level, V gene clusters for V_H and V_L are located on different chromosomes. On the other hand, primitive antibodies must have started with combining sites made up from a single polypeptide. Apparently, during the course of V gene expansion, chromosomal duplication or gene translocation created a condition whereby V_H–V_L recombination became established. Since each binding specificity can now be created by the random combination of two polypeptides, the repertoire and the discriminatory ability increased dramatically. More importantly, however, we should note that the element of randomness in terms of specificities was also introduced. For instance, each time new V genes arise, many new specificities can be created. The point is that the immune system does not know whether they are useful or not. Nevertheless, this method of random expansion of antibody repertoire has proven to be powerful, and probably the only way to counterbalance the equally forceful strategy of antigenic mutation used by microorganisms. This point was discussed [6] and is emphasized in figure 1.

Actually, 100 each of V_H and V_L genes in the germ-line (as is suggested for the number for mouse antibody genes) plus newly found J and D gene segments may be capable of generating as many as a million antibodies of different specificities through a random combinatorial process. On the other hand, for having such an enrichment in repertoire, the immune system had to pay certain costs. As the specificity is created randomly, anti-self antibodies are bound to arise which might cause an autoimmune attack. Because the problem of self-nonself discrimination could no longer be handled at the genetic level, as was done in the early stage, new mechanisms of avoiding autoimmune response had to be devised at the phenotypic level. Thus, self-nonself discrimination for self-tolerance became a learned property, and mechanisms of clonal deletion and suppression were born [5].

Fig. 1. The foresight of the immune system is due to the random generation of numerous antigen-binding sites (V genes) clonally expressed: by chance, some are bound to be directed against antigens not yet in existence.

In this line of conjecture, we would like to understand the emergence of the idiotype network interaction. Idiotype and anti-idiotype interactions involving various lymphocyte subsets and antibodies are a kind of autoimmune response without any harm to the body. Although they were brought into being by the evolutional process of random enrichment of the repertoire, the immune system was quick to use the newly emerged idiotype network for its own advantage. Thus anti-idiotypic suppressor T cells, by checking continually arising anti-self clones, would provide a fail-safe mechanism for self-tolerance.

The same kind of suppression should also be powerful in regulating ongoing immune responses as a feedback force. An activation signal would be more efficiently delivered when anti-idiotypic helper T cells join the interaction for B cell triggering.

For the moment we do not know whether the present immune system can control the immune response without an idiotype network or not. One way to look at this question would be to study the repertoire in lower vertebrates. There is the interesting observation that the antibody repertoire in one of the amphibian species might be smaller than that of mice [2]. If in the future we can experimentally manipulate the size of the immune repertoire, it would become possible to answer a question of this kind.

So far in our argument we have avoided the possibility that somatic mutation might play a role in forming the idiotype network. But after all, this mechanism is another way to increase the size of repertoire in a random fashion. Together with the combinatorial mechanism we have discussed, somatic mutation could constitute a major mechanism of diversity generation. If so, there is little doubt that the immune system could exploit somatic mutations to perfect the idiotype network. If, for the sake of argument, somatic mutation for antibody diversity were to appear at a very early stage of evolution, and if the size of repertoire were large enough at that time, it follows that the idiotype network could have appeared at the same time. Thus, in this case, the GOD (generator of diversity) could save the whole network.

References

1 Augustin, A.A.; Julius, M.H.; Cosenza, H.; Matsunaga, T.: in Pernis, Vogel, Regulatory T lymphocytes, pp. 171–184 (Academic Press, New York 1980).
2 Du Pasquier, L.; Wabl, M.R.: Eur. J. Immunol. *8:* 428–433 (1978).
3 Eichmann, K.; Rajewsky, K.: Eur. J. Immunol. *5:* 661–666 (1975).
4 Jerne, N.K.: Annls Immunol. Inst. Pasteur *125:* 373–389 (1974).
5 Matsunaga, T.; Simpson, E.; Meo, T.: Transplantation *30:* 34–39 (1980).
6 Ohno, S.; Matsunaga, T.; Epplen, J.T.; Hozumi, T.: in Fougereau, Dausset, Immunology 80 (Academic Press, New York, 1980).

Dr. T. Matsunaga, City of Hope Research Institute, 1500 E. Duarte Road, Duarte, CA 91010 (USA)

The Immune System, vol. 1, pp. 81–88 (Karger, Basel 1981)

Analysis of B Cell Precursors in Fetal and Adult Mice

Christopher J. Paige[1], *Paul W. Kincade*

Basel Institute for Immunology, Basel, Switzerland; Sloan-Kettering Institute for Cancer Research, Rye, N.Y., USA

During the last several years B lymphocyte populations which differ in maturity and/or function have been distinguished. The resolution of these populations was achieved through the development of mitogenic and antigen-specific assay systems as well as the discovery of cell surface components which these cells possess. Cells which may be precursors of B Lymphocytes have also been characterized primarily as assay systems which allow them to express their B cell-generating potential [1, 6, 11, 13, 18–24]. Recently, cell surface markers present on these progenitors have also been recognized [5, 8, 22]. Nonetheless, the properties and differentiation potential of these precursors, their relationship to other hematopoietic stem cells, and even the definitive relationship between these cells and the various populations of B lymphocytes which subsequently arise remains to be established. Furthermore, it is not yet certain that B cell development during adult life proceeds in a manner identical to fetal differentiation. The resolution of these questions should provide useful information for a more thorough understanding of the developmental relationship between the lymphoid lineage and other hematopoietic cells, the generation and maintenance of self-tolerance and the underlying mechanism of immunodeficiency states.

[1] *Christopher Paige* has been a member of the Basel Institute for Immunology since 1980.

Assay for B Cell Progenitors

Evaluation of the differentiative and proliferative potential of various B cell progenitors was undertaken using several systems in which the emergence of B lymphocytes could be closely monitored. For these studies we have taken advantage of the finding that partially immunodeficient CBA/N mice completely lack B cells capable of forming colonies in soft-agar (colony forming unit-B cell, CFU-B) but nevertheless provide the necessary environment in which normal B cell progenitors can develop into colony-forming B cells [3, 9, 21, 22]. Thus, irradiated or unirradiated CBA/N mice are ideal recipients for adoptive transfer experiments in which the development of very small numbers of potential B cell precursors can be assessed. Furthermore, CBA/N cells are useful in liquid culture assays where they contribute important filler cells but do not interfere with the assessment of B cell generation. This protocol has proven quite useful because the CFU-B assay detects a functional B cell which is heterogeneous by other criteria, the cloning efficiency in agar is not dependent on accessory cells, the assay is linear under appropriate conditions, and the emergence of very small numbers of B cells can be unequivocally detected due to the sensitivity of the B cell cloning assay as well as the background-free environment provided by the CBA/N cells [3–10, 15, 16, 21, 22]. An additional advantage is that the normal cells used in theses studies are derived from CBA/H-T6T6 mice which possess an easily identifiable chromosome marker that permits us to compare B cell reconstitution to reconstitution in other cellular compartments. It should be noted that while the absence of background makes the CBA/N mouse convenient for these studies, it is not essential, and similar data have been obtained using normal CBA/H mice [21]. A detailed description of these methods has been provided previously [21, 22].

Emergence of B Cells in Reconstituted CBA/N Mice

It has been postulated that the primary sites of murine B cell differentiation are the fetal liver and adult bone marrow, as cells which may be intermediates in the differentiation of pluripotent stem cells into B lymphocytes have been detected in these tissues [1, 19, 24]. Thus, we initiated our studies by examining the ability of cells from these sites to generate clonable B cells after adoptive transfer into irradiated (1,050 rad) CBA/N recipients. We found detectable levels of clonable B lymphocytes in the spleen and bone marrow of recipient mice 1 week after transfer of 10^5 donor cells, and these reached normal levels in approximately 1 month [21]. The fetal liver in these experiments was obtained from 13-day-old embryos. Since the ability to clone in soft agar does not arise until day 16.5 of gestation, concomitant with the appearance of surface immunoglobulin-positive (sIg^+) cells, these data indicate that B cell progenitors present in the fetal liver were able to differentiate into clonable B cells in the CBA/N recipients. In contrast, the adult bone marrow contains cells which already possess sIg and are capable of proliferation in soft agar. To determine the relative contribution of these mature cells in our assay system we depleted the bone marrow of sIg^+ B lymphocytes before transplantation. This was accomplished either by in vivo suppression of B cells induced by treatment with anti-μ antibodies from birth or by B cell removal on anti-μ-coated plastic Petri dishes [1, 28]. We found that the presence of sIg^+ cells had absolutely no effect on the number of CFU-B subsequently detected in recipient mice [21]. This indicates that, similar to our findings with the fetal liver, progenitor cells in the bone marrow contributed to the generation of the majority of detected CFU-B. This was confirmed by sedimentation velocity at unit gravity which separates hematopoietic cells primarily on the basis of size [17]. Our analysis revealed that the bone marrow cells capable of immediate cloning in agar sedimented on average at 3 mm/h (as do sIg^+ B cells) while the cells which restore B cell function in irradiated mice sedimented at 5 mm/h [22]. The fetal liver cells responsible for B cell generation sedimented even more rapidly with an average rate of 6.1 mm/h [22].

Distinctions between B Cell Progenitors and Myeloid Stem Cells (CFU-s)

While sedimentation velocity clearly separated the B cell progenitors from B lymphocytes it did not allow a clear separation between B cell progenitors and the class of multipotent hematopoietic stem cell which forms foci of proliferating cells in the spleen of irradiated/reconstituted recipients [27]. These are termed colony forming units-spleen (CFU-s), and while it has been demonstrated that this cell type can give rise to several hematopoietic lineages, and thus is clearly multipotent, the relationship between these cells and the lymphoid lineage remains controversial. It has been suggested, for example, that clonable B cells are derived from CFU-s and can be found within splenic foci [12, 14]. We examined this question by analyzing splenic foci which arose in irradiated recipients injected with a 50:50

mixture of CBA/H-T6T6 + CBA/N bone marrow cells. CBA/N mice, though lacking clonable B cells, have normal numbers of CFU-s and we reasoned that half of the spleen foci would be generated from CBA/N stem cells. If clonable B cells detected in a splenic focus did indeed arise from the CFU-s which generated that focus, we predicted that none of the CBA/N derived foci would contain CFU-B. Individual splenic foci were thus analyzed for chromosome markers to determine whether they were derived from CBA/H-T6T6 stem cells and for the presence of colony-forming B cells. We found that while half of the splenic foci were CBA/N-derived, over 90% of the foci, regardless of origin, contained detectable numbers of clonable B cells [21]. We confirmed that all of the clonable B cells generated from the 50:50 mixture were T6T6, indicating that the majority of CBA/N-derived spleen foci contained CBA/H-T6T6 B cells. Furthermore, we consistently found more CFU-B between foci than within them. These data do not disprove the possibility that CFU-s are capable of generating some CFU-B but they strongly suggest that other cell types, also present in the bone marrow inoculum, randomly seeded throughout the host spleen, expanded contiguously with the CFU-s, were often incorporated into the developing splenic foci, and generated the majority of CFU-B detected.

A major prediction of this suggestion is that conditions can be found under which the number of CFU-s would not be correlated with the ability to restore B cell function in recipient mice. We have subsequently found several situations in which this prediction is fulfilled. CBA/N mice were reconstituted with 10-day yolk sac cells, since components of the B lineage have not been detected in that site even after extensive organ culture [20]. While 12-day fetal liver cells generated clonable B cells in 6 days, we found that 10-day yolk sac cells, containing an equivalent number of CFU-s, generated low numbers of CFU-B only after 2 months [21]. The fact that even a few colony forming B cells were in fact generated reveals that the entire developmental sequence from stem cell to B lymphocyte probably occurs in an irradiated adult recipient. The length of time involved however suggests that the differentiation of stem cells into B cells may not be a very efficient process under these conditions.

Because organ or age specific differences among CFU-s could be postulated to explain this discrepancy, we performed a similar comparison between 13- and 16-day fetal liver. While the incidence of CFU-s was the same in these populations, we found that 16-day fetal liver generated 10 times the number of B cells as did 13-day fetal liver and that these were detected at least 2 days sooner [22]. In contrast, the number of granulocyte/macrophage progenitors (CFU-c) detected in the same irradiated recipients always reflected the input number of CFU-s. Since it could still be argued that age differences in stem cells accounted for these data, we next utilized hybridoma antisera to distinguish between CFU-s and B cell progenitors found in age- and organ-matched populations. A cytotoxic rat anti-mouse brain antisera was found which detected an antigen on pre-B lymphoma cell lines but had no effect on CFU-s number [26]. We therefore treated both 13-day fetal liver and adult bone marrow with this antisera + complement and compared the B cell reconstitution potential of treated cells to complement-treated controls. We found that B cell reconstitution was significantly reduced (60–95%) while CFU-s number remained unchanged [11]. We are thus left with the possiblity that the entry of stem cells into the B lineage is minimal compared to the expansive potential of pre-committed cells and/or that the relevant stem cells for the humoral immune system are not well enumerated by the CFU-s assay. Preliminary support for the first suggestion comes from cytogenic analysis of long-term (8–16 months) recipents of 10^5 13-day fetal liver cells in which the reconstitution of the B cell compartment (LPS blasts) was compared to that of the T cell (PHA blasts), granulocyte/macrophage (CSA blasts), or CFU-s (spleen foci) compartments [21]. This analysis revealed that donor cells (T6T6) accounted for 95% of the B cell compartment while only 40–60% in other hematopoietic compartments were of donor origin [21]. In several individual mice this difference was quite dramatic since donor cells accounted for $>$ 90% of the B cells while $<$ 10% in other compartments. This may indicate that the cell responsible for the B cell reconstituion observed is indeed an already committed B cell progenitor, although we have not yet ruled out the possiblity that the irradiated host played a selective role in regulation of the final degree of reconstitution achieved in various hematopoietic compartments.

Heterogeneity among Different Populations of B Cell Precursors

An additional level of complexity in B cell differentiation is suggested by a comparison of B cell progenitors during fetal and adult life. As noted previously, the cells in adult bone marrow which are responsible for the emergence of colony-forming B lymphocytes in the spleen of irradiated-reconstituted CBA/N recipients sediment less rapidly than a similar cell in 14-day fetal liver [22]. We have also found that these populations significantly differ in their ability to repopulate unirradiated CBA/N recipients [21]. When 14-day fetal liver cells were injected into unirradiated CBA/N mice, B cells were not detected for 14 days, after which they were found in increasing quantities and ultimately accounted for more than 60% of some B cell populations in the recipient mice. In contrast, adult bone marrow cells more rapidly generated detectable numbers of CFU-B but these never increased above the input levels of 10–20% of normal controls. Differences between fetal and adult B cell progenitors were also evident in in vitro maturation experiments. For example, B cell-depleted adult bone marrow cells held in liquid culture, generating CFU-B in 12–24 h, were significantly influenced by the presence or removal of adherent cells, and were not responsive to LPS [6]. However, 14-day fetal liver cells requiring 4 days of in vitro maturation, were not affected by adherent cell removal, and are inhibited by the presence of LPS [6]. We must point out that it is not yet certain that these differences detected in in vitro maturation are intrinsic to pre-B cells rather than accessory cells. Finally, pre-B cells in fetal and adult life differ in their sensitivity to lysis by cytotoxic antisera directed against either Lyb-2 or Qa-2 antigens [2, 8, 25]. In the latter case, this is caused by an absence of the relevant target antigen from the fetal pre-B cells. Since Lyb-2 can be found by other methods, the former case must be due to either a difference in the sensitivity of the cells to antibody + complement-mediated lysis or a difference in the surface presentation of this antigen.

Conclusions

These studies reveal that B cell progenitors found during both fetal and adult life have the capacity to generate large numbers of B lymphocytes in a relatively short time. That this progenitor is already committed to the B lineage is suggested by: (1) chromosome analysis of long-term recipients which show a preponderance of B cell reconstitution when compared to reconstitution of other compartments; (2) monoclonal antisera which selectively reduce B cell reconstitution in the presence of normal myeloid reconstitution, and (3) our failure to observe any significant contribution to the B cell pool by CFU-s stem cells. The latter observation may in fact be irrelevant if most CFU-s are already restricted to non-lymphoid hematopoiesis. B cell progenitors are quite heterogeneous however and the earliest of these, found in 12- to 14-day fetal liver, have significantly more proliferative potential under some conditions than do older adult progenitors. They also differ in their requirements for in vitro maturation and in cell surface components [6, 8]. This may indicate that there exists in the fetal liver a unique phase of pre-B generation which results in a large precursor pool of committed B cells in adult life. If this model is correct it will be important to determine whether the generation of the long-lived precursor pool involves selection of variable region genes and to what differentiation signals these cells are responsive.

Acknowledgement. The work at Rye was supported by Grants AI-12741, CA-17404, CA-08748, T32CA09149, and RCDA AI-00265.

References

1 Burrows, P.D.; Kearney, J.F.; Lawton, A.R.; Cooper, M.D.: J. Immun. *120:* 1526–1531 (1978).
2 Flaherty, L.: Immunogenetics *3:* 533–539 (1976).
3 Kincade, P.W.: J. exp. Med. *145:* 149–263 (1977).
4 Kincade, P.W.; Flaherty, L.; Lee, G.; Watanabe, T.; Michaelson, J.: J. Immun. *124:* 2879–2885 (1980).
5 Kincade, P.W.; Lee, G.; Paige, C.J.: in Seligmann, The primary immunodeficiencies (Elsevier/North-Holland, Amsterdam, in press, 1980).
6 Kincade, P.W.; Lee, G.; Paige, C.J.; Scheid, M.P.: Submitted for publication.
7 Kincade, P.W.; Lee, G.; Scheid, M.P.; Blum, M.D.: J. Immun. *124:* 947–953 (1980).
8 Kincade, P.W.; Lee, G.; Watanabe, T.; Scheid, M.P.: Submitted for publication.
9 Kincade, P.W.; Moore, M.A.S.; Lee, G.; Paige, C.J.: Cell. Immunol. *40:* 294–302 (1978).
10 Kincade, P.W.; Paige, C.J.; Parkhouse, R.M.E.; Lee, G.: J. Immun. *120:* 1289–1296 (1978).
11 Lafleur, L.; Miller, R.G.; Phillips, R.A.: J. exp. Med. *135:* 1363–1374 (1972).
12 Lala, P.K.; Johnson, G.R.: J. exp. Med. *148:* 1468–1477 (1978).
13 Melchers, F.: Eur. J. Immunol. *7:* 476–481 (1977).
14 Metcalf, D.; Johnson, G.: in Battisto, Strodein, Immunoaspects of the spleen, pp. 27–36 (Elsevier/North-Holland, Amsterdam 1977).
15 Metcalf, D.; Nossal, G.J.V.; Warner, N.L.; Miller, J.F.A.P.; Mandel, T.E.; Layton, J.E.; Gutman, G.A.: J. exp. Med. *142:* 1534–1549 (1975).
16 Metcalf, D.; Warner, N.L.; Nossal, G.J.V.; Miller, J.F.A.P.; Shortman, K.; Rabellino, E.: Nature, Lond. *255:* 630–632 (1975).
17 Miller, R.G.; Phillips, R.A.: J. cell. Physiol. *73:* 191–205 (1970).
18 Nossal, G.J.V.; Pike, B.L.: Immunology *25:* 33–45 (1974).
19 Osmond, D.G.; Nossal, G.J.V.: Cell. Immunol. *13:* 132–145 (1974).
20 Owen, J.J.T.; Raff, M.C.; Cooper, M.D.: Eur. J. Immunol. *5:* 468–473 (1975).
21 Paige, C.J.; Kincade, P.W.; Moore, M.A.S.; Lee, G.: J. exp. Med. *150:* 548–563 (1979).

22 Paige, C.J.; Kincade, P.W.; Shinefeld, L.A.; Sato, V.L.: J. exp. Med. *153:* 154–165 (1981).
23 Phillips, R.A.; Melchers, F.J.: Immunology *117:* 1099–1103 (1976).
24 Raff, M.C.; Megson, M.; Owen, J.J.T.; Cooper, M.D.: Nature, Lond. *259:* 224–226 (1976).
25 Sato, H.; Boyse, E.A.: Immunogenetics *3:* 565–572 1976).
26 Shinefeld, L.A.; Sato, V.L.; Rosenberg, N.E.: Cell *20:* 11–17 (1980).
27 Till, J.E.; McCulloch, E.A.: Radiat. Res. *14:* 213–222 (1961).
28 Wysocki, L.J.; Sato, V.L.: Proc. natn. Acad. Sci. USA *75:* 2844–2848 (1978).

Dr. Christopher Paige, Basel Institute for Immunology, Postfach, CH–4005 Basel (Switzerland)

The Immune System, vol. 1, pp. 89–94 (Karger, Basel 1981)

Development of B Cell Function in the Chicken

Auli Toivanen[1], *Paavo Toivanen*[1], *Olli Vainio, Olli Lassila, Yoshikazu Hirota*

Departments of Medicine and Medical Microbiology, University of Turku, Turku, Finland; Basel Institute for Immunology, Basel, Switzerland

With this paper we wish to present our cordial congratulations to *Niels K. Jerne.* We hope it will remind him of the time when two of us (A.T. and P.T.) worked with chickens in Basel and enjoyed the stimulating atmosphere of the Institute.

The chicken has served extensively as a model for studies on the differentiation of lymphoid cells. In an avian model it is possible to experimentally manipulate both the bursal and the thymic system, and the availability of sufficient numbers of embryos matched in age and histocompatibility permits developmental studies during the early stages of ontogeny. Although several basic questions still remain unanswered and several new problems have arisen, a general concept of the differentiation process of chicken B cells is emerging. It has become clear that the same basic biological rules that control lymphocyte functions and interactions apply for mammals as well as for birds or even amphibians.

In this article we give a short summary of the present knowledge about B cell differentiation in the chicken and discuss two aspects of special interest: the ability of B precursor cells to mature in an allogeneic bursal environment and the role of T cells in the early B cell maturation.

[1] *Auli* and *Paavo Toivanen* were members of the Basel Institute for Immunology in 1979 and 1980.

General Scheme of B Cell Differentiation

Three different degrees of maturity characterize the differentiation process of the chicken B cells: the prebursal, bursal and postbursal phase.

The first *prebursal* (and prethymic) *stem cells* originate within the hemopoietic foci of the axial mesenchyme of the early embryo. This has been demonstrated using chimeras constructed of the *area pellucida* (giving rise to the embryo proper, allantois and amnion) of one chicken embryo and of the *area vasculosa* (giving rise to the yolk sac) of another embryo at the age of 2 days, when circulation has not yet been established. Using either chromosomal or immunoglobulin allotype markers it has been possible to trace the origin of the lymphoid cells of the developing chicken after hatching. In these experiments, relying on functional criteria, the lymphocytes of the grown-up chimera always proved to be of the same type as the original *area pellucida,* i.e. of the embryo type and not of the yolk sac-type [9, 11, 15]. With an established circulation the stem cells move not only within the embryo but also appear in the yolk sac and can be demonstrated there in a fair number on the 7th day of incubation [12, 16].

The exact localization of the prebursal stem cells within the embryo has not yet been established, but most probably it is related to the hematopoietic foci described in 1909 by *Danchakoff* [2, 17]. A mixed cell population prepared from the general axial mesenchyme of 7-day-old embryos is able to induce morphological and functional reconstitution of the immunological apparatus when transplanted into 18-day-old cyclophosphamide-treated embryonic recipients [10].

The bursal phase of development is characterized by strict bursa-dependence. During the late embryonic development and around the time of hatching, the bursa contains *bursal stem cells.* These immunologically incompetent pre-B cells need to be able to enter the bursa and to proliferate there in order to give rise to mature B cells later in life. After the chick hatches both the cells present in the bursa and those already seeded out become increasingly independent of the bursa. After the 6th week of life, all lymphoid organs of the chicken contain *postbursal cells* which no longer can or need to enter the bursa for further maturation. During a short period around the age of 3–5 weeks the bursa, spleen and even the bone marrow and thymus contain *early postbursal cells* which have retained the ability to enter the bursa. However, this is no longer necessary for their further differentiation [21–24].

Table I. Characteristics of chicken B cells representing different degrees of maturity

	Stem cells		Postbursal cells	
	prebursal	bursal	early	mature
Need for bursal microenvironment	?	+	–	–
Restoration of:				
Bursal structures	+	+	±	–
Germinal centers in the spleen	+	+	±	–
Antibody formation	+	+	+	+

The characteristics of the various degrees of maturity of B cells are summarized in table I; for more details the reader is referred to a recent review [20].

Maturation of Bursal Stem Cells in an Allogeneic Bursa

When bursal stem cells are transplanted into a newly hatched allogeneic host with cyclophosphamide-induced B cell deficiency [13, 14], they home to the bursa and proliferate there, induce tolerance, and no graft-versus-host reaction takes place [1, 25]. The resulting B cells are seeded into the periphery, and they are able to express thymus-independent antibody production. However, the immune functions requiring collaboration between the donor-derived B cells and the host-derived T cells remain absent in an allogeneic combination; no thymus-dependent antibodies and no germinal centers are formed [18, 25, 26].

It could therefore be suspected that bursal stem cells would not undergo full differentiation in the histoincompatible bursal microenvironment. However, all available evidence indicates that this is not the case. The cells undergo a normal switch from IgM to IgG, they acquire postbursal characteristics, and they express normal function when transferred into histocompatible secondary hosts. The explanation for the absence of collaboration between the donor-derived B cells and the host-derived T cells is rather to be found in the fact that B cells in the allogeneic host express their original, genetically determined Ia-like antigens (table II) [27, 28].

Table II. Occurrence of Ia-like antigen-positive cells in the bursa of cyclophosphamide-treated chicks 2–4 weeks after transplantation of syngeneic or allogeneic bursal stem cells

Donor genotype	Host genotype	% cells reacting with antiserum against	
		Ia^2	Ia^{15}
$B^{15}B^{15}$	$B^{15}B^{15}$	<2	65.4
$B^{15}B^{15}$	B^2B^2	<2	67.7
–	$B^{15}B^{15}$	<2	67.9
–	B^2B^2	70.4	<2

Ia-like antigens were demonstrated by immunofluorescence. Mean values for 9–10 birds are given.

In other words, bursal stem cells are able to enter an allogeneic bursal microenvironment, proliferate there and acquire full functional maturity, but they express their own genetically determined Ia-like surface antigens – without any influence from the allogeneic bursa. Therefore, the host-derived T cells will recognize them as foreign and fail to collaborate with them.

The Role of T Cells in the Early Maturation of B Cells

Evidence is accumulating that T cells or their precursors are essential in early B cell differentiation. Thymectomy in ovo or at hatching also induces lymphocyte depletion in the bursa [6, 7, 29]. *Hirota and Bito* [4] observed that bursal stem cells were not able to induce reconstitution of young cyclophosphamide-treated birds, if the birds had also been thymectomized at hatching. Later *Hirota* et al. [5] and *Toivanen* et al. [19] extended these observations; the birds were depleted of both T and B cells by thymectomy, X-irradiation and cyclophosphamide in the newly hatched period. When such birds were transplanted with bursal stem cells alone, reconstitution of the bursal structures and of the germinal center formation was poor, and antibody formation against both *Brucella* and SRBC remained deficient. However, if the bursal stem cells were supplemented with thymus cells from

10-day-old chickens, complete reconstitution was observed. Similarly in birds that had one or two thymic lobes left, complete reconstitution was achieved with bursal stem cells.

As discussed earlier, bursal stem cells are able to enter the bursa of an allogeneic host and induce full reconstitution of its histological structure and of thymus-independent antibody formation. Taken together these two experimental series therefore indicate that the interaction between the primitive T cells and the bursal stem cells is not restricted by histoincompatibility in the same manner as the interaction between mature T and B cells.

Concluding Remarks

During the bursal phase of maturation the genes coding for the variable region are activated and the B cells become able to respond to different antigens. At the same time, the cells start to express Ia-like surface antigens on their surface. It has been suggested that lymphoid cell precursors during early differentiation learn the relevant compatibilities required for effective cell-cell interactions and that this learning process would be dictated by the major histocompatibility genotype of the host. This would also apply to B cells, which would preferably interact with T cells and macrophages representing the maturation environment [3, 8]. Our observations do not support this view. On the contrary, our results clearly indicate that the surface characteristics of the B cells are not genetically modulated by the allogeneic environment. Furthermore, in the experiments where bursal stem cells have been permitted to differentiate within an allogeneic host, all conditions for adaptation between the host-derived T cells and donor-derived B cells would be ideal. Yet we could never demonstrate any collaboration between these two cell populations.

The rules and mechanisms by which the differentiation environment determines the interaction preferences of macrophages and T cells remain to be established. Regarding the B cells the situation seems more clear: the bursal microenvironment serves as an activator of genetically determined factors but does not dictate appearance of new or modified cell surface determinants.

Acknowledgements. This work was supported in part by a grant from the Sigrid Jusèlius Foundation and by contracts with NIH (N01-CB-74177) and the Association of Finnish Life Insurance Companies.

References

1 Cain, W.A.; Cooper, M.D.; Good, R.A.: Nature, Lond. *217:* 87–89 (1968).
2 Danchakoff, V.: Arch. mikrosk. Anat. EntwMech. *73:* 117–181 (1909).
3 Gorczynski, R.M.; Kennedy, M.J.; MacRae, S.; Steele, E.J.; Cunningham, A.J.: J. Immun. *124:* 590–596 (1980).
4 Hirota, Y.; Bito, Y.: Immunology *35:* 889–899 (1978).
5 Hirota, Y.; Vainio, O.; Toivanen, P.: Acta pathol. microbiol. scand., C, Immunol. (in press, 1981).
6 Janković, B.D.; Isaković, K.: Int. Archs Allergy appl. Immun. *24:* 278–295 (1964).
7 Janković, B.D.; Knezević, Z.; Isaković, K.: Folia biol., Praha *25:* 301–302 (1979).
8 Katz, D.H.: in Kunkel, Dixon, Advances in immunology, vol. 29, pp. 137–207 (Academic Press, New York 1980).
9 Lassila, O.; Eskola, J.; Toivanen, P.; Martin, C.; Dieterlen-Lièvre, F.: Nature, Lond. *272:* 353–354 (1978).
10 Lassila, O.; Eskola, J.; Toivanen, P.: J. Immun. *123:* 2091–2094 (1979).
11 Lassila, O.; Martin, C.; Dieterlen-Lièvre, F.; Nurmi, T.E.I.; Eskola, J.; Toivanen, P.: Transplant. Proc. *11:* 1085–1088 (1979).
12 Lassila, O.; Eskola, J.; Toivanen, P.; Dieterlen-Lièvre, F.: Scand. J. Immunol. *11:* 445–448 (1980).
13 Lerman, S.P.; Weidanz, W.P.: J. Immun. *105:* 614–619 (1970).
14 Linna, T.J.; Frommel, D.; Good, R.A.: Int. Archs Allergy appl. Immun. *42:* 20–39 (1972).
15 Martin, C.; Lassila, O.; Nurmi, T.; Eskola, J.; Dieterlen-Lièvre, F.; Toivanen, P.: Scand. J. Immunol. *10:* 333–338 (1979).
16 Moore, M.A.S.; Owen, J.J.T.: Nature, Lond. *215:* 1081–1082 (1967).
17 Romanoff, A.L.: The avian embryo. Structural and functional development (Macmillan, New York 1960).
18 Toivanen, A.; Toivanen, P.: Transplant. Proc. *7:* 165–167 (1975).
19 Toivanen, A.; Toivanen, P.; Lassila, O.; Vainio, O.; Hirota, Y.: in Seligmann, Hitzig, Primary immunodeficiencies, pp. 189–196 (Elsevier/North-Holland, Amsterdam 1980).
20 Toivanen, A.; Toivanen, P.; Eskola, J.; Lassila, O.: in Rose, Payne, Freeman, Avian immunology (British Poultry Science Ltd., Edinburgh, in press).
21 Toivanen, P.; Toivanen, A.: Eur. J. Immunol. *3:* 585–595 (1973).
22 Toivanen, P.; Toivanen, A.; Good, R.A.: J. Immun. *109:* 1058–1070 (1972).
23 Toivanen, P.; Toivanen, A.; Good, R.A.: J. exp. Med. *136:* 816–831 (1972).
24 Toivanen, P.; Toivanen, A.; Linna, T.J.; Good, R.A.: J. Immun. *109:* 1071–1080 (1972).
25 Toivanen, P.; Toivanen, A.; Sorvari, T.: Proc. natn. Acad. Sci. USA *71:* 957–961 (1974).
26 Toivanen, P.; Toivanen, A.; Vainio, O.: J. exp. Med. *139:* 1344–1349 (1974).
27 Vainio, O.: Scand. J. Immunol. *10:* 517–523 (1979).
28 Vainio, O.; Toivanen, A.: J. Immun. *123:* 1960–1964 (1979).
29 Weber, W.T.: Transplant. Rev. *24:* 113–158 (1975).

Dr. Auli Toivanen, Department of Medicine, University of Turku,
SF–20520 Turku 52 (Finland)

The Immune System, vol. 1, pp. 95–101 (Karger, Basel 1981)

Peyer's Patches and the Ontogeny of B Lymphocytes in Sheep

John Reynolds[1]

Basel Institute for Immunology, Basel, Switzerland

Just as there is one particular area in an animal where T lymphocytes are produced (the thymus), so there is one main site of B lymphocyte production, but this site differs from species to species. In birds, it is the bursa of Fabricius that has this role [10], but in mice, guinea pigs and some other mammals, the bone marrow is the most important site [12]. However, the bone marrow is not the main site of B lymphogenesis in all mammals because it is now known that, in sheep at least, the Peyer's patches have this function [7, 16, 17]. In both birds and sheep, therefore, the lymphoid follicles in the wall of the intestine have a function that is crucial for the normal differentiation of B lymphocytes. These morphological and functional similarities, together with evidence from studies of ontogeny, have led to the conclusion that Peyer's patches in sheep are a mammalian equivalent of the avian bursa of Fabricius [16, 17].

Clear evidence of the association between Peyer's patches and the production of B lymphocytes was obtained following the removal of 60–70% of the Peyer's patches from lambs a few days before birth [7, 16, 17]. This was done by surgically resecting the last 10% of the small intestine, where most Peyer's patches are concentrated in lambs [15]. During the 18 months after birth, surface immunoglobulin was never found on more than 5% of the cells in the lymph of these lambs, whereas in normal lambs, the proportion of B lymphocytes present increases from about 5% at birth to adult values of 20–30% by 3 months after birth. The shortage of B lymphocytes after excision of most Peyer's patches is not only restricted to the lymph but is

[1] *John Reynolds* has been a member of the Basel Institute for Immunology since 1979.

Table I. Percentage of B lymphocytes in the lymph and in cell suspensions of various lymphoid tissues from 3 lambs

Source of cells	1 month before birth; normal	10 months after birth; normal	10 months after birth; PP excised[1]
Lymph	3	24	2
Peyer's patches	55	35	–
Thymus	0.1	0.2	0.1
Spleen	8	11	4
Mesenteric node	2	17	7
Prescapular node	5	28	9

Surface immunoglobulin was detected with rabbit anti-sheep immunoglobulin coupled to FITC (Nordic Laboratories, Tilburg).

[1] The last 10% of the small intestine was surgically resected a few days before birth, making the lamb 60–70% deficient in Peyer's patches (PP).

also detected in cell suspensions of the lymph nodes and spleens of these animals (table I). Therefore, lambs deficient in Peyer's patches are also severely deficient in B lymphocytes.

Studies of the types of lymphocyte and their rates of proliferation in different lymphoid tissues also confirm that Peyer's patches contribute greatly to the ontogeny of B lymphocytes in lambs. Analysis of cell suspensions prepared from various lymphoid tissues of sheep aged from 1 month before birth to 12 months after birth showed, in each of 12 sheep examined, that Peyer's patches had the highest proportion of B lymphocytes compared with the other tissues (table I). The differences were most striking before birth: B lymphocytes accounted for 55% of the cells in Peyer's patches, but for less than 10% of the cells in the spleen, lymph nodes and lymph. Peyer's patches are relatively small in fetal lambs, but the lymphoid follicles they contain expand rapidly in size to become one of the largest constituents of the lymphoid system by 2–3 months after birth [16]. Considering their size and their content of cells with surface immunoglobulin, Peyer's patches must have one of the largest pools of B lymphocytes in lambs.

The most logical interpretation of the high proportion of B lymphocytes present in Peyer's patches, in conjunction with the widespread shortage of B lymphocytes after removal of most Peyer's patches, would be that B lymphocytes are generated within the Peyer's patches. Although it has not yet been formally established, it is most likely that B lymphocytes do originate within the Peyer's patch follicles, where cells are produced at rates great enough to satisfy a lamb's B lymphocyte requirements. The follicles contain many large transforming cells (fig. 1) that become labelled after intravenous injection of ^{3}H-thymidine (fig. 2), which shows that they are in the S phase of the cell cycle. In smears prepared from cell suspensions of Peyer's patches biopsied 1 h after ^{3}H-thymidine injection, about 40% of the cells were labelled [17]. 3 days of repeated injections caused almost all of the cells in Peyer's patches to become labelled, showing that cells in the follicles are newly produced [*Reynolds,* unpublished].

An estimate was made of the rate at which the cells divide by injecting the drug colcemid into a number of 3-month-old lambs and counting the proportion of cells arrested in metaphase in smears prepared from Peyer's patches. From this it was calculated that 2–3% of all cells in Peyer's patches enter mitosis each hour [17]. Assuming that each mitosis produces 2 daughter cells, the total number of cells would increase by 2–3% per hour. The Peyer's patches in a 3-month-old lamb contain a total of about 1.0×10^{11} cells and it has been calculated that $2–3 \times 10^9$ cells would need to leave the Peyer's patches each hour to maintain steady-state conditions. (It is valid to assume steady-state conditions because the size of the follicles remains constant in lambs between 2 and 5 months of age [15].) There are a number of possibilities that could account for the disappearance of $2–3 \times 10^9$ cells per hour: (1) Some cells may enter directly into the bloodstream or into the lumen of the intestine, but there is controversy about the existence of such pathways [5, 6]. (2) The most likely route for cells leaving Peyer's patches is to enter the lymphatics which drain the intestine. To examine this possibility the mesenteric lymph nodes were excised from neonatal lambs and 2–3 months later the main intestinal lymphatic was cannulated; lymph collected in this way drained directly, without modification, from the intestine. It was found that the lymph from the whole small intestine transports $5–8 \times 10^8$ cells per hour [15]. However, only a small proportion of these cells could be newly produced in Peyer's patches because most cells in the lymph are long-lived recirculating lymphocytes [8]. These results demonstrate that only a small proportion of the $2–3 \times 10^9$ cells produced each hour in Peyer's patches eventually enter the lymph. (3) It is evident that many lymphocytes never leave the Peyer's patches but die in the follicles. Fragments of cell nuclei are a normal feature in the follicles and are seen in large numbers (fig. 1). Some of the fragments appear to be in the interstitial space, whereas others are clearly within macrophages (fig. 1). Ten hours after an intravenous injection of ^{3}H-thymidine many of the cell fragments

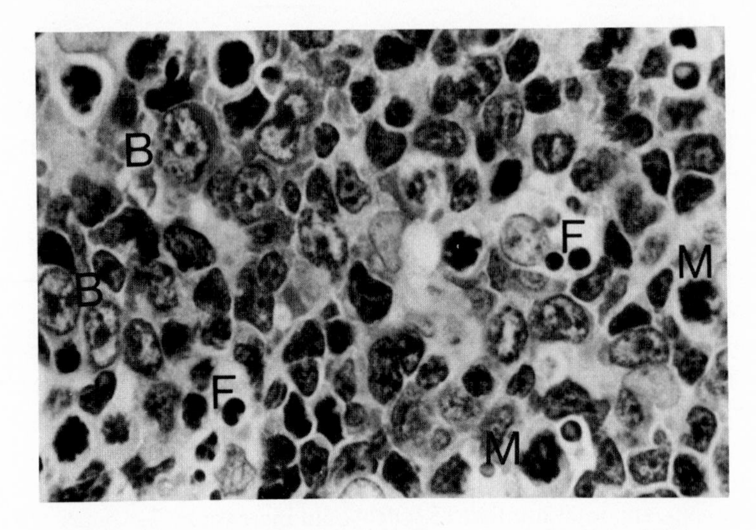

Fig. 1. Cells in a Peyer's patch follicle from a 2-month-old lamb. There are many blast-like cells (B), cells in mitosis (M) and fragments of cell nuclei (F). Methacrylate. HE. × 460.

were labelled, as shown by examining tissue sections by autoradiography [*Reynolds,* unpublished]; this suggests that some cells disintegrate either in the S phase, during mitosis, or after the cells divide. It has also been proposed that there is a high rate of cell death in the thymus, based on comparisons of the rate of cell production and the rate of cell export [19]. However, no confirmation of this has been obtained from morphological studies, because the thymus, in contrast to the Peyer's patches, contains little evidence of cell destruction [2].

One major piece of information lacking is the nature of the cell that is induced to divide in the Peyer's patch follicles; the precursor B lymphocyte, as defined in other species [11], seems to be the most likely candidate. The nature of the stimulus which causes the cells to divide is also unknown, but the possibility that it is simply in response to antigen can be excluded for a number of reasons. Firstly, there is a high rate of cell proliferation in the follicles of fetal lambs during the last third of gestation, when the placenta protects the fetus from extrinsic antigen (fig. 2). This is confirmed by the morphological appearance of the lymph nodes and spleen, both of which change rapidly when antigen is present although in a normal fetus they

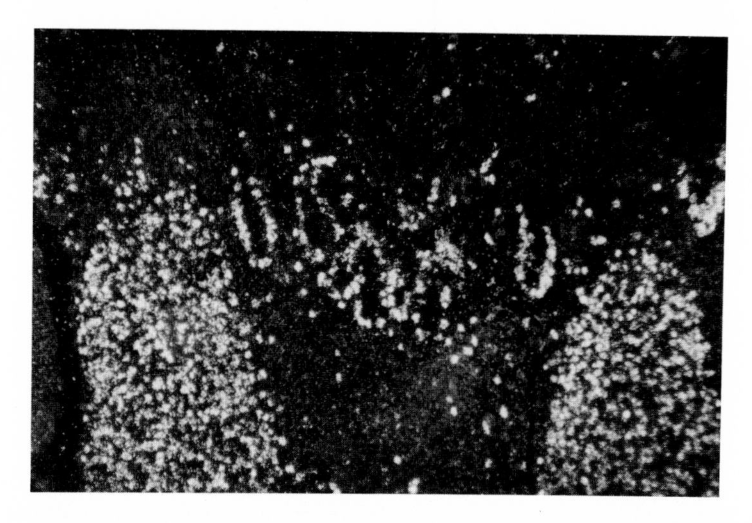

Fig. 2. Distribution of labelled cells in a Peyer's patch of a fetal lamb 10 days before birth and 1 h after an intravenous injection of ^{3}H-thymidine. The dark-field autoradiograph shows that extensive cell proliferation is restricted to the follicles; there are few proliferating cells in the lymphoid region between the follicles. × 60.

appear completely unreactive [15, 16]. In contrast, Peyer's patches and the thymus mature before birth and their development cannot be modified by injecting antigen [15, 16]. Another observation which is evidence against antigen being the main cause of the proliferation is that Peyer's patches begin to involute at 6 months of age, despite the continual presence of antigen [15–17]. Figure 2 shows that the proliferative stimulus is restricted to within the clearly defined boundary of the follicle and has no influence on the surrounding lymphoid tissue. It may be that the epithelial reticular cells, or some other non-lymphoid cells, help provide the microenvironment appropriate for the proliferation in the follicles. Similar cells are thought to have a crucial function in the differentiation of T lymphocytes in the thymus [20].

Since B lymphocytes are produced in Peyer's patches in sheep, the question arises whether this could be a site where antibody diversity is generated. Current concepts of how antibody diversity is established have come from the recent advances in immunoglobulin genetics. It has been shown that the 100–1,000 V segments of DNA and the 5 J segments which code for the variable region of the immunoglobulin light chain are separated in the germ-line genome, but during B lymphocyte differentiation one V segment is joined to one J segment from each group by a process of DNA translocation [1, 18]. The process is similar for the immunoglobulin heavy chain [3]. These gene rearrangements, together with variations in the exact position in the DNA where the junction is formed as well as point mutations in the hypervariable regions, are thought to contribute to antibody diversity [18]. However, there is evidence that abortive gene rearrangements are very common [14, 18] and if defective B lymphocytes are rapidly eliminated this could account for some of the cell wastage seen at sites of B lymphocyte differentiation, such as the bone marrow in guinea pigs [13] and the Peyer's patches in sheep. The importance of mutations in germ line V genes was predicted by *Jerne* [9] in 1971 in his theory on the generation of antibody diversity. He proposed that the primary lymphoid organs, such as the thymus and bursa of Fabricius, are mutant breeding sites and that mechanisms exist for recognizing and destroying any deleterious mutations. Now the Peyer's patches in sheep can also be classified as a primary lymphoid organ [16, 17] and should therefore be considered as a possible site where mutant B lymphocytes arise. Assuming that mutations occur once in every 10^5 cell divisions [9] it can be calculated that the 2–3 $\times$ 10^9 cells which divide in Peyer's patches could, in theory, produce 2–3 $\times$ 10^4 mutations each hour. These figures are based on Peyer's patches from 2- to 3-month-old lambs which are fully immunocompetent but it cannot be discounted, even at this stage of maturity, that Peyer's patches are still producing mutant B lymphocytes. Fetal lambs are able to respond to most antigens by half way through gestation [4] and although Peyer's patches have developed at this stage [15] it is not known if they are already producing lymphocytes. Therefore, confirmation, or otherwise, of the possible involvement of Peyer's patches during the early phase of the development of antibody diversity must await information on lymphogenesis in the earlier stages of ontogeny.

It is not known why the predominant site of B lymphocyte production is the bone marrow in some species and the wall of the intestine in others, but one advantage offered by the demonstration of the importance of Peyer's patches in sheep in this role is that it is now possible to use a new range of techniques to study the ontogeny of B lymphocytes.

Acknowledgements. I am grateful to the following people for their help and advice: *Bede Morris, Ross Cahill, Zdenek Trnka, Gerard Bordmann, Lisbeth Dudler* and *Birgit Kugelberg.*

References

1 Bernard, O.; Hozumi, N.; Tonegawa, S.: Cell *15:* 1133–1144 (1978).
2 Claësson, M.H.; Hartmann, N.R.: Cell Tiss. Kinet. *9:* 273–291 (1976).
3 Early, P.W.; Huang, H.V.; Davis, M.M.; Calame, K.; Hood, L.: Cell *19:* 981–992 (1980).
4 Fahey, K.J.; Morris, B.: Immunology *35:* 651–661 (1978).
5 Fahy, V.A.; Gerber, H.A.; Morris, B.; Trevella, W.; Zukoski, C.F.: Monogr. Allergy, vol. 16, pp. 82–99 (Karger, Basel 1980).
6 Ferguson, A.: Gut *18:* 921–937 (1977).
7 Gerber, H.A.: Functional studies on gut associated lymphoid tissues; thesis, Canberra (1979).
8 Hall, J.G.; Morris, B.: J. exp. Med. *121:* 901–910 (1965).
9 Jerne, N.K.: Eur. J. Immunol. *1:* 1–9 (1971).
10 Lydyard, P.M.; Grossi, C.E.; Cooper, M.D.: J. exp. Med. *144:* 79–97 (1976).
11 Melchers, F.: in Le Douarin, Cell lineage, stem cells and cell determination, pp. 281–288 (Elsevier/North-Holland, Oxford 1979).
12 Osmond, D.G.: J. reticuloendoth. Soc. *17:* 99–114 (1975).
13 Osmond, D.G.; Everett, N.B.: Nature, Lond. *196:* 488–489 (1962).
14 Perry, R.P.; Kelley, D.E.; Coleclough, C.; Seidman, J.G.; Leder, P.; Tonegawa, S.; Matthyssens, G.; Weigert, M.: Proc. natn. Acad. Sci. USA *77:* 1937–1941 (1980).
15 Reynolds, J.D.: The development and physiology of the gut-associated lymphoid system in lambs; thesis Canberra (1976).
16 Reynolds, J.: Gut-associated lymphoid tissues in lambs before and after birth. Monogr. Allergy, vol. 16, pp. 187–202 (Karger, Basel 1980).
17 Reynolds, J.D.; Cahill, R.N.P.; Trnka, Z.: in Solomon, Developmental and comparative immunology, vol. 1 (Pergamon Press, Oxford, in press, 1981).
18 Robertson, M.: Nature, Lond. *287:* 390–392 (1980).
19 Scollay, R.G.; Butcher, E.C.; Weissman, I.L.: Eur. J. Immunol. *10:* 210–218 (1980).
20 Zinkernagel, R.M.; Callahan, G.N.; Althage, S.; Cooper, P.; Klein, P.A.; Klein, J.: J. exp. Med. *147:* 882–896 (1978).

Dr. John Reynolds, Basel Institute for Immunology, Postfach, CH–4005 Basel (Switzerland)

The Immune System, vol. 1, pp. 102–109 (Karger, Basel 1981)

Monoclonal Antibodies as a Tool for the Study of Embryonic Development

R. Kemler, P. Brûlet, F. Jacob[1]

Service de Génetique cellulaire du Collège de France et de l'Institut Pasteur, Paris, France

Although immunology and developmental biology have ignored each other, it was *Niels Jerne's* opinion that they would progressively increase their mutual interest. Analysis of the differentiation process at any stage of development requires the use of cell markers. Immunological techniques can provide a source of such markers insofar as specific reagents can be obtained. In this respect, the introduction by *Köhler and Milstein* [11] of the myeloma fusion technique for the production of monoclonal antibodies (mAbs) has proved invaluable. Not only do monoclonal antibodies constitute especially reliable reagents, they also provide a new way of dissecting molecular components of complex structures.

Mouse Embryo and Teratocarcinoma

In the past years, our laboratory has been interested in the early stages of development of the mouse embryo. In addition to the analysis of embryos, embryonal carcinoma (EC) cells and some of their differentiated derivatives were isolated from mouse teratocarcinomas and used as a material for in vitro studies [6]. A large part of this work was focussed on the cell membrane of EC cells and early embryonic cells. The aim was to define some of the cell surface antigens for two reasons. First, to obtain specific cell markers for early stages of development, and second to define some of the

[1] *François Jacob* was a member of the Board of Advisors of the Basel Institute for Immunology in 1976.

structures involved in cell-cell interaction and eventually in the differentiation process [7].

A step in the latter direction was realized with the use of a xenogenic rabbit anti-EC serum. Monovalent Fab fragments prepared from IgG of such sera reversibly inhibit compaction and differentiation, both in morula and aggregates of EC cells [9]. A gp 84 glycoprotein has thus been purified which appears to take part in the compaction process of these cells [5].

Detection of specific cellular markers at early stages of mouse development was attempted with the use of allogeneic and syngeneic sera. EC cells are devoid of H-2 histocompatibility antigens [2]. Syngeneic anti-EC cell antisera have allowed the detection of a cell surface-marker called F9 [1]. This F9 antigen is present on all EC cells but disappears when they differentiate. It is absent from adult somatic cells but present at all stages of spermatogenesis. Finally, it can be detected on multipotential cells of the mouse embryo. It turned out, however, that the anti-F9 syngeneic antisera contain an heterogeneous population of antibodies that detect several distinct surface antigens [4]. Hybridomas were then prepared in order to obtain more specific reagents.

Monoclonal Antibodies against EC Cells

With conventional antisera, syngeneic immunization had been used in order to obtain a restricted immune response and avoid as much as possible polyabsorptions of the sera. In contrast, to prepare specific monoclonal antibodies, xenogeneic immunizations were first used in order to obtain a broader immune response. As long as cell surface determinants are not present on all cell types but exhibit a specific distribution, many of them could be useful as markers for a study of development. Rats were, therefore, immunized with mouse EC cells and their splenocytes were fused with mouse myeloma cells. A high number of hybridomas was thus obtained and maintained, 50% of which secreted some mAbs reacting with EC cells. Among these, a variety of different specificities was found and the clones were in general stable [10].

Despite these properties, however, the mAbs thus obtained did not prove very useful for two main reasons. (1) The presence of the target antigens could not be correlated with particular cell types – rather these antigens corresponded to what *Milstein and Lennox* [12] called 'jumping antigenic specificities', and (2) most antigens turned out to be probably of glycolipidic nature and difficult to characterize biochemically.

Monoclonal antibodies against EC cells were then prepared after immunization of allogeneic BALB/c mice. Although hybridomas producing anti-EC cell specific mAbs are about 100 times less frequent with mouse than with rat splenocytes, these antibodies turned out to detect surface antigens with more restricted cell distribution. For instance one of the hybridomas produced using splenocytes of a BALB/c mouse immunized with EC cells PCC4 secreted a mAb reacting only with EC cells – but not with differentiated derivatives – as well as with mouse preimplantation embryos and ectodermal cells of 6- to 8-day embroys [8]. The chemical nature of this surface antigen has not yet been elucidated. Actually, similar difficulties in the biochemical characterization of antigenic structures detected by mAbs against EC cells have been reported from other laboratories [13, 14].

Monoclonal Antibodies against a Protein Marker of Trophectoderm

In the development of the mouse embryo, the first morphological differentiation occurs during the transition morula → blastocyst. While all the blastomeres that compose a morula look alike, the blastocyst is made of two cell types: an outer layer of trophectoderm cells surrounds a small inner cell mass (ICM). Using two-dimensional gel electrophoresis, the patterns of the proteins synthesized by trophectoderm and ICM cells were analyzed and compared with those of various teratocarcinoma-derived cells [3]. Three of the proteins that characterize trophectoderm cells were found to copurify with preparations of intermediate filaments from these cells. These proteins are also present in parietal yolk sac cells as well as in a trophoblastoma cell line derived from teratocarcinoma, but not in ICM or in EC cells. Monoclonal antibodies were prepared by fusing mouse myeloma cells with splenocytes from rats immunized against a preparation of intermediate filaments from trophoblastoma cells. Out of a series of 92 independent hybridomas, several were selected that secreted mAbs reacting with trophoblastoma and parietal yolk sac cells, but not with EC cells. In the embryo, they were found to react with trophectoderm but not with ICM cells. In indirect immunofluorescence tests, these mAbs decorate an intracellular network in attached trophoblastoma and trophectoderm cells.

A number of monoclonal antibodies secreted by independently isolated hybridomas were thus found to give a similar pattern of reactions. When the reactivity of these mAbs towards the protein preparation of intermediate filaments was tested, different patterns of reactivity were observed (fig. 1). It

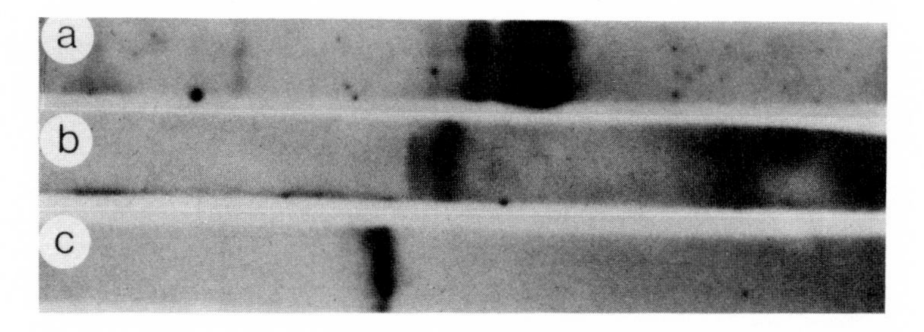

Fig. 1. Biochemical identification of the intermediate filament proteins reacting with mAbs, revealed by a protein transfer as described in ref. 3. *a* TROMA 1; *b* TROMA 2; *c* TROMA 3.

seems, therefore, that mAbs directed against different antigenic determinants react with the same intracellular network. Three of these mAbs, each detecting a different protein pattern, were used for further investigations. They are called TROMA 1, 2 and 3 (TROMA for trophectoderm monoclonal antibodies).

Reactivity Pattern of Three Anti-Intermediate Filament Monoclonal Antibodies during Embryonic Development

It has proved difficult to analyze the reactivity of conventional antisera with sections of embryos at different ages because of frequent non-specific labeling. In contrast, clearcut differential staining could be obtained with the mAbs just described, using an indirect immunofluorescence test with a second layer of antibodies purified by affinity column and conjugated with fluorescein isothiocyanate. At various times during development, cryostat sections of embryos were prepared and their reactivity with the three TROMAs was investigated.

Up to day 10, the three mAbs exhibit the same reactivity: they stain exclusively trophectoderm as well as parietal and visceral endoderm. At day 12, the three mAbs react with most of the epithelial cells – gut, lung, pancreas, kidney, thymus (fig. 2), uterus and skin – whatever their embryonic origin, while neuroepithelium remains always negative. Liver cells, however, are strongly stained with TROMA 1 and 2, but not with 3. After day 14, the three mAbs no longer stain skin epithelium while they still react with mesoderm and endoderm-derived epithelia.

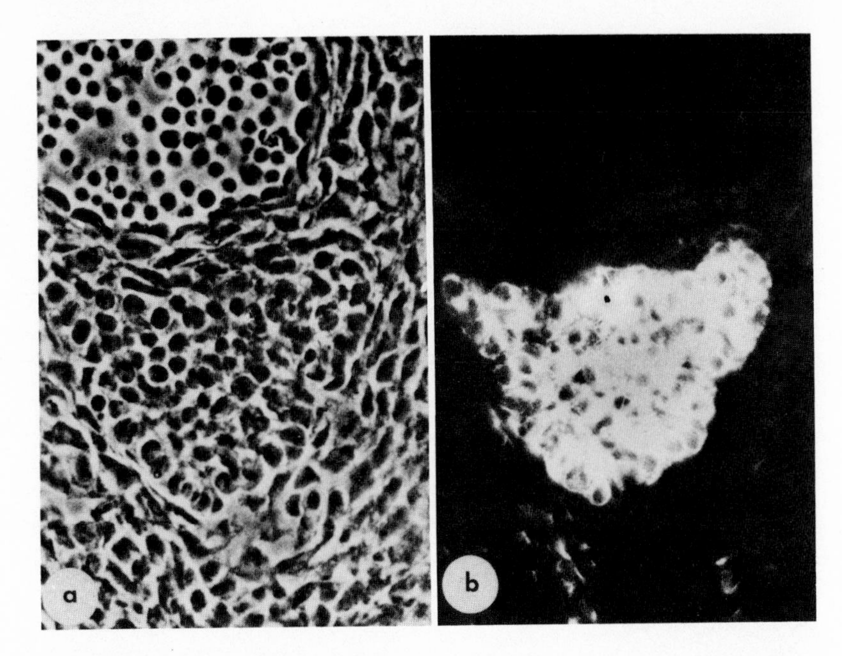

Fig. 2. Indirect immunofluorescence test with TROMA 1 on cryostat sections of thymus epithelial cells of a 12-day-old embryo. *a* Phase, note the embryonic aorta in the left upper part; *b* immunofluorescence. × 880.

Adult tissues were then investigated in a similar way. Skin epithelium was not stained by the three mAbs which were again found to react preferentially with epithelia of endodermal and mesodermal origin. Differences could, however, be detected in the reactivity of epithelia with the three TROMAs as shown by three examples. (1) In gut sections (fig. 3), TROMA 1 and 3 react strongly with all epithelial cells while TROMA 2 reacts strongly with goblet cells and only very weakly with epithelial cells. (2) In salivary glands (fig. 4), TROMA 1 and 2 react with salivary duct cells,

---

Fig. 3. Indirect immunofluorescence test with TROMA 1 and 2 on cryostat sections of adult gut epithelium. *a* TROMA 1 labels uniformly all epithelial cells; *b* TROMA 2 reacts heavily with goblet cells. × 880.

Fig. 4. Indirect immunofluorescence test with TROMA 1 and 3 on cryostat section of adult salivary gland. *a* TROMA 1 reacts with duct and acini epithelial cells; *b* TROMA 3 labels duct epithelial but not acini cells. × 560.

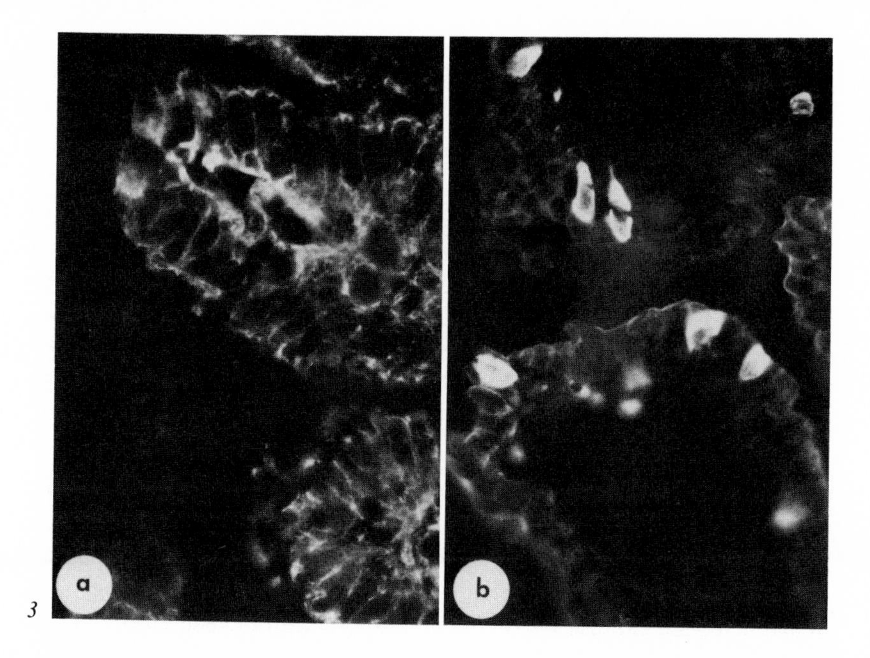

3

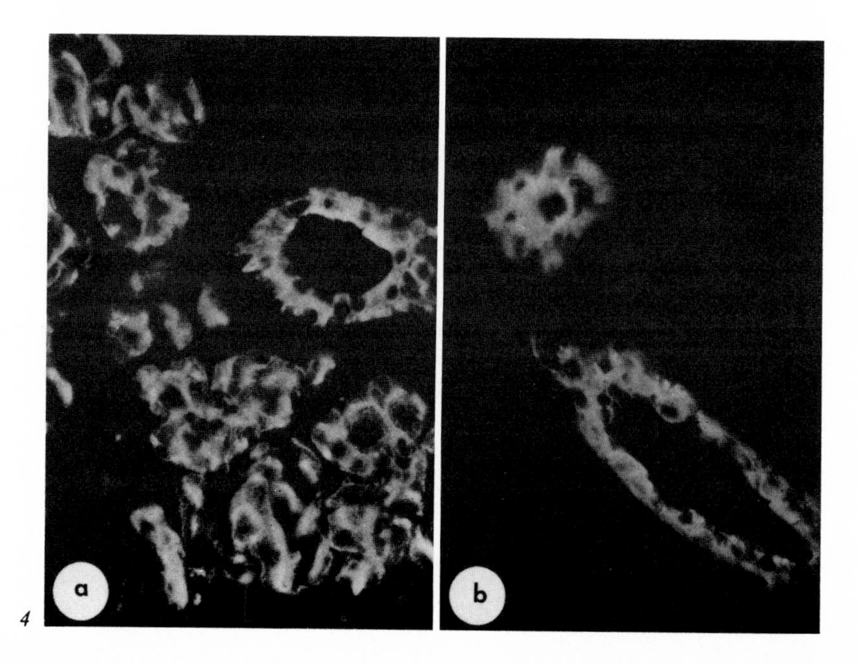

4

intercalated duct and acini cells, while TROMA 3 stains only intercalated and salivary duct cells. (3) In bladder sections, TROMA 3 reacts only with the internal epithelial layer while 1 and 2 react with the 3 layers.

Conclusion

A few years ago, it seemed possible to obtain markers specific for a particular cell state at a certain stage of embryonic development. The use of such markers would, therefore, allow unambiguous characterization of the developmental cell state. This, however, appears now more and more unlikely for early stages of development.

There exist some markers which define unique cell states of terminal differentiation, such as hemoglobin, immunoglobulin and so forth. Such specific markers, however, have not yet been found, and actually one should perhaps not expect to find as specific a marker as hemoglobin, to characterize, say, early mesodermal cells. For early states, it seems as though a particular cell state expresses a particular combination of gene products which are not unique for that state, but are expressed in different places and at different times in development. What is unique is probably not some particular product, but a particular *combination* of products also used elsewhere. In French restaurants, there are two possible types of meal: one may eat a *menu,* in which one gets a fixed, unchangeable course of dishes, or one can eat *à la carte* and just choose any combination, in quality and quantity, of those dishes which best fit one's present mood or appetite. Gene products expressed in early stages of development seem to represent a combination 'à la carte' rather than a fixed menu.

Acknowledgements. This work was supported by grants from the Centre National de la Recherche Scientifique, the Délégation Général à la Recherche Scientifique et Technique, the Fondation pour la Recherche Médicale Française, the Institut National de la Santé et de la Recherche Médicale, the Fondation André Meyer, and the Ligue Française contre le Cancer.

References

1 Artzt, K.; Dubois, P.; Bennett, D.; Condamine, H.; Babinet, C.; Jacob, F.: Proc. natn. Acad. Sci. USA *70:* 2988–2992 (1973).
2 Artzt, K.; Jacob, F.: Transplantation *17:* 633–634 (1974).

3 Brûlet, P.; Babinet, C.; Kemler, R.; Jacob, F.: Proc. natn. Acad. Sci. USA *77:* 4113–4117 (1980).
4 Damonneville, M.; Morello, D.; Gachelin, G.; Stanislawski, M.: Eur. J. Immunol. *9:* 932–937 (1979).
5 Hyafil, F.; Morello, D.; Babinet, C.; Jacob, F.: Cell *21:* 927–934 (1980).
6 Jacob, F.: Imm. Rev. *33:* 3–32 (1977).
7 Jacob, F.: Proc. R. Soc. Lond. B *201:* 249–270 (1978).
8 Kemler, R.: in Lindauer, Fortschritte der Zoologie, Würzburg Symp.: Progress in Developmental Biology (1980).
9 Kemler, R.; Babinet, C.; Eisen, H.; Jacob, F.: Proc. natn. Acad. Sci. USA *79:* 4449–4452 (1977).
10 Kemler, R.; Morello, D.; Jacob, F.: in Le Douarin, Cell lineage, stem cells and cell determination. Inserm Symp. No. 10 (Elsevier/North-Holland/Biomedical Press, Amsterdam 1979).
11 Köhler, G.; Milstein, C.: Nature, London *256:* 495–497 (1975).
12 Milstein, C.; Lennox, E.: Curr. Top. devl Biol. *14:* 1–32 (1980).
13 Solter, D.; Knowles, B.B.: Proc. natn. Acad. Sci. USA *75:* 5565–5569 (1978).
14 Stern, P.; Willinson, K.; Lennox, E.; Gafre, G.; Milstein, C.; Secher, D.; Ziegler, A.: Cell *14:* 775–783 (1978).

Prof. F. Jacob, Unité de Génétique cellulaire, Département de Biologie Moléculaire, Institut Pasteur, 25, rue du Dr-Roux, 75724 Paris Cédex 15 (France)

The Immune System, vol. 1, pp. 110–115 (Karger, Basel 1981)

Monoclonal Antibodies in the Study of the Differentiation of Mammary Cells

M. Unger, H. Battifora, R. Dulbecco[1]

The Salk Institute, La Jolla, Calif., USA

This article is the continuation of a conversation that one of the authors *(R.D.)* and *Niels Jerne* initiated 30 years ago. The work described here was initiated as part of an effort to develop new approaches to understanding, diagnosing, and treating human cancer. It is based on the recognition that such developments must be centered on cellular specificity. Specificity may be the result of several processes such as normal differentiation, abnormal differentiation due to an alteration of the normal pattern (e.g. retrograde differentiation), or abnormal differentiation due to the expression of abnormal genes (e.g. viral genes, or structural cellular genes altered by mutation). The specificity of a cell is reflected in the set of gene products it produces.

Determination of the components produced by a cell type presents two technical difficulties. The first difficulty is that animal tissues, both normal and neoplastic, contain a multiplicity of cell types. This difficulty can be overcome by cultivating cells in vitro, generating clonal lines. These lines may in turn undergo variation, but if a cell with a well-defined set of characteristics is utilized as the paradigm, it can be maintained by subsequent cloning. The other difficulty concerns the large number of cellular products, which are generated in different quantities. The more abundant products are not necessarily the most characteristic for the cell types. This difficulty can be overcome by the preparation of monoclonal antibodies to a cell type, and by their use to isolate the corresponding antigens.

[1] *Renato Dulbecco* was a colleague of *Niels Jerne* at Caltech in Pasadena and was a member of the Board of Advisors of the Basel Institute for Immunology from 1970 to 1975.

The work described conforms to the above strategy. It utilizes the identification of cell types in the mammary gland of the young female rat. It is based on the production and characterization of monoclonal antibodies to two clonal cell lines derived from a rat mammary carcinoma. The two cell lines are related to each other because one (fusiform cells) is derived from the other (cuboidal cells) by in vitro differentiation. Therefore, this system permits an insight into the type of changes of gene expression in differentiation.

Materials and Methods

Cell Lines. The cells employed are derived from the RAMA 25 line which was obtained from a DMBA-induced mammary carcinoma in a Sprague-Dawley rat [1]. The RAMA 4 line has fusiform cell morphology and is derived from the same tumor. The LA7 line is a clonal derivative of the RAMA 25 line with cubical morphology. Both the RAMA 25 and the LA7 lines continue to generate fusiform cells with the characteristics of the RAMA 4 cells at low frequency.

Production of Monoclonal Antibodies (McAb). Balb/c mice were inoculated intraperitoneally three times (every 14 days) with $5 \cdot 10^7$ cells, and a month later with $2\text{–}3 \cdot 10^6$ cells intravenously. 4 days later spleen cells were fused with the NS1 myeloma cell line [6] and hybrid cells were isolated in selective medium [5]. After cloning, the antibody released into the culture medium was characterized by a binding assay using the cells of the line used in the immunization and a ^{125}I-labeled second antibody; an increase of fourfold over background was regarded as significant. The antigen identified by a given McAb was determined by labeling cells with ^{125}I by the lactoperoxidase-glucose oxidase reaction [2] and solubilizing the membrane protein with NP-40. The extract was immunoprecipitated with the McAb; the proteins in the immunoprecipitates were separated and identified by SDS-PAGE electrophoresis followed by radioautography. The McAbs were also characterized for the ability to bind to *S. aureus* protein A [3]. The heavy chain classes were determined by biosynthetic labeling with ^{3}H-leucine followed by SDS-PAGE and by immunoprecipitation with specific antisera. McAbs are characterized for their ability to bind to both LA7 and RAMA 4 cells by measuring binding at antibody saturation using an excess of both the McAb and the second antibody.

The ability of McAb to bind to cells in the rat mammary gland or in other organs was determined by indirect immunofluorescence of frozen sections prepared in a cryostat. The second antibody (goat anti-mouse IgG) was purchased from Cappel Labs or Antibodies, Inc.

Results

Many antibody-secreting hybridomas have been isolated and are at various stages of characterization. Several generalities can be made regarding their properties.

Table I. Histochemical specificity of hybridomas

Hybrid	Eliciting cell type	Relative binding[1]	Antigen[2]	Immunofluorescence of sections[3]			
				LA7	RAMA 4	mammary epithelium	other sites
57B15	RAMA 4	6	H 1	++++	++++	+++	endothelial cells
57B23	RAMA 4	6	±[5]	+	+++	++	basement membrane[4] in kidney glomeruli and other organs
57B33	RAMA 4	8.5	> 100kd	±	++++	++	renal brush borders *liver canaliculae*
57B38	RAMA 4	3	< 50kd	++++	++++	+++	myelin
57B41	RAMA 4	4	25kd	++	+++	ND	ND
50A35	LA7	36	> 120kd	++++	±	+ (and stroma)	ND
50A4	LA7	12	94kd	+++	–	ND	ND
SF9A9	LA7	9	120kd	++++	++	++ (and stroma)	basement membranes[4]
SF9A12	LA7	6	50kd	++++	++++	++	negative
SF9A16	LA7	11	100kd	++++	–	±	negative

[1] Increase in binding of ^{125}I-labeled goat anti-mouse immunoglobulin over background.
[2] Immunoprecipitation of ^{125}I-labeled cell surface protein as revealed by SDS-PAGE.
[3] Indirect immunofluorescence of cell or tissue sections relative intensity of fluorescence is indicated by –, +, up to four +s.
[4] Different distribution of label in kidney glomeruli.
[5] Ambiguous results.

Specificity of McAbs for Cuboidal and Fusiform Cells. Out of 178 hybridomas that have been tested, 152 produce McAbs each able to bind both LA7 and RAMA 4 cells. Of them, 74 were elicited by immunization with LA7 cells, 78 by RAMA 4 cells. Of the remaining 26 McAbs, 18 (elicited by LA7 cells) bind to LA7 cells (counts bound more than five-fold background) but not to RAMA 4 cells (count bound not significantly greater than background). The remaining 8 McAbs (elicited by RAMA 4 cells) bind only to RAMA 4 cells.

Out of a set of 178 McAbs, 54 immunoprecipitated an antigen, whereas 124 could not. Presence or absence of an immunoprecipitated antigen does not correlate with the extent of binding. For McAbs that recognize both

LA7 and RAMA 4 cells the immunoprecipitated antigens (if any) derived from the two cell types migrate at the same rate in PAGE-SDS and are probably identical.

Specificity of McAbs for Antigens of the Adult Rat. For a set of McAbs the ability to recognize cells or constituents of various rat tissues was determined by immunofluorescence of frozen sections. The results are reported in table I, which also reports the results of the interactions of these antibodies with LA7 or RAMA 4 cells, such as binding and immunoprecipitation of surface antigens and immunofluorescence in sectioned cell pellets. The immunofluorescence data with LA7 or RAMA 4 cells show reasonable agreement with the binding data. Observations with mammary gland (50- to 60-day-old virgin females) show little correlation with the specificity detected with the cultured cells; moreover, none of the antibodies of this set distinguishes between epithelial and myoepithelial cells. Observations with other tissues (brain, liver, kidney, thymus, intestine) show some crossreactions with cells of other unrelated tissue. Cross-reactions appear to be mainly directed toward surface structures (basement membrane, brush border, myelin). The two basement membrane McAbs recognize different components; in fact the distribution of immunofluorescence in sections of kidney glomeruli and the immunoprecipitated antigens are different for the two antibodies.

Discussion

The results show that McAbs can clearly distinguish between two cell types related by a single differentiation step, such as the cuboidal LA7 and the fusiform RAMA 4 cells. When this differentiation from the cuboidal to the fusiform type takes place, it involves many surface changes, both acquisition and loss of antigens. The fact that in a sample of 178, 26 McAbs distinguish the two cell types suggests that the actual number of surface changes is much larger. The binding data do not indicate whether the differences between the two cells are absolute (i.e. one antigen absent on a cell type) or quantitative. They do not say whether an antigen absent from the surface of a cell type is present internally – as is the case for the Thy-1 antigen, which is absent from the surface of the cuboidal RAMA 25 cells but present internally, and appears at the surface in the fusiform RAMA 4 cells [4].

The results obtained with the small set of McAbs of table I show also the converse effect, i.e. that the same antigen can be present on cells of different tissues. Several of these antigens are components of specialized structures present at the surface of many cell types.

These two sets of results taken together suggest that differentiation involves a change in the expression of many surface components, but that some of these components are utilized in cells at distant stages of differentiation. The individuality of differentiated cells rests therefore on patterns of antigens. This does not exclude that some antigens are unique to a given cell type (at least in a quantitative sense as ascertained by immunofluorescence).

The cross-reactions of certain McAbs between the two types of cultured cells and cells of various organs may be useful for characterizing the cell lines. The fact that both McAbs reacting (in different ways) with basement membrane are present in large amounts on the surface of RAMA 4 cells may support the hypothesis that these cells are equivalent to the myoepithelial cells present in the gland, as has been suggested [1]. However, LA7 cells also produce one of the two constituents in large quantity; and they produce collagen IV, which is a component of basement membranes [unpublished data]. Hence they also display similarities to myoepithelial cells. On the other hand, if the RAMA 4 cells correspond to myoepithelial cells, they should not produce constituents present at the luminal surface in other tissues (such as kidney brush border or liver bile canaliculi). In either cell line, the presence of abnormal constituents may result from a perturbation of differentiation due to the neoplastic state of the cells.

Whether the McAbs described so far, or others, will be useful for distinguishing cell types within the mammary gland, remains to be seen. The absence of McAbs able to distinguish between various cell types in the adult mammary gland may be due to the very small number so far examined. It may also result from the limitation of our study to a single developmental stage of the mammary gland, if the tumor from which the lines we used as immunogens derived from a precursor cell sparsely represented in the test material.

The numerous questions raised by the results obtained so far are being pursued in the hope of attaining the goals stated in the introduction.

Acknowledgements. This work was supported by Grant 1-R01CA21993 of the National Cancer Institute, and by grants from the Hammer Foundation, the Educational Foundation of America, the Pardee Foundation, and the V. Samuel Roberts Noble Foundation, Inc. This research was conducted in part by the Clayton Foundation for Research – California Division. *M. Unger* is a Clayton Foundation Investigator; *R. Dulbecco* is a Senior Foundation Investigator. The permanent address of *H. Battifora* is Northwestern Memorial Hospital, Surgical Pathology Department, Chicago, Ill.

References

1 Bennett, D.C.; Peachey, L.A.; Durbin, H.; Rudland, P.S.: Cell *15:* 283–298 (1978).
2 Hynes, R.O.: Proc. natn. Acad. Sci. USA *70:* 3170–3174 (1973).
3 Kessler, S.W.: J. Immun. *115:* 1617–1624 (1975).
4 Lennon, V.A.; Unger, M.; Dulbecco, R.: Proc. natn. Acad. Sci. USA *75:* 6093–6097 (1978).
5 Littlefield, J.W.: Science *145:* 709–710 (1964).
6 Williams, A.F.; Galfre, G.; Milstein, C.: Cell *12:* 663–673 (1977).

Dr. Renato Dulbecco, The Salk Institute, PO Box 85800, San Diego, CA 92138 (USA)

Antibodies and B Cell Function

The Immune System, vol. 1, pp. 116–123 (Karger, Basel 1981)

Remembrance of Plaques Past

Claudia Henry[1]

Department of Microbiology and Immunology, University of California, Berkeley, Calif., USA

With the latitude to write in this 'Festschrift' of anything 'that you think *Niels* would enjoy reading', I have opted to dust off old records and write of recollections of his years in Pittsburgh and Frankfurt from 1962 to 1968. These years contained within them the development of the practical technique that has been the basis of quantitative cellular immunology – the plaque technique.

When *Niels* left the WHO in Geneva in 1962 to assume a Chair in Pittsburgh, he came with the intention of developing a simple and accurate technique for quantitating antibody-secreting cells, even when they were a minority population as in a primary response. His appreciation of the need for such a cellular assay and the requirements it must meet were born of his talent for quantitative thinking, while experience in the bacteriophage world indicated the way. He found himself among colleagues of other disciplines who knew of his natural selection theory and had heard his plea for simplified immunological terminology, coining such words as epitope, paratope, etc. [6]. No one, save *Al Nordin,* knew anything of practical immunology. Realizing this, *Niels* enticed *Ole Rostock* from Copenhagen for a short period to help set up a laboratory. *Ole's* sense of alienation reached its peak during a trip to the Allegheny mountains and he left with relief, bequeathing us wormwood and a recipe for absinthe. *Niels* also pronounced the Pennsylvanian woods as unacceptably random and advanced the opinion that the

[1] *Claudia Henry* is a long-time friend and colleague of *Niels Jerne* and was a member of the Basel Institute for Immunology in 1979.

belching stacks of the antiquated J & L mills were artistically superior. *Al*, choosing to look for hemolysin-producing cells, had meanwhile injected rabbits with sheep red cell stroma. In the phage manner, he and *Niels* poured soft agar containing various concentrations of lymph node cells and sheep red cells (SRC) into Petri dishes and added guinea pig serum as a source of complement. A proselyte from an adjacent corridor, I was invited to share the excitement of the first plaques [9]. For my PhD I had exploited the techniques of the phage workers and alarming levels of ^{32}P to study the structure and replication of polio virus RNA. The complexity and loneliness of this endeavor, *Niels'* assertion that 'RNA replication can hold no novelties', and above all the enthusiasm of the two plaque makers, all combined in my decision to join them. There were few improvements [10] to the original method. To eliminate the unreliability of the complement, I drew on my animal virus heritage to propose that DEAE dextran might neutralize the anti-complementary properties of the agar. *Aaron Stock* recognized the need for better definition and, calling on a clinical past, suggested that the plates be stained with benzidine. Holding the dark blue plates at arm's length and looking like his compatriot, *Tycho Brahe, Niels* delivered his verdict: 'like stars in the heavens'. We made some attempts to disclose cells making anti-phage antibodies as 'reverse' plaques, but the system of dividing bacteria, replicating virus and secreting lymphocytes was clearly too complex. We therefore decided to use the hemolytic plaque assay to unravel the mysteries of the primary immune response. With the arrival of *Hiroshi Fuji* with *Kiyoko* and their 'twin princes', *Aurelia Koros* to do her graduate work, and *Art Park*, the Pittsburgh nucleus was now complete.

For the next 3 years considerable energy was invested to establish the kinetics of the primary plaque-forming cell (PFC) response of mice to increasing doses of SRC, and to disclose the significance of the specific PFC which we invariably found in the spleens of unimmunized mice. We hoped that the dose-response relationships would enable us to quantify the number of cells that could respond to a single antigen, and that they might also help to identify the immediate target of the antigen. Influenced by the natural selection theory [5], we entertained the possibility that a specific event involving 'natural' antibody might precede stimulation of the specific precursor cells predicted by clonal selection [1]. Establishment of the curves required only hard work and the mass slaughter of mice. We were immediately impressed by the variability between mice, both in their response to SRC and the PFC of unstimulated mice. We found these distributions to be approximately log normal or, more precisely, negative binomial, a distribution that arises from a simple process with constant probabilities of immigration, birth, and death. To establish the experimental points with reasonable confidence, we used a minimum of 30 mice for each experimental point. When snowfalls blocked the streets, responsibility fell on those who could walk or ski to the microscopes, and *Niels* proved to be a bulwark, trudging down an avenue of snow-laden elms.

The PFC responses per spleen to i.v. doses of 4×10^4 to 4×10^9 SRC in tenfold increments are the solid lines in figure 1. As a source of challenge and torment to us for at least 6 years, I feel they have earned the right now to public display. We aired them in an immunological information bulletin [11], deferring 'real' publication until we had satisfactorily interpreted all their salient features, a goal that perpetually eluded us. We proceeded to establish that the kinetic curves as illustrated were true measurements of the splenic response to tenfold increments of antigen. Using ^{51}Cr-labeled SRC, we found that within 4 h about 1 % of the antigen arrived in the spleen over the dose range of 4×10^4 to 4×10^7 and somewhat more at higher doses. Eyeballs throbbed as we measured plaque diameters under different plating conditions and at different stages of the response. Since methods that increased plaque diameters did not increase the number of small plaques, we concluded that all cells secreting anti-SRC antibodies did so in amounts above the threshold of detection. We also established that departure of PFC from the spleen did not materially change the curves. The dotted line in figure 1, which represents the PFC recovered from an estimated blood volume of 2 ml, shows that the number of PFC in the blood is 15% of the number in the spleen during the rise of the response curve. By splenectomizing mice before injection, we established that the spleen contributed the majority of blood PFC. By splenectomizing mice with large numbers of blood PFC, we determined that about 30% left the blood in 24 h. We thus calculated that loss to the blood represented about 23% of the PFC remaining in the spleen and would not significantly alter the curves. *Aurelia* and *Fuji* established that the exponential increases represented cell multiplication. Autoradiographs of spleen suspensions obtained from injected mice or chickens and exposed to pulses of ^{3}H-thymidine revealed that about half the PFC were heavily labeled. *Fuji* concentrated on the responses of chickens, where the use of chicken serum as a complement source maintained the PFC in a condition permitting cytological studies. Delicately adding three dyes, he proclaimed all plaque formers to be members of the plasma cell series with immature plasma cells and blasts predominating early after immunization and mature plasma cells at later stages and in unimmunized

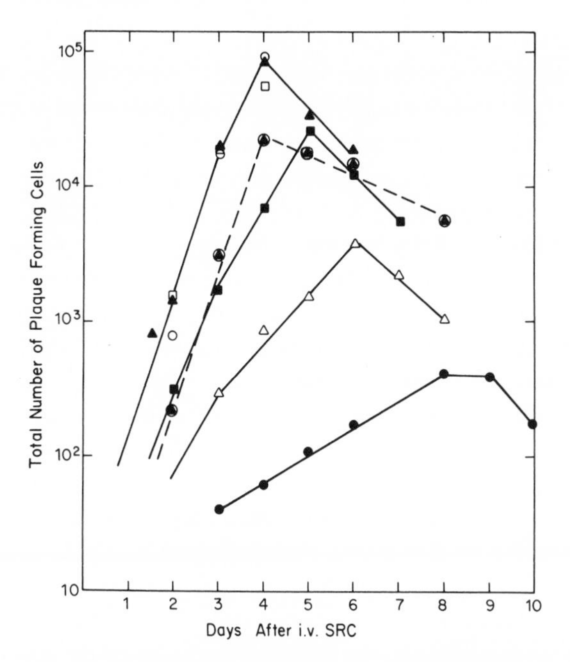

Fig. 1. The solid lines are the PFC responses per spleen after an i.v. injection of SRC of the indicated doses: ● = 4×10^4; △ = 4×10^5; ■ = 4×10^6; ○ = 4×10^7; ▲ = 4×10^8; □ = 4×10^9. Each point is the arithmetic mean of 40–80 mice after deduction of the mean normal PFC. The dotted line is the mean PFC response in peripheral blood (2 ml) after an i.v. injection of 4×10^8 SRC. Each circle containing a triangle is the arithmetic mean of 10–30 mice.

chicks. Administration of colcemide 3 h before killing arrested about one third of the immature cells in metaphase. *Al* and I, whose initial attempts to remove a thymus had yielded a salivary gland, had now been reassured by new-found surgical skills and the resistance of mice, and expanded our techniques to include hemisplenectomy. Though we established that the PFC levels in the two halves of a normal spleen were approximately the same, we found no correlation between the number of preexisting plaques in one half-spleen and the response elicited by antigen from the remaining half. Thus, the combined evidence indicated that the average of about 100 PFC per unimmunized spleen were terminal cells that arose spontaneously and were *not* the targets of the antigen.

In proposing models to account for the dose-response curves we began with the simplest one, essentially a restatement of the clonal selection theory. We assumed that there were a limited number of precommitted precursor cells, *n,* each of which would respond to a single hit by a red cell with development of a clone of PFC. This postulate accounted for the phenomenon of exhaustion of the response by 4×10^7 SRC, and our finding that saturation with one antigen did not affect the capacity to respond to another antigen which aims at a different target. We could also estimate the number of target cells responsive to SRC. If *h* was the probability that a red cell effectively hit a given target cell, the probability that a given target cell escaped a hit from a dose of D red cells would be e^{-hD}. A mouse with n target cells would remain unhit with probability e^{-hDn}. The dose 4×10^6 SRC, which gave 25% of the maximal response, presumably engaged only one fourth of the available target cells ($e^{-h4 \times 10^6} = 0.75$). We ascertained that one eighth of the mice given 4×10^4 SRC were indistinguishable from unimmunized mice, both in their response to that dose or in their behavior to reinjection ($e^{-hn4 \times 10^4} = 0.125$). Thus, $h \simeq 7 \times 10^{-8}$ and $n \simeq 700$. We assumed that differentiation limited the clonal yield of a hit target cell to 100–200. In accord with expectation, the peak levels attained were approximately proportional to dose. The model, however, had one disqualifying deficiency. It failed to explain why the curves were not parallel and reached their maxima at different times. Figure 1 shows that the apparent rate of multiplication of PFC was dose-dependent, ranging from an apparent doubling time of 7 h for the dose of 4×10^7 to 44 h for 4×10^4 SRC. The incorporation into the model by *Dankward Kodlin* of the assumptions of a variable target with a negative binomial distribution and a distribution of hit times gave a slightly different estimate of *n,* but did not materially affect the slopes. We concluded that we must abandon the simple notion that antigen acts independently or only once. The postulate that the rate of division of the offspring of the target cells was dependent on further antigenic encounters was quickly discredited by our finding that such a model was incapable of predicting a response over a 1,000-fold dose range. It seemed most plausible that the decisive element missing from our assumptions was a system of feedback control, and it seemed clear that this must be provided by antibodies. Though we had some evidence of the regulatory role of antibody, we deferred further consideration of this issue to another geographical location.

In 1966, in recognition of his contributions to the field of biological assay, e.g. [12], *Niels* was offered the directorship of the Paul Ehrlich Institute in Frankfurt, an institute that specialized in biologics standardization. It was appropriate that he should head the institute that bore the name of the forerunner of selective theories. Current immunological journals soon joined the records of *Paul Ehrlich's* experiment in the 'memorial room', and the beautiful copper and porcelain incubators of the 'Chemische Abteilung' were rapidly impressed into plaque service. The long evenings of immunological discussion were delayed while *Alexandra* and *Niels* restored a survivor of the bombings at Schumann Strasse to its former patrician splendor and helped me attempt the reconstruction of another high-ceilinged relic in a tawdrier section of the West End. I recall with some surprise that these efforts were done with entertainment from *Harry Belafonte* and the *Smothers Brothers* and recognize with sadness that both houses finally fell victim to the wreckers' ball and the encroachments of insurance companies. The *Fuji* family arrived, *Dankward Kodlin* joined us, and we renewed our efforts to clarify the essential components of the primary stimulus.

We found that the anti-SRC response was inhibited by injection of specific IgG antibodies, but enhanced by administration of specific IgM antibodies [3]. The quantities that were effective in influencing the response, especially to low doses of antigen, were very small, and in fact were amounts that might be expected in the sera of normal mice. These findings did not fit into current immunological thought. Quantitative considerations precluded that IgG antibody was merely masking antigenic determinants. Since inhibition was many more times effective when the IgG was given in multiple small doses over a 3-week period before antigen than when given in a single dose 1 day previously, it appeared that antibody alone might have a direct effect on precursor cells. Furthermore, when antibody was given concurrently with antigen, there was always a delay between the injection of antibody and an effect – 1–2 days for IgG inhibition and 3 days for IgM enhancement.

Dankward finally succeeded in simulating the decreasing slopes of the Pittsburgh dose-response curves by incorporating into the single-hit theory an antigen availability function, D(t), through which 'available' antigen rose to a maximum 1 day after injection and then slowly declined. Thus, the probability of a hit at time t is $1\text{-}e^{-cD(t)}$, and this varies with time. Furthermore, the number of cells first hit at various time intervals was considered, giving the total yield at a given time as the sum of the clonal yields at that time from all cells that had been hit. At high doses, all target cells receive their first hit simultaneously and early, while for lower doses the time of first hit is distributed over a considerable period. Choice of parameters c

and n was based on estimates discussed above, as was a clonal yield of 2^8 and a constant division time of 7 h. We imagined that 'availability' was due to modification of the administered antigen, possibly by antibodies in concert with macrophages. Appreciation of the essential role of T cell help in B cell triggering did not come till 1968. By then all of us, save *Niels,* had recrossed the Atlantic, and *Niels* was preparing to move to Basel to realize his dream of an international institute devoted entirely to basic immunology.

Our results and interpretation were consistent with clonal selection, but indicated that considering it in an immune system of independent lymphocyte clones unregulated by circulating immunoglobulin was oversimplification. Crucial elements were missing from our consideration, and a leap in intuitive thought was needed. *Niels* later provided this with the concept that the immune system is a network of paratope-idiotope interactions, regulated by its own elements which are internal images of foreign antigens [7]. Certain of our Frankfurt findings with specific antibodies are now loosely explainable in terms of this concept [2, 8]. Though general notions of immune network have been explored mathematically [4, 13], a formulation that would account quantitatively for the kinds of findings discussed here remains a challenge.

As I write this, *Niels* is retiring as director of the Basel Institute, and one has cause to ponder how he will use his creatitivity and remarkable instinct for rightness. Within 30 years he has contributed three seminal ideas [5, 7, 8], an experimental technique and a way of thinking to direct the course of immunology. It is difficult to imagine him content now with merely viewing the Provençal heavens.

References

1 Burnet, F.M.: The clonal selection theory of acquired immunity (Cambridge University Press, London 1959).
2 Forni, L.; Coutinho, A.; Köhler, G.; Jerne, N.K.: Proc. natn. Acad. Sci. USA *77:* 1125–1128 (1980).
3 Henry, C.; Jerne, N.K.: J. exp. Med. *128:* 133–152 (1968).
4 Hoffman, G.W.: Eur. J. Immunol. *5:* 638–647 (1975).
5 Jerne, N.K.: Proc. natn. Acad. Sci. USA *41:* 849–857 (1955).
6 Jerne, N.K.: A. Rev. Microbiol. *14:* 341–358 (1960).
7 Jerne, N.K.: Eur. J. Immunol. *1:* 1–9 (1971).
8 Jerne, N.K.: Annls Immunol. *125C:* 373–389 (1974).
9 Jerne, N.K.; Nordin, A.A.: Science *140:* 405 (1963).

10 Jerne, N.K.; Nordin, A.A.; Henry, C.: in Amos, Koprowski, Cell-bound antibodies, pp. 109–122 (Wistar Institute Press, Philadelphia 1963).
11 Jerne, N.; Nordin, A.; Henry, C.; Fuji, H.; Koros, A.: NIH Information Exchange Group 5, Memo 46, pp. 1–41 (1965).
12 Jerne, N.K.; Wood, E.C.: Biometrics *5:* 273–299 (1949).
13 Richter, P.H.: Eur. J. Immunol. *5:* 350–354 (1975).

Prof. Claudia Henry, University of California, Berkeley, Department of Microbiology and Immunology, Berkeley, CA 94720 (USA)

The Immune System, vol. 1, pp. 124–131 (Karger, Basel 1981)

Exciting B Cells

Fritz Melchers[1], *Jan Andersson*[1]

Basel Institute for Immunology, Basel, Switzerland; Biomedicum, University of Uppsala, Uppsala, Sweden

In 1969 *Niels Kaj Jerne* [13] paid tribute to the 70th birthday of Sir *MacFarlane Burnet* by proposing 20 questions which, when answered, would constitute the complete solution of immunology. He concluded then, one year before the opening of the Basel Institute for Immunology, that even before these questions were answered, 'in principle, immunology was solved in 1957 when *Burnet* published his Clonal Selection Theory of Acquired Immunity.' So, in 1980, we ask ourselves whether we wasted our time on trying to answer questions 16, 17, and 20:

16. 'By what molecular mechanisms does an antigen-sensitive cell respond to antigenic and hormonal stimuli?' 17. 'In how many different ways can a cell react to such stimuli, and what are the locations and the characteristics of the different stimulatory signals?' 20. 'How is antibody formation related to cellular immunity?'

The central dogma of immunology states that one cell makes one antibody. The foundations for this dogma were laid by *Jerne* [11] in 1955 and expanded by *Burnet* [7]:

'It is assumed that when an antigen enters the blood or tissue fluids it will attach to the surface of any lymphocyte carrying reactive sites which correspond to one of its antigenic determinants ... It is postulated that when antigen-natural antibody contact takes place on the surface of a lymphocyte the cell is activated to ... undergo proliferation to produce a variety of descendants ... The descendants will include plasmacytoid forms capable of active liberation of soluble antibody ...'

[1] *Fritz Melchers* has been a member of the Basel Institute for Immunology since its founding; *Jan Andersson* was a member of the Institute from 1971 to 1973 and has been an intermittent visitor ever since.

The bias of those times was clear: binding of antigen alone to specific lymphocytes would suffice to trigger clonal proliferation and amplification of that specific antibody synthesis. This simple concept of antigen recognition was no longer tenable when it was found that different types of cells had to cooperate in immune responses. *Jerne* himself was puzzled by the finding that two determinants were needed on an antigen to induce lymphocytes to produce specific antibody against one of the determinants [24]. Initially macrophages had been suspected of cooperating with B cells, but *Jerne* [14] never liked the idea very much, since macrophages do not produce specific antibody but only cytophilically borrow it from the serum. When it became clear that T cells cooperate with B cells [21] and that T cells were clonally specific for antigen, confidence in the exquisite specificity of individual members of the immune system was restored. *Mosier's* [22] finding that, in fact, all three types of cells – T cells, B cells, *and* macrophages – were needed could be ignored if binding of two different antigenic determinants to two different lymphocytes were all that was needed for triggering.

When, 'waiting for the end', *Jerne* [12] wrote in 1967:

> 'The antibody problem ... has a beginning and an end, and the people approaching this problem can, accordingly, be divided into two groups: (1) the *trans*-immunologists that start at the end with the structure of antibody molecules, hoping to work their way backwards, and (2) the *cis*-immunologists that start at the beginning, with the effects of antigenic exposure, hoping to work their way forwards ... The precise dividing point between *cis* and *trans* is the receptor molecule on the antigen-sensitive lymphocyte. The interest of the *trans*-immunologist wanes when confronted with tales about the vicissitudes of the antigen in the tissues before reaching the triggering point, whereas the *cis*-immunologist becomes sceptical at stories of inexorable mechanisms leading from the origin of the vertebrates to the endowment of a lymphocyte with genes for producing only a single foreordained antibody. The result is that the two hardly speak to each other. Or rather, a *cis*-immunologist will sometimes speak to a *trans*-immunologist; but the latter rarely answers.'

The History of the Cis-*Immunologist*

The simple concept that binding antigen to cell-bound antibody was the only thing that was needed for triggering antibody synthesis was also questioned by *Andersson* et al. [6] after the discovery of B lymphocyte mitogens. Such mitogens could activate, by circumventing the step of antigen binding to surface Ig molecules, a large part of all B cells in a polyclonal fashion. This finding called for a series of experiments that demanded working hours characteristic of a *trans*-immunologist. Thus, *cis* and *trans* started talking to each other.

The History of the Trans-*Immunologist*

'... and an analysis by *Melchers and Knopf* [20] indicate(s) that carbohydrate attachment proceeds by several steps requiring more than 8 min and is probably requisite to secretion' [12]. These experiments, although somewhat removed from studies of the structure of antibody molecules, clearly started at the end of an immune response, with the simple-minded hope of a biochemist that understanding of the biochemical reactivities controlling synthesis, transport, and secretion of antibody molecules would lead to an understanding of the control of humoral antibody responses.

Basel's Roaring Seventies

It did not take long after arriving in Basel to discover that question 18 [13] could not be answered by studying antibody-secreting plasma cells. Mitogen-stimulated normal B cells synthesize and secrete Ig as myeloma cells do, with no control by environmental influences.

Working our way backwards, we decided to study antigen-sensitive, resting B cells, the synthesis and deposition of Ig in their surface membrane, the turnover of the surface-deposited Ig, and the changes which occurred when these resting B cells were stimulated by mitogens. We asked whether proliferation and maturation to Ig secretion were two processes which were separable, possibly even antagonistic in the clonal development of B cells. But since the use of mitogens circumvented the step of antigen binding to surface Ig, our experiments were often regarded as 'unphysiological', not the 'normal' response of B cells: 'It is like approaching archeology with bulldozers' [16].

The importance of mitogens was not popularized by the discovery of more mitogens [18], nor by the determination of the active principle of two of the mitogens [5], nor even by the finding that anti-Ig antibodies inhibited the mitogen-induced maturation of resting B cells [1]. We concluded at that time:

'... that a complex of structures on the surface of Ig-positive, small lymphocytes is involved in lymphocyte stimulation ... The complex, probably exposed to the outside of the cell on the surface membrane, represents the initial portion of a series of molecules which are involved in reactions leading to DNA synthesis, cell division and differentiation into secreting plasma cells ... Surface-bound Ig molecules in these proposed "receptor complexes" appear to modulate reactions leading to stimulation. Binding of antigen or anti-Ig antibodies alone appears not to be sufficient for stimulation. Additional factors have to act. T cells, adherent cells and/or serum factors may be sources for such natural stimulating factors in vitro. Modulations of B cells by antigen could render these cells susceptible to the action of antigen-unspecific, "mitogenic" factors. This may be achieved either by a change of the capacity of the cells to bind these factors or by more intricate conformational changes in the complex of molecules involved in stimulation from a conformation in which binding of factors does not lead to stimulation to a conformation (favored by antigen) in which binding results in stimulation. The modulating role of surface-bound Ig molecules may be positive or negative for stimulation ...'

But still *Jerne* would not hear of additional signals aside from specific antibody. When we told him of our results, he would counter with results of experiments with sheep erythrocytes as specific antigen done by *Fuji and Jerne* [9]. Soon the discussion would drift to the specificity of antibody molecules, to their repertoire of recognition, to idiotypes and networks.

Although *Niels Jerne,* in his 10 years at the Basel Institute, came into our laboratory only once, he was very concerned about our experimental technique. He disapproved of our way of doing his hemolytic plaque assay [15], mixing erythrocytes, lymphocytes, developing antibody and complement all at once. Even when we modified the erythrocytes by coupling protein A onto them [10], so that now all Ig-secreting cells of a given type or class could be detected irrespective of the antigen specificity of the secreted Ig, he still wasn't overly excited.

Three advances, and the arrival of *Antonio Coutinho* in Basel, finally made mitogens acceptable: (1) the protein A plaque assay [10]; (2) mercaptoethanol in lymphocyte cultures [8, 19], and (3) 'filler' thymus cells in lymphocyte cultures [2, 26]. Now we could count cells which exponentially increased in number, in order to quantitatively measure lymphocyte stimulation in vitro. With the addition of thymus 'filter' cells [26] we could finally use limiting dilution analyses to determine the frequencies of mitogen-reactive B cells. Since now one third of all B cells were found to be reactive [3], all arguments of cross-reactivity of mitogens with antigens suddenly disappeared. This type of analysis also allowed us to estimate frequencies of antigen-specific B cells within the mitogen-stimulable set of cells. The results were greeted, inside and outside the Institute, with well-controlled enthusiasm, since the frequencies were 20- to 100-fold higher than had been estimated previously by others [23].

The Never-Ending Love for Natural Antibodies

The 'physiology' of polyclonal stimulation became popular. *Jerne* was now fascinated that an ongoing polyclonal stimulation was a way to explain natural antibody as the result of stimulation of every B cell in the system, once in a while. We went to Freiburg into a 'Nebenzimmer' in 'Oberkirch's Weinstuben' to argue all night long with ever increasing levels of alcohol in our blood stream, provided by Kaiserstühler wines, what the original repertoire of germ line-encoded antibody genes could be, and what would somatically develop from that – and whether we now finally had ways to measure it all.

Is 'natural antibody' derived from 'background'-activated normal B cells that secrete the total repertoire of B cells? *Jerne* [11] wrote in 1955:

'In a rabbit the half-life of circulating globulin is about 5 days ... normal mammalian serum contains more than 10^{19} globulin molecules per milliliter ... This means that the rabbit must daily synthesize about 10^{18} globulin molecules ... Lymphocytes have a life span of not more than a few days. It can be calculated that the daily output in the rabbit is of the order of 10^{10} of these cells ... For 10^{10} lymphocytes to accomplish this task (i.e. to synthesize 10^{18} globulin molecules daily), each would have to synthesize about a thousand globulin molecules per second.'

By 1976, when we sat in 'Oberkirch's Weinstuben', the numbers had been slightly corrected (fig. 1). Now, 10 mg/ml of normal Ig in serum corresponded to 5×10^{16} Ig molecules/ml, and a mouse with approximately 2 ml serum was estimated to have approximately 10^9 B cells. From our experiments we also knew that only 1% of all B cells were 'background'-activated B cells secreting 10^3–10^4 molecules per second per cell. The other 99% were found to release less than 10^3 molecules per hour per cell [4]. By all quantitative analyses we had accumulated so far, it appeared that 10^7 'background-secreting' cells in a mouse, in 5 days, therefore, could secrete at best 10^{15}, not 5×10^{16}, Ig molecules. It is remarkable that the apparently simple question of the origin, turnover and specificity of 'natural antibodies' is still unanswered.

Jerne [12] also wrote, while 'waiting for the end' in 1967:

'If, as a mathematician has said, five parameters suffice for the description of an elephant, then surely five types of antigen-receptor molecules, each with a wide range of affinities for a given antigenic determinant, and all competing for the antigen, would suffice to explain regulatory phenomena at the antigen level.'

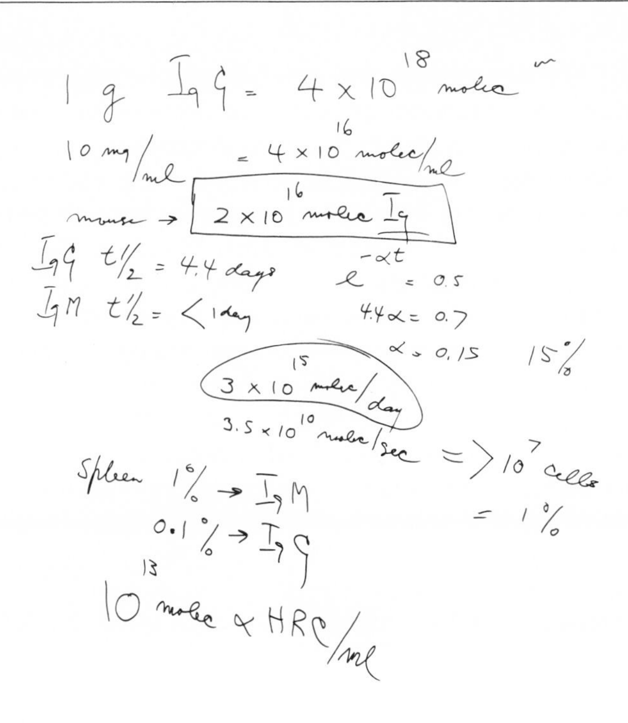

Fig. 1. Quantitative immunology in 'Oberkirch's Weinstuben'.

Cis- *and* Trans-*Immunologists on Their Way to Describe an Elephant*

We returned to the problem of how mature, resting lymphocytes are triggered. The problem of B cell stimulation had become even more complicated when it was recognized that the interactions between the three cells – T cells, B cells and macrophages – were H-2-restricted. And we still had no way to study T cell-dependent B cell stimulation under conditions which would be as efficient as mitogenic stimulation.

'Help' finally came from *Max Schreier. Max* had improved conditions of in vitro stimulation by antigen-specific helper T cell lines and clones so much that the frequency of B cells responding to erythrocyte antigens was as high as that obtained under polyclonal stimulation by mitogens [25]. This started a short but fruitful cooperation which perished when we began to publish the results on T-B collaboration. There was disagreement on whether T cells or B cells were more important in this collaboration.

At the end we now know that T cell-dependent B cell stimulation to clonal proliferation and antibody production proceeds in two steps. In the first step, antigen-specific helper T cells require antigen and I region-compatible macrophages to produce 'help' for B cells. 'Help' constitutes antigen-unspecific B cell growth and maturation factors (BRMF) which can be found released into the supernatant medium of such cultures of helper T cells, macrophages and antigen. In the second step the same helper T cells interact with I region-encoded determinants on those resting B cells which bind antigen. Thus, dual recognition of antigen (via Ig) and of helper T cells ('anti-Ia', via Ia) excites the resting B cells so that they become susceptible to stimulation by BRMF to clonally proliferate and mature to antibody secretion. Exposure of resting B cells to BRMF in the absence of antigen leads to polyclonal and H-2-unrestricted maturation without proliferation. H-2 restriction and antigen specificity in T-B collaboration is, therefore, controlled at the level of the resting cells. Three elements are involved in this control: Ig, Ia, and receptors for BRMF. Mitogens, such as LPS and lipoprotein, break such H-2 restrictions and antigen specificities of resting B cells [17].

Our elephant [12] so far has three dimensions and is, therefore, well on its way to being defined. If all that interested you were the specificity of the immune system, we have not added anything to the understanding of the repertoire of B cells. If, however, you wanted to know of the molecular and cellular mechanisms that excite and stimulate a B cell, then the lonesome role of antigen-specific Ig receptors has been challenged by growth factors and histocompatibility interactions.

In all the excitement we have failed to describe *Niels Jerne's* crucial influence on our work. The logic of his understanding of the immune system provoked us, his clarity in defining what he did not understand of it inspired us. That made us respect, admire and love him, he who gave us all complete freedom to think and to do what we wanted.

References

1 Andersson, J.; Bullock, W.W.; Melchers, F.: Eur. J. Immunol. *4:* 715 (1974).
2 Andersson, J.; Coutinho, A.; Lernhardt, W.; Melchers, F.: Cell *10:* 27 (1977).
3 Andersson, J.; Coutinho, A.; Melchers, F.; Watanabe, T.: Cold Spring Harb. Symp. quant. Biol. *41:* 227 (1976).

4 Andersson, J.; Lafleur, L.; Melchers, F.: Eur. J. Immunol. *4:* 170 (1974).
5 Andersson, J.; Melchers, F.; Galanos, C.; Lüderitz, O.: J. exp. Med. *137:* 943 (1973).
6 Andersson, J.; Sjöberg, O.; Möller, G.: Eur. J. Immunol. *2:* 349 (1972).
7 Burnet, F.M.: Aust. J. Sci. *20:* 67 (1957).
8 Click, R.E.; Benck, L.; Alter, B.J.: Cell. Immunol. *3:* 156 (1972).
9 Fuji, H.; Jerne, N.K.: Annls Inst. Pasteur, Paris *117:* 801 (1969).
10 Gronowicz, E.; Coutinho, A.; Melchers, F.: Eur. J. Immunol. *6:* 588 (1976).
11 Jerne, N.K.: Proc. natn. Acad. Sci. USA *41:* 849 (1955).
12 Jerne, N.K.: Cold Spring Harb. Symp. quant. Biol. *32:* 591 (1967).
13 Jerne, N.K.: Australas. Ann. Med. *4:* 345 (1969).
14 Jerne, N.K.: Cold Spring Harb. Symp. quant. Biol. *41:* 1 (1977).
15 Jerne, N.K.; Nordin, A.A.: Science *140:* 405 (1963).
16 Loor, F.: Eur. J. Immunol. *4:* 210 (1974).
17 Melchers, F.; Andersson, J.; Lernhardt, W.; Schreier, M.H.: Immunol. Rev. *52:* 89 (1980).
18 Melchers, F.; Braun, V.; Galanos, C.: J. exp. Med. *142:* 473 (1975).
19 Melchers, F.; Coutinho, A.; Heinrich, G.; Andersson, J.: Scand. J. Immunol. *4:* 853 (1975).
20 Melchers, F.; Knopf, P.M.: Cold Spring Harb. Symp. quant. Biol. *32:* 255 (1967).
21 Miller, J.F.A.P.; Mitchell, G.F.: Nature, Lond. *216:* 659 (1967).
22 Mosier, D.E.: Science *158:* 1573 (1967).
23 Quintans, J.; Lefkovits, I.: Eur. J. Immunol. *3:* 392 (1973).
24 Rajewsky, K.; Schirrmacher, V.; Nase, S.; Jerne, N.K.: J. exp. Med. *129:* 1131 (1969).
25 Schreier, M.H.: J. exp. Med. *148:* 1612 (1978).
26 Stocker, J.W.: Immunology *30:* 181 (1976).

Dr. Fritz Melchers, Basel Institute for Immunology, Postfach, CH–4005 Basel (Switzerland)

The Immune System, vol. 1, pp. 132–138 (Karger, Basel 1981)

Structure and Arrangement of Human Heavy Chain Variable Region Genes

G. Matthyssens[1], *T.H. Rabbitts*

Vrije Universiteit Brussel, Institute of Molecular Biology, Sint-Genesius-Rode, Belgium; MRC Laboratory of Molecular Biology, Cambridge, England

The organization of gene sequences in the mouse has been studied extensively, and this for both λ and κ light chains [3, 4, 10, 14, 23–26]. As is evident from these references, a major part of this work has been performed at the Basel Institute for Immunology under the directorship of Prof. *Jerne.* The first immunoglobulin gene isolation realized there by *Tonegawa's* [26] group established that in the embryonic genome the conventionally defined variable (V) region is encoded in a DNA segment coding for amino acid 1–95 which is located upstream (5′ with respect to the transcription direction) from a DNA segment (J DNA) coding for the rest of the V region. In the cell expressing this particular V region, the V DNA is contiguous with J DNA due to a recombinational event. This event is accompanied by deletion of the DNA sequences between the V and J DNA segments, with the preservation of the intron between the V-J and C DNA segments [15, 19].

The 5′-flanking regions of the mouse light chain J segments and the 3′-non-coding region of embryonic V genes have been implicated in the V-J joining step which is considered to be the important event in the activation of the immunoglobulin gene [15, 19].

For a better understanding of the structural features which might be implicated in the recombination sites between V and J segments, we decided to study the germ line organization of heavy chain genes in humans, and more precisely those belonging to the V_HIII subgroup to which the majority of the variable region sequences studied until now belong [6].

[1] *Gaston Matthyssens* was a member of the Basel Institute for Immunology from 1974 to 1977.

Materials and Methods

Recombinant DNA Techniques. The internal *Eco*RI fragments of the Charon 4 phage [2] were replaced by DNA inserts of approximately 15–20 kilobases (kb). These were prepared from human fetal liver by a non-limit digestion with restriction endonucleases *Hae* III + *Alu* I followed by the addition of *Eco*RI linker molecules [9, 11].

The amplified library was screened according to the in situ plaque hybridization technique of *Benton and Davis* [1] using the nick-translated mouse plasmid pμ/107 [12].

Sequencing Technique. Nucleotide sequencing was carried out in M13 vectors [16, 22] using the dideoxy chain termination method [21].

Results

Complete Nucleotide Sequence of a Human Germ-Line V Gene. In a first score of about 2×10^5 individual phage clones (which corresponds roughly to ¼ of a total library), we detected 28 positives which hybridized to our mouse cDNA plasmid pμ/107. This probe contains a full copy of a V_H sequence belonging to the V_HIII subgroup [12] which is very similar to the V_H subgroup in humans [6]. It also contains the complete constant region μ sequence of mouse, which is known to have a striking gradient of increasing homology with the human μ chain [7] (CH1: 48% → C-terminus: 89%).

However, none of the positives selected contained the human μ gene, as they failed to hybridize with the pμ/118 cDNA clone, which contains only a copy of the constant μ chain and the 3′-untranslated region of the mouse μ mRNA [12].

The average size of human DNA in our clones is 18 kb. Two of the clones analysed possessed two V_H segments; the spacing between these two segments varied between approximately 12.5 and 15 kb [13].

In order to determine the structure of the human V_H genes, we have subcloned one of our positives designated λV_H26 in M13 phage. The sequence of this gene (fig. 1) readily fits into the V_HIII subgroup as defined by *Kabat* et al. [6]. It includes a 46-base pair exon which codes presumably for part of the hydrophobic signal peptide, a 103-base pair intron, the entire V-gene exon containing at the 5′ side 11 bases coding for the remainder of the signal peptide, and the 3′-flanking region. The coding region corresponds to the published amino acids only up to residue 97. This was confirmed by the sequence of two other human V_H genes (λV_H32 and λV_H52, see [13]). It thus seems that the 20-residue peptide comprising the third hypervariable region is encoded separately in the human genome.

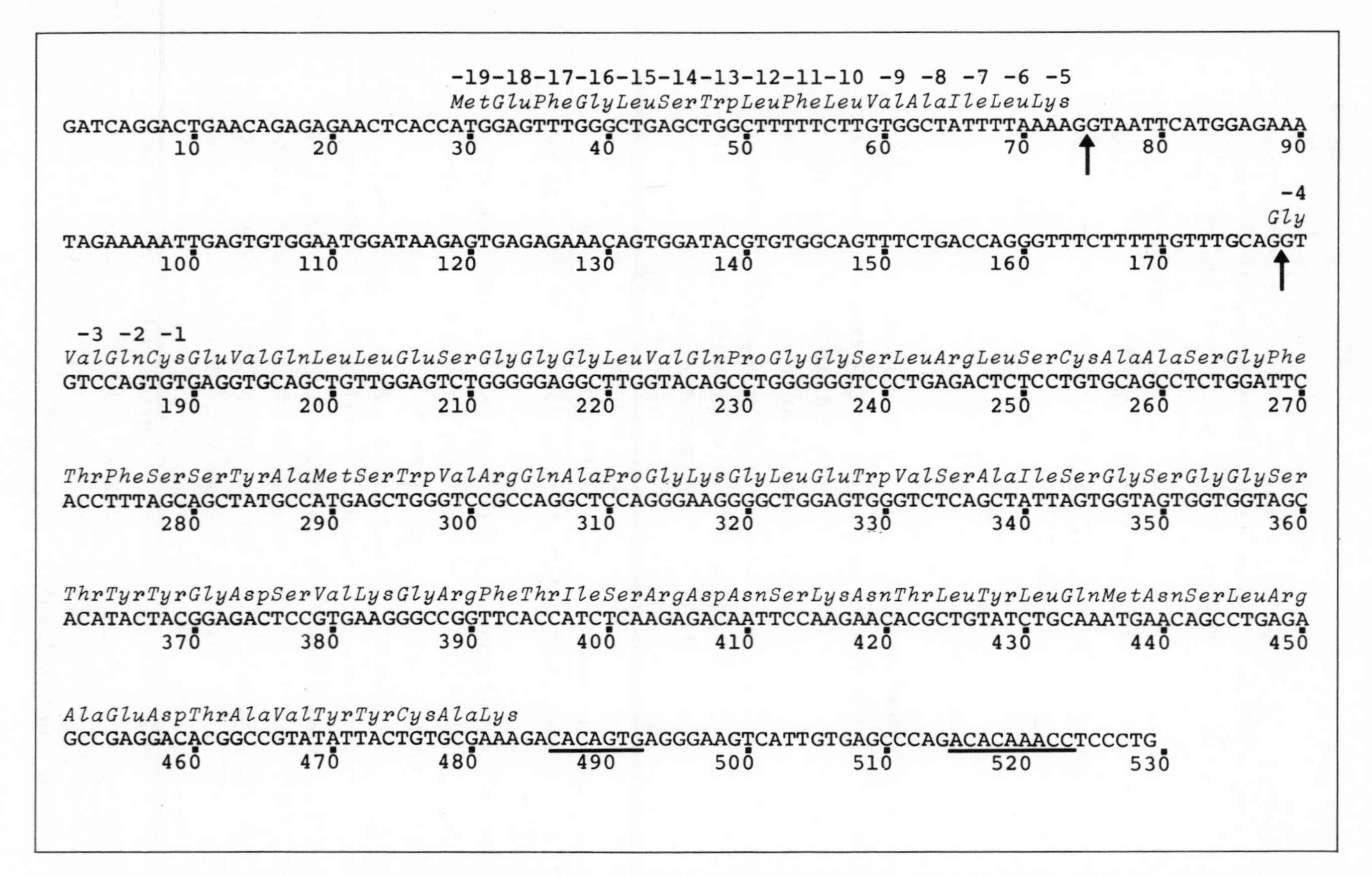

Fig. 1. Nucleotide sequence of an isolated V_H gene cloned from human fetal liver DNA. A 1.9 kb *Eco*RI fragment from clone λV_H26 was subcloned in M13 phage and the nucleotide sequence determined using the dideoxy method [21]. Putative splicing positions are shown by vertical arrows. The underlined sequences are thought to be involved in the recombination event leading to a complete V gene.

Table I. Conserved sequences flanking V genes of human and mouse

V DNA segments		3′-Flanking region of germ-line V gene			Ref.
V_H26	(H)	C A C A G T G	A G G G A A G T C A T T G T G A G C C C A G	A C A C A A A C C	13
V_H52	(H)	C A C A G T G	A G G G A A G T C A G T G T G A G C C C A G	A C A C A A A C C	13
V_H107	(M)	C A C A G T G	A GAG G A C G T C A T T G T G A G C C C A G	A C A C A A A C C	5
V_H141	(M)	C A C A G T G	A G G G A A G T CCA A T G T G A G C C T G C	A C A **A** A **T** A C C	20
$V_{\lambda I}$	(M)	C A C A **A** T G	A A T G T G T A G A T G G G G A A G T A G A	**T** C A **A** **G** A A C **A**	3
$V_{\lambda II}$	(M)	C A C A **A** T G	ACA T G T G T A G A T G G G G A A G T A G A	A C A **A** **G** A A C **A**	26
V_K101	(H)	C A C A G T G	T T A C A C A C C C A A	A C A **T** A A-A C C	18
$V_\kappa 2$	(M)	C A C A G T G	A T T C A A G C C A T G	A C A **T** A A A C C	23
$V_\kappa 21C$	(M)	C A C A G T G	C T C A G G G C T G A	A C A **A** A A A C C	19
V_K41	(M)	C A C A G T G	A T A C A A A T C A T A	A C A **T** A A A C C	24
Basic sequence		C A C A G T G		A C A C A A A C C	

Two blocks of conserved sequences found 3′ to all germ-line V genes are boxed. Bases different from those of our basic sequence are boldfaced. H = Human; M = mouse.

Two Blocks of Conserved Sequences Adjacent to the V_H Gene Segment. Table I shows the sequence of part of the 3′-flanking region of 2 V_H gene segments (λV_H26 and λV_H52). This shows that these two germ-line genes have very conserved flanking regions, in addition to their homologous coding sequences [13]. Furthermore, we have compared our human sequences to these found flanking the V_H, V_κ, and V_λ genes in mouse: table I shows that a heptameric and nanomeric block of sequences in the 3′-flanking region of all these germ-line V DNA segments is conserved.

Discussion

We have previously shown that the genes detectable in human DNA with a homologous probe (such as λV_H26) and the heterologous mouse probe (pμ/107) are almost identical [13]. It has also been shown that these experiments detect all members of the V_HIII subgroup (roughly 20–23 genes) [18]. The genes we have studied here, and which therefore belong to this subgroup, are contained within a gene cluster with spacings between 12 and 16 kb, a situation very similar to the V_H-gene cluster of mouse [8].

This is direct evidence for the presence of multiple related V_H genes in the human germ line which contribute to immunoglobulin diversity. One of these genes has been sequenced completely. Compared to the most commonly occurring amino acids of the human V_HIII proteins, it differs in 14 codon positions, of which one occurs in the first and nine in the second hypervariable region [17]. As in all κ, λ, and heavy chain genes studied so far in mouse [3, 5, 20, 24, 26], the V coding begins at amino acid positions –4 within the signal peptide. Also, the coding by this human germ-line gene ends prematurely with Ala 97 (or Lys 98). This implies that the whole third hypervariable region is encoded in a separate DNA segment (J_H) or segments (D and J_H, see [5, 20]) which contrasts markedly with the mouse V_L chains where part of this region occurs within the V segment. The presence however of the conserved heptameric and nanomeric sequences in similar positions of both light- and heavy-chain germ-line V genes (table I) suggests that the recombination event generating a complete V gene is carried out by a similar or identical mechanism. The fact that these sequences are found in humans is further evidence for their involvement in this site-specific recombination. An additional feature is the conserved length of the spacer between these two blocks: the V_H spacer in humans is 22 nucleotides long whereas in the mouse it is 23 nucleotides long. This highly conserved spacing corresponds to two turns of the DNA helix and therefore results in the maintenance of precise spatial orientations between blocks of conserved nucleotides [5]. It is also striking that the sequences of the V_H human and mouse spacer are well conserved, and this in contrast to the V_κ spacers which have diverged widely.

In conclusion, the present study predicts that the somatic recombination which is required to generate a complete heavy chain gene is very similar or identical in humans and the mouse. Final proof of this mechanism awaits identification and characterization of the J_H (and D) cluster in humans.

Acknowledgements. We thank Dr. *T. Maniatis* for kindly providing the human phage library and *A. Forster* for expert technical assistance. *G. Matthyssens* thanks the Royal Society and the Nationaal Fonds voor Wetenschappelijk Onderzoek for a fellowship.

References

1 Benton, W.D.; Davis, R.W.: Science *196:* 180–182 (1977).
2 Blattner, F.R., et al.: Science *196:* 161–169 (1977).
3 Bernard, O.; Hozumi, N.; Tonegawa, S.: Cell *15:* 1133–1144 (1978).
4 Brack, C.; Hirama, M.; Lenhard-Schuller, R.; Tonegawa, S.: Cell *15:* 1–14 (1978).
5 Early, P.; Huang, H.; Davis, M.; Calame, K.; Hood, L.: Cell *19:* 981–992 (1980).
6 Kabat, E.A.; Wu, T.T.; Bilofsky, H.: NIH Publication No. 80–2008 (1979).
7 Kehry, M.; Sibley, C.; Fuhrman, J.; Shilling, J.; Hood, L.E.: Proc. natn. Acad. Sci. USA *76:* 2932–2936 (1979).
8 Kemp, D.J.; Cory, S.; Adams, J.M.: Proc. natn. Acad. Sci. USA *76:* 4627–4631 (1979).
9 Lawn, R.M.; Fritsch, E.F.; Parker, R.C.; Blake, G.; Maniatis, T.: Cell *15:* 1157–1174 (1978).
10 Lenhard-Schuller, R.; Hohn, B.; Brack, C.; Hirama, M.; Tonegawa, S.: Proc. natn. Acad. Sci. USA *75:* 4709–4713 (1978).
11 Maniatis, T.; Hardison, R.C.; Lacy, E.; Lauer, J.; O'Connell, C.; Quon, D.: Cell *15:* 687–701 (1978).
12 Matthyssens, G.; Rabbitts, T.H.: Nucl. Acids Res. *8:* 703–713 (1980).
13 Matthyssens, G.; Rabbitts, T.H.: Proc. natn. Acad. Sci. USA *77:* 6561–6565 (1980).
14 Matthyssens, G.; Tonegawa, S.: Nature, Lond. *273:* 763–765 (1978).
15 Max, E.E.; Seidman, J.G.; Leder, P.: Proc. natn. Acad. Sci. USA *76:* 3450–3454 (1979).
16 Messing, J.; Gronenborn, B.; Mueller-Hill, B.; Hofschneider, P.H.: Proc. natn. Acad. Sci. USA *74:* 3642–3646 (1977).
17 Rabbitts, T.H.; Bentley, D.L.; Dunnick, W.; Hobart, M.; Matthyssens, G.; Milstein, C.: Cold Spring Harb. Symp. quant. Biol. (in press).
18 Rabbitts, T.H.; Matthyssens, G.; Hamlyn, P.H.: Nature, Lond. *284:* 238–243 (1980).

19 Sakano, H.; Hüppi, K.; Heinrich, G.; Tonegawa, S.: Nature, Lond. *280:* 288–294 (1979).
20 Sakano, H.; Maki, R.; Kurosawa, Y.; Roeder, W.; Tonegawa, S.: Nature, Lond. *286:* 676–683 (1980).
21 Sanger, F.; Nicklen, S.; Coulson, A.R.: Proc. natn. Acad. Sci. USA *74:* 5463–5467 (1977).
22 Schreier, P.H.; Cortese, R.: J. molec. Biol. *129:* 169–172 (1979).
23 Seidman, J.G.; Leder, A.; Edgell, M.H.; Polsky, F.; Tilghman, S.M.; Tiemeier, D.C.; Leder, P.: Proc. natn. Acad. Sci. USA *75:* 3881–3885 (1978).
24 Seidman, J.G.; Max, E.E.; Leder, P.: Nature, Lond. *280:* 370–375 (1979).
25 Tonegawa, S.; Hozumi, N.; Matthyssens, G.; Schuller, R.: Cold Spring Harb. Symp. quant. Biol. *41:* 877–889 (1976).
26 Tonegawa, S.; Maxam, A.M.; Tizard, R.; Bernard, O.; Gilbert, W.: Proc. natn. Acad. Sci. USA *75:* 1485–1489 (1978).

Dr. G. Matthyssens, Institute of Molecular Biology, Vrije Universiteit Brussel, Paardestraat 65, B–1640 Sint-Genesius-Rode (Belgium)

The Immune System, vol. 1, pp. 139–148 (Karger, Basel 1981)

Electron Microscopy of Particles Related to the Immune System and Use of IgG for Specific Labelling: Review, Summary, and Perspectives

Edward Kellenberger[1], *Werner Villiger*

Microbiology Department, Biozentrum, University of Basel, Basel, Switzerland

With Contributions by *E. Delain* (Paris), *J. Engel* (Basel), *E.R. Podack* (La Jolla, Calif.), *J. Roth* (Geneva), *M. Wurtz* (Basel) and *M. Yanagida* (Kyoto)

The purpose of this paper is to provide a short appreciation of the role played by electron microscopy with respect to both the structure of particles of the immune system and in problems of specific identification of antigens in cytology.

Structure of Antibodies and of Complement

It was not very long ago that IgGs were considered to be cylindrical or cigar-shaped proteins with interaction sites on both ends. The first micrographs suggesting the Y shape of IgG and the presence of hinges between the Fab pieces and Fc were provided in 1965 by *Feinstein and Rowe* [5]. 2 years later the fully convincing images by *Valentine and Green* [36; see also 10] were published. This typical form was later confirmed by X-ray crystallography [2, 8, 29]. Despite the highly increased instrumental resolving power of modern electron microscopes, the demands involved in specimen preparation and imaging (electron beam-induced damage) have prevented much more information from being obtained.

[1] *Edward Kellenberger* is a long-time friend and associate of *Niels Jerne*. He organized the electron microscopy facility at the Basel Institute for Immunology in 1970–1971, and he has been a member of the Board of Consultants of the Institute since its founding.

The first convincing images of IgM were provided by *Parkhouse* et al. [21] and *Feinstein* et al. [4]. The latter authors showed it as a sort of spider having its radial legs bound to bacterial flagellae chosen as antigens.

IgA has been studied by *Munn* et al. [18] and found to consist of different types of aggregates of Y-shaped structures resembling IgG.

Figures 1a and b provide a gallery of IgG-dimers and of IgM, prepared by surrounding the particles by heavy metal containing non-crystallising salts (negative stain). In figure 1c we see IgM and IgG molecules absorbed to double-stranded RNA (Rotavirus) prepared by air drying and bidirectional metal deposit and then observed by dark field electron microscopy [19].

For IgG the antigen-binding sites are located at the upper parts of the Fab arms (upper arms of the Y). This was concluded from electron micrographs of complexes of IgG with a bifunctional antigen [36] and later confirmed by X-ray crystallography [20; further references in 2]. The binding site for C1q, which is the binding protein of the first component of complement (C1), is located at the Fc stem [24]. It was recently demonstrated by electron microscopy that IgG-dimers bind to the heads of C1q [34].

IgM resembles a cog-wheel (fig. 1b, c) in which the summits of the cogs carry the antigen-binding sites. IgM bound to flagellae shows different conformations, suggesting that this immunoglobulin is also highly flexible. There are indications that a specific conformation must be reached for the binding of complement to IgM [4].

Fig. 1. Electron micrographs of macromolecules and biological structures (except for *d* and *f*). *a, b* IgG-dimers *(a)* and IgM molecules *(b)* prepared by negative staining. Bar = 20 nm and 30 nm, respectively. *c* IgM and IgG molecules absorbed to double-stranded RNA (Rotavirus) prepared by air drying, bidirectional heavy metal shadowing and observed by dark-field electron microscopy (courtesy of *E. Delain*). Bar = 40 nm. *d* Scheme of the protein A-gold (pAg) technique. 1 = antigen, 2 = IgG, 3 = colloidal gold particle coated with protein A. *e, f* C1q molecule, side view, negatively stained. Bar = 10 nm. Structure model *(f)* (courtesy of *J. Engel*). *g* C1r C1s-complexes, subcomponents of C1, prepared by negative staining. Bar = 30 nm. *h, i* T4 phage *(h)* with antibodies coated fibers after reaction with specific antiserum (courtesy of *M. Yanagida*). T4 phage *(i)* from a control preparation (courtesy of *M. Wurtz*). Samples *h* and *i* prepared by negative staining. Bar = 50 nm. *j* Erythrocyte membrane with typical lesions after treatment with the membrane attack complex (MAC) of the complement system. Negative stain preparation (courtesy of *E.R. Podack*). Bar = 40 nm.

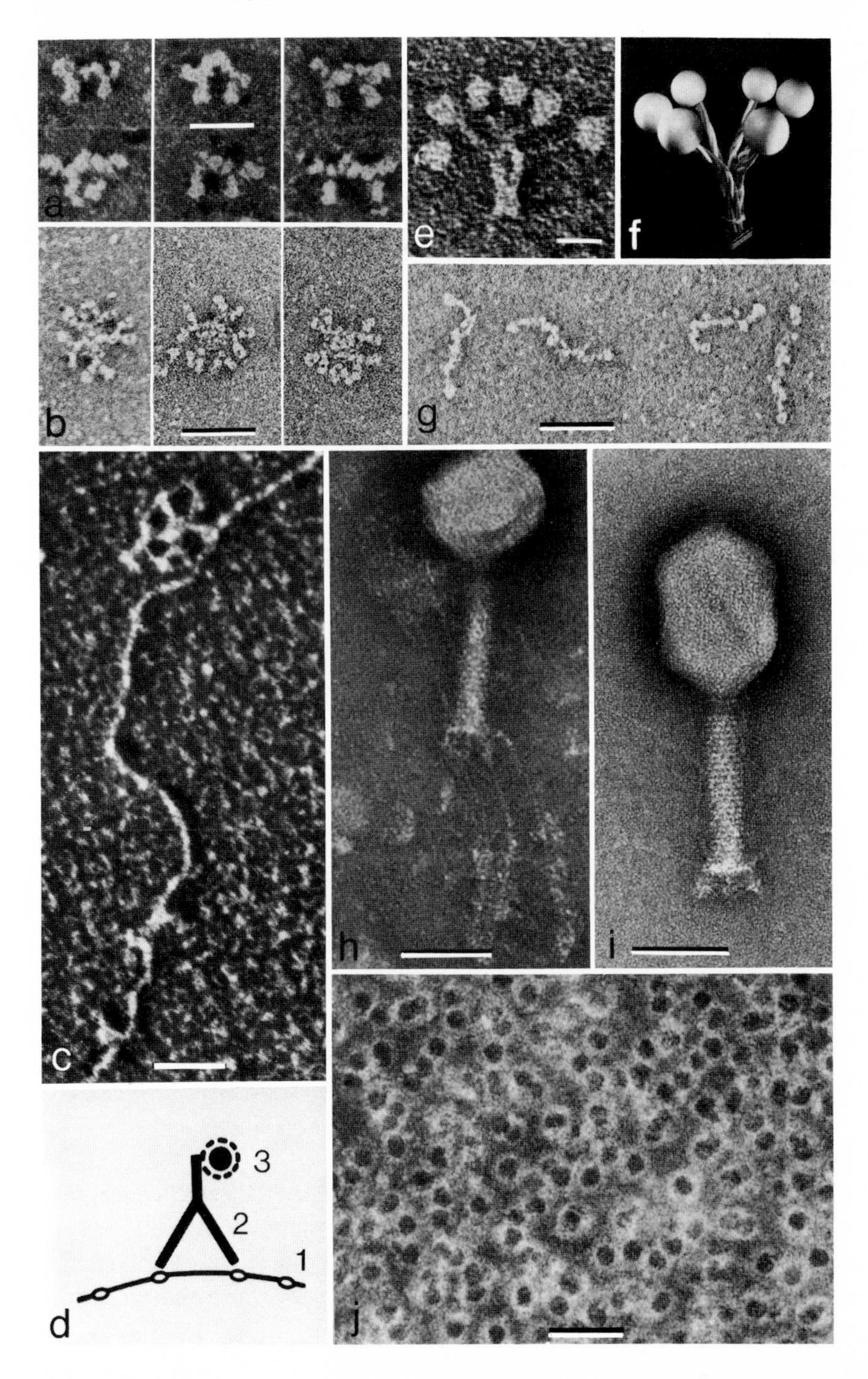
a
b
c
d
e
f
g
h
i
j
1
2
3

Attempts at gathering more information by observing altered IgG, e.g. after opening of the disulphide bridges, have only shown differences in behaviour of the structure that are compatible with an increased plasticity of the antibody molecules [28].

X-Ray crystallography of IgG and of Fab and Fc fragments has revealed the multidomain structure of IgG as well as the detailed folding of the four peptide chains in these domains [reviewed in 2]. Processed micrographs of negatively stained microcrystals show an excellent correspondence of domains with those of X-ray data [32], although they are far from atomic resolution. Micrographs of single molecules can obviously never compete with the structural information obtained through the redundancy of crystals.

Electron microscopy has a high potential for the elucidation of the larger multidomain proteins of the complement system [for references see 23, 34] and of the complexes of IgG with them. Recently, progress has been achieved for the morphology of two of the constituents of the complement [16, 24]. Figures 1e, g show C1q and C1r C1s complexes, obtained by negative stain. The structural information on these components (fig. 1f) provides a stimulating basis for biochemical and biophysical studies [34, 35], which should elucidate the mechanism of complement activation.

Eventually, structures have to be related to functions. We illustrate in figure 1j by *Podack* erythrocyte membranes lysed by the membrane attack complex (MAC) of complement. This 1.7×10^6 molecular weight complex is assembled from five precursor proteins – C5, C6, C7, C8 and C9 – on the membrane to form the ring-like structures (complement lesions). It has been shown that the MAC is composed of a C5b-9 dimer and causes membrane perturbation by the formation of micellar domains in the lipid bilayer [22a, b]. The most challenging questions about how a lymphocyte, after binding with an antigen, is stimulated to produce antibodies is far from being understood. Some perspectives, further discussed in this paper, might also help towards elucidating this problem by electron microscopy.

Use of IgGs for Purposes of Biochemical Identifications

It is well known that the direct methods of specific cytochemical staining for electron microscopy present severe limitations. Indirect methods, like autoradiography, and particularly immunolabelling, however, have provided numerous very successful applications of which we can summarize here only the technical principles.

For supramolecular structures composed of identical subunits, which for this reason show many identical antigenic sites, *Anderson* et al. [3] found very early that the coating by IgGs becomes easily visible in negatively stained preparations, even though individual IgGs are not neatly resolved [1]. In figures 1h, i, we show an application of this method to the identification of proteins composing the bacterial virus T4. Specific IgGs were produced by absorbing a complete antiserum with protein mixtures lacking only one of them. With this technique it was possible to locate all the major proteins constituting the outside of phage head and tail [39].

By using site-specific antibodies it is also possible to follow the movement of a given antigenic site on a protein during physiologically controlled conformational changes like, for example, those which the shell of a virus undergoes when the latter matures from a previrion into the very stable virus [15]. In this particular case, labelling with Fab pieces provided evidence that the observed conformational changes are not due to a simple rearrangement of rigid subunits within a quarternary structure, but that they are correlated with tertiary structural changes (movement of domains) of the subunit itself.

The divalent nature of IgG allows for specific pairing of supramolecular structures by cross-linking identical sites. With this technique it was, for example, possible to locate a large number of the different protein subunits of ribosomes [17, 33; reviewed in 38].

The most important questions asked of electron microscopy are, however, those concerning the localization of biochemically defined elements on or within a cell, because here no other methods are available. In situ information is obviously crucial for understanding vital processes of the cell. Two techniques are available:

(a) *Freeze-fracturing* is most suitable for the observation of membrane surfaces, because the fracture follows them preferentially; lipid bilayers are always and biological membranes mostly split in the median. Cytochemical identifications are, however, extremely difficult to achieve with this technique because only a replica of the still deep-frozen fractured surfaces can be observed. Although this technique is very important in membrane research in general, we will not treat it further in this paper.

(b) *Sectioning techniques* provide very thin tissue slices for direct observation. They are obtained either from frozen material (cryotomy) or from resin embeddings in which the cellular water has become replaced by a solid resin.

While cryotomy is still extremely difficult and hazardous, the second technique has become a cytological routine. Sectioning is complementary to freeze fracturing because it is particularly suited for obtaining cross-sections through membranes and for observing intracellular organization, e.g. membrane systems and cytoskeleton [37]. Antibody labelling is possible in the following ways: (1) Cellular surfaces can become labelled before embedding and sectioning [reviewed in 22]. (2) Antibodies do not penetrate into intact cells. In order to label intracellular structures, the cells have to be permeabilized as is usual in immunofluorescence studies [6, 37]. The penetration of IgGs is obviously also accompanied by leakage of small intracellular components. Besides ions and small organic components, soluble proteins (of sizes up to that if IgGs) are also prone to leave the cell. Such leakage is not just deleterious, but it can also present advantages; this will be discussed further below. (3) Cryosections can be labelled by antibodies after thawing. Thawing and drying – which is required for observation – involve physical effects related to changes of interfacial energies [13] which might strongly alter the structures. (4) By sectioning of embedded cells, antigenic sites become revealed on the section surfaces and are accessible for interaction with IgG. Antibody labelling is limited here by the low number of available sites at the surface; antibodies are indeed too large to penetrate into the resin embedded tissue. The affinity of the accessible sites might, in addition, be reduced, or even destroyed, by denaturation induced by the organic fluids involved in embedding and by the heat normally used for curing (polymerization) of the resin. Improvements in the technique will be discussed below.

When using conventional transmission electron microscopy the antibodies involved in the above-described methods must be visualized by the use of heavy metals. Most commonly ferritin is coupled to antibodies, but peroxydase-coupling, which leads to a deposit of heavy metals by an enzymatic reaction, is also used. This tagging of IgG is performed either by covalent linkage or by so-called sandwiching techniques involving one to several additional antibodies, which are specific for the first antibody and/or to the tag [reviewed in 31].

The resolution of such techniques can obviously never be better than the distance of the tag from the structure-specific site of the first IgG and thus, at best, is some 25 nm. The ferritin technique is, in addition, severely hampered by non-specific interactions of ferritin, which frequently are as strong as, or even stronger than, the specific binding of the IgG. This severe limitation seems to be reduced to substantially lower values when ferritin is replaced by collodial gold coated with protein A [7, 25, 27]. This protein

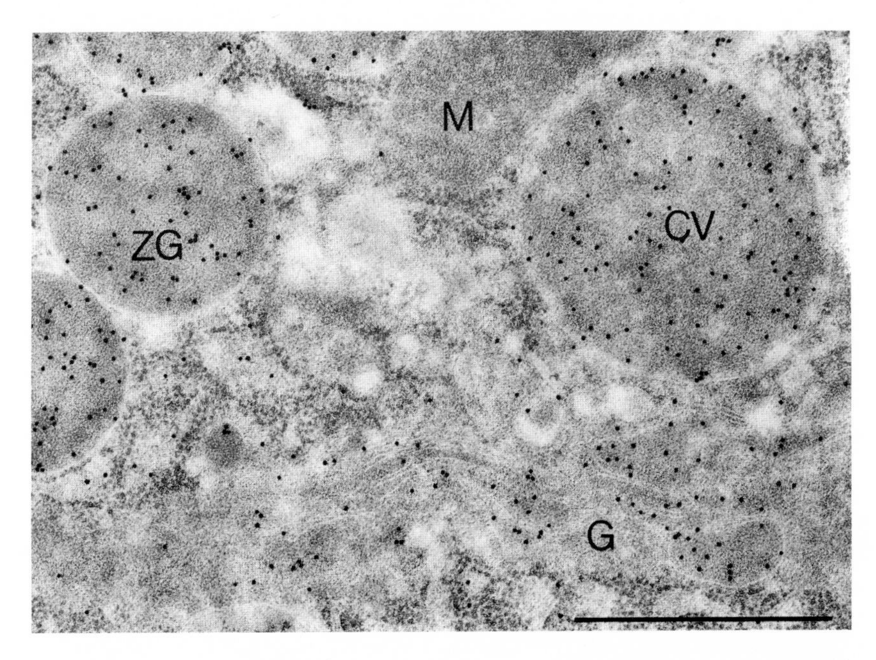

Fig. 2. A rat exocrine pancreas, low temperature embedded with Lowicryl K4M, thin sectioned and stained for amylase by the pAg technique (see fig. 1d). G = Golgy apparatus, CV = condensing vacuoles, ZG = zymogen granules and M = mitochondrion (courtesy of *J. Roth*). Bar = 1,000 nm.

from staphylococcus is highly specific for Fc (fig. 1d), and has shown extremely little non-specific adsorption [26].

Conventional embedding involves dehydration by organic liquids (at room temperature) and curing of the resin at 50–70 °C; both actions are known to denature proteins. For this reason many authors have tried low temperature procedures which included UV curing of the resin. Particularly suitable resins have been developed in our laboratory and improved preservation has been demonstrated recently [14]. One of these resins has been found to be particularly suitable for immunolabelling with the gold protein A technique (fig. 2) [26].

Comparison of gold protein A antibody labelling on sections from low denaturation embeddings with that on cryosections indicates that the sensitivity of these methods is now comparable [*Roth,* personal communication]. The first method is obviously easier to perform and hence more reproducible.

Perspectives for Improvements of Immunolabelling Techniques

There is hope for improvement of sectioning techniques. Sections of resin-embedded tissues are visible in conventional transmission electron microscopy (CTEM) only when the material is massively impregnated ('stained') with heavy metals like Os, U, Pb. It is the heavy metal deposit which is observed and not the protein itself. There is no reason to think that this deposit is inside the proteins rather than irregularly deposited on their surfaces. Recent observations in our laboratory confirm the latter view. In order to observe proteins directly, other stains have to be discovered or/and other means for providing contrast have to be explored. Unstained material in sections can be observed with very high contrast in the dark-field mode of CTEM [30]. It involves, unfortunately, very high doses of electrons which destroy matter by radiochemical effects (beam induced destruction). In the scanning transmission electron microscopy (STEM), dark-field and related types of high contrast imaging are achieved with much lower doses and the observations of unstained sections have lead to very promising results [9]. In our laboratory unstained protein fibers of 1–2 nm were observed incidentally by this technique [11]. This provides hope that the much larger IgGs can be visualized without additional tag and that their attachment sites might thus become more precisely defined; antigens might then be identified as molecules and localised within 2–3 nm. It is obvious that intracellularly this will be possible only on permeabilized cells out of which the soluble proteins should have leaked out before they are precipitated onto structural elements by the ensuing fixation and dehydration. It is predictable that such non-specific precipitates will not be easily distinguishable from specific IgGs. Low temperature embedding, as mentioned above, is also here a promising method [reviewed in 12].

Other developments in electron microscopy, such as observation of frozen-hydrated specimens at 10 °K [11, 12], will, in the future, certainly lead to applications in the broad field of immunology.

Acknowledgements. The authors are grateful to Prof. *J. Engel* for his help and critical reading of the manuscript, to all who contributed pictures (mentioned in the figure legends) and *E. Amstutz* for typing the manuscript. They also wish to thank *M. Steiner, L. Müller* and *M. Zoller* for excellent photographic work and *M. Jaeggi* for the drawing.

References

1 Almeida, J.; Cinader, B.; Howatson, A.: J. exp. Med. *118:* 327–340 (1963).
2 Amzel, L.M.; Poljak, R.J.: A. Rev. Biochem. *48:* 961–997 (1979).
3 Anderson, T.F.; Yamamoto, N.; Hummeler, K.: J. appl. Physiol. *32:* 1639 (1961).
4 Feinstein, A.; Munn, E.A.; Richardson, N.E.: Ann. N.Y. Acad. Sci. *90:* 104–121 (1971).
5 Feinstein, A.; Rowe, A.J.: Nature, Lond. *205:* 147–151 (1965).
6 Franke, W.W.; Schmid, E.; Kartenbeck, J.; Mayer, D.; Hacker, H.J.; Bannasch, P.; Osborn, M.; Weber, K.; Denk, H.; Wanson, J.-C.; Drochmans, P.: Biologie cell. *34:* 99–110 (1979).
7 Horisberger, M.: Biologie cell. *36:* 253–258 (1979).
8 Huber, R.: Trends biochem. Sci. *4:* 271–276 (1979).
9 Jones, A.V.; Leonard, K.R.: Nature, Lond. *271:* 659–660 (1978).
10 Kabat, E.A.: Structural concepts in immunology and immunochemistry; 2nd ed. (Holt, Rinehart & Winston, New York 1968).
11 Kellenberger, E.: J. Electron. Microscopy *28:* 549–556 (1979).
12 Kellenberger, E.: in Brederoo, Priester, Electron microscopy, vol. 2, pp. 632–637. 7th Eur. Congr. on Electron Microscopy Foundation, Leiden 1980.
13 Kellenberger, E.; Kistler, J.: in Hoppe, Mason, Advances in structure research by diffraction methods. Unconventional electron microscopy for molecular structure determination, pp. 49–79 (Vieweg & Sohn, Braunschweig/Wiesbaden 1979).
14 Kellenberger, E.; Carlemalm, E.; Villiger, W.; Roth, J.; Garavito, R.M.: Low denaturation embedding for electron microscopy of thin sections (Chemische Werke Lowi, Waldkraiburg 1980).
15 Kistler, J.; Aebi, U.; Onorato, L.; Ten Heggeler, B.; Showe, M.K.: J. molec. Biol. *126:* 571–589 (1978).
16 Knobel, M.R.; Villiger, W.; Isliker, H.: Eur. J. Immunol. *5:* 78 (1975).
17 Lake, J.A.; Pendergast, M.; Kahan, L.; Nomura, M.: Proc. natn. Acad. Sci. USA *71:* 4688–4692 (1974).
18 Munn, E.A.; Feinstein, A.; Munro, A.J.: Nature, Lond. *231:* 527–529 (1971).
19 Nahon-Merlin, E.; Delain, E.; Couland, D.; Lacour, F.: Nucleic Acid Res. *8:* 1805–1822 (1980).
20 Padlan, E.A.; Davies, D.R.; Rudikoff, S.; Potter, M.: Immunochemistry *13:* 945–949 (1976).
21 Parkhouse, R.M.E.; Askonas, B.A.; Dourmashkin, R.R.: Immunology *18:* 575–584 (1970).
22 De Petris, S.: in Korn, Methods in membrane biology, vol. 8, pp. 1–198 (Plenum Press, New York 1978).
22a Podack, E.R.; Biesecker, G.; Müller-Eberhard, H.J.: Proc. natn. Acad. Sci. USA *76:* 897–901 (1979).
22b Podack, E.R.; Esser, A.F.; Biesecker, G.; Müller-Eberhard, H.J.: J. exp. Med. *151:* 301–313 (1980).
23 Porter, R.R.: Fed. Proc. *36:* 2191–2196 (1977).
24 Reid, K.B.M.; Porter, R.R.: in Inmann, Mandy, Contemporary topics in molecular immunology, vol. 4, pp. 1–22 (Plenum Press, New York 1975).
25 Romano, E.L.; Romano, M.: Immunochemistry *14:* 711–715 (1977).

26 Roth, J.; Bendayan, M.; Carlemalm, E.; Villiger, W.; Garavito, M.: J. Histochem. Cytochem. (in press).
27 Roth, J.; Bendayan, M.; Orci, L.: J. Histochem. Cytochem. *26:* 1074–1081 (1978).
28 Seegan, G.W.; Smith, C.A.; Schumaker, V.N.: Proc. natn. Acad. Sci. USA *76:* 907–911 (1979).
29 Silverton, E.W.; Navia, M.A.; Davies, D.R.: Proc. natn. Acad. Sci. USA *74:* 5140–5144 (1977).
30 Sjöstrand, F.S.; Dubochet, J.; Wurtz, M.; Kellenberger, E.: J. Ultrastruct. Res. *65:* 23–29 (1978).
31 Sternberger, L.A.: Immunocytochemistry; 2nd ed. (Wiley & Sons, New York 1979).
32 Steven, A.C.; Navia, M.A.: Proc. natn. Acad. Sci. USA *77:* 4721–4725 (1980).
33 Tischendorf, G.W.; Zeichhardt, H.; Stöffler, G.: Proc. natn. Acad. Sci. USA *72:* 4820–4824 (1975).
34 Tschopp, J.; Villiger, W.; Lustig A.; Jaton, J.-C.; Engel, J.: Eur. J. Immunol. *10:* 529–535 (1980).
35 Tschopp, J.; Villiger, W.; Fuchs, H.; Kilchherr, E.; Engel, J.: Proc. natn. Acad. Sci. USA *77:* 7014–7018 (1980).
36 Valentine, R.C.; Green, W.M.: J. molec. Biol. *27:* 615–617 (1967).
37 Webster, R.E.; Henderson, D.; Osborn, M.; Weber, K.: Proc. natn. Acad. Sci. USA *75:* 5511–5515 (1978).
38 Wittman, H.G.; Nierhaus, K.H.; Sieber, G.; Stoffler, B.; Tesche, G.W.; Tischendorf, B.; Wittman-Liebold, B.: in Sturgess, Electron microscopy 1978, vol. III, pp. 459–469 (The Microscopical Society of Canada, Toronto 1978).
39 Yanagida, M.; Ahmad-Zadeh, C.: J. molec. Biol. *51:* 411–421 (1970).

Prof. E. Kellenberger, Biozentrum, Klingelberstrasse 70, CH–4056 Basel (Switzerland)

The Immune System, vol. 1, pp. 149–154 (Karger, Basel 1981)

Specificity – or Why Immunology Is Unlikely to Become Boring Soon

Helmut M. Pohlit[1]

Institut Pasteur, Paris, France

Many immunologists feel that the immune system poses a problem which, even though it cannot easily be separated from much of biology, nevertheless seems to be special enough to make the idea of a characteristic solution reasonable. However, the ever-increasing mass and complexity of biochemical and molecular biological data about the immune system, together with ever more sophisticated technology, instills in some the fear that this would tend to inhibit the posing and facing of fundamental questions and that, thus, boredom is surely approaching. In many other immunologists it instills the hope that the solution indeed is nothing but the enumeration and detailed description of the molecular and cellular ramifications of structures and functions. I believe this hope is false. I believe a characteristic solution must be in terms of evolutionary theory on the one side, and on the other side, in terms of the objects of this theory, namely biological adaptive systems. The search for these terms, the need for embedding in a wider framework, and the accessibility of the immune system to experimentation are why immunology will remain fascinating.

Nevertheless, we need detailed molecular data. The situation may be likened to the centuries-old ingenious search for perpetual motion machines. The invention of the system state function entropy, compatible with

[1] *Helmut Pohlit* was a member of the Basel Institute for Immunology from 1970 to 1978.

observation, made this search illusory, but created at the same time completely new avenues for theory and application. Entropy that could not have been deduced from molecular data, no matter how detailed, was thus, on the one hand essential for the understanding of the (thermic) behavior of large ensembles of molecules; on the other hand, it soon began to play a significant role outside thermodynamics, i.e. in the theory of probability and its applications, into which thermodynamics could be imbedded. Immunology is probably in a similar situation. For whatever purpose we do research in this field, the problem of characteristic terms, applicable to the immune system and generalizable, is a real one and not just philosophical. To explain how this is meant, in the context of some experimental observations, is the intention of this article, which I dedicate to the immunologist who has in a conspicuous manner demonstrated where at different stages of development of the art the line between speculation and theory has to be drawn so that theory be practial – *Niels K. Jerne.*

The Completeness Problem

Several different epithets describe different manifestations of the same central enigma of immunology. Diversity and specificity (of combining sites) are perhaps most frequently used. The experiments to be described below suggest another term – fidelity (with respect to clonal specificity). 'Completeness' (with respect to antigen recognition) has been considered rather little. However, it was with respect to completeness that we did an experiment which of course did not solve the completeness problem but led to the realization of the high degree of fidelity of antigen recognition. The reasoning was as follows.

To account for the high degree with which the respective antisera distinguish between the members of a given set of antigens, supposedly chosen at random, a combinatorial consideration yields a lower limit for the number of combining sites available to the animal. Whether this number is 10^5 or even 10^7, it does not follow that there should always be an antibody to a given antigen. If the generation of combining sites is assumed to be a random process, and if the antigenic universe is considered a random sample of all possible determinants, the probability of finding a matching pair of epitope and paratope in a given animal should be practically zero. This kind of problem is apparent, in a heightened fashion, for instance in the obviously continuous recognition capability of sound frequency in the auditory sensory system. If we were to assume a (necessarily) finite set of neurons, each tuned to recognize a random frequency, the probability of recognizing a given, arbitrary frequency would be zero. In reality, it has been found that there exists a finite number of brain cell groups which seem to be tuned rather sharply to 'their' frequency [1]. Now, to account for the absence of holes in the perception of pitch one has to assume a rather even distribution of tuning curves over the audible range, each with a certain width to generate sufficient overlap and thus a seemingly continuous recognition spectrum. If, however, tuning is a random event (through the randomness of some detection element) by the same process the range should be a random event as well, and one would have to ask: (1) Why is there tuning? (2) Why are the tuning ranges so well organized that continuous recognition is possible within a certain frequency range? In considering only the frequency we have, of course, greatly oversimplified the multi-parameter recognition of sound. Returning now to the immune system, recognition of antigenic structures is also not describable in terms of a single parameter. But the two questions above may be rephrased and asked again: If the combining site structures are random: (1) Why are they distributed in such a way? (2) Why are the individual recognition ranges such that they permit completeness in antigen recognition?

Actually, rather little is known about the distribution of combining sites and their recognition ranges, because in order to answer such questions one would have to somehow classify the recognition properties of the different clonal antibody combining sites. One way of obtaining such a classification for a subset of combining sites, namely those reacting with a given antigen, is as follows: Let us assume an antiserum is raised against antigen A, and when tested against a series of related antigens (B, C, D, etc.) it is found that it reacts with these antigens to a lesser and lesser extent, in the order given. Two extreme answers are imaginable when one asks what cross reaction one should expect for each of the clonal antibody species of which the antiserum is composed: (hypothesis A) all antibody species exhibit the same order of reactivity as the whole antiserum; (hypothesis B) the order varies among the antibody species without limit. The specificity of the whole antiserum is due to an opportune mixture of these antibodies.

Hypothesis B would be more what one would expect for random generation of combining-site diversity. Hypothesis A implies, if verified, either an even greater repertoire of combining sites than assumed for random generation, or some plasticity on the part of the primary recognition mechanism.

The Little/Eisen Experiment

Little and Eisen [2] approached such an analysis more than 10 years ago. They came to the conclusion that hypothesis A was probably correct. The antigens were TNP and DNP, and from each corresponding antiserum they withdrew about 1% of the specific antibodies, namely those complexing best, under conditions of reaction equilibrium, with the cross-reacting antigen. For this small portion of antibodies they showed that it exhibited the same parameters of cross-reactivity as the unfractionated antiserum. Clearly, since the investigated antibody fraction had been selected for high affinity towards the cross-reacting antigen, affinity and specificity do not seem correlated. In addition to affinity, this fraction must also have differed in clonal composition from that of the whole antiserum, although it is impossible to estimate the extent without knowledge of the affinity distribution. Thus, one might also cautiously conclude that the specificity was not strongly different among different clones. Thus, hypothesis A seems correct.

Experiments with Three Cross-Reaction Bacteriophages

By taking the investigation from two to three cross-reacting antigens, and by making use of isoelectric focusing chromatography (IEF) of antisera in slab gels as a means of clonal fractionation, we hoped to sharpen the analysis. The following experiments were done at the Basel Institute for Immunology, together with *Reinhard Schulze:* Rabbits were immunized with any of the three bacteriophages f_2, f_r, and f_4. For all practical, immunological purposes the shell of these phages may be considered to consist only of 180 identical polypeptide chains whose amino acid composition varies by up to 14 (of 130) residues among the three phages. After a booster with the homologous antigen the sera were fractionated on slab gels under conditions of IEF within a pH range from 4 to 9. After the run the gel was cut into 40 parallel strips and the antibody eluted. Subsequently, the cross-reaction pattern was determined for each fraction against each phage, and this pattern was compared with that of the corresponding whole antiserum.

The following results were obtained: (1) All IEF fractions, varying widely in concentration of antibody, exhibited the same cross-reaction pattern as the whole antiserum. (2) The immunizing antigen always reacted best. The cross-reactivity patterns depended, naturally, on the immunizing phage. For example, with f_2 as immunizing phage, the relative cross-reactivities with f_2, f_r, and f_4 were about 100, 10, and 1, respectively. With f_r as immunizing phage the result was about 100, 25, and 10 for f_r, f_2 and f_4, respectively. (3) This pattern of cross-reactivity differed very little among identically immunized animals. (4) When the boosted animals had been rested for several months a further booster with a cross-reacting phage induced essentially the same type of antiserum as had been obtained after the first (homologous) boost.

Again, it cannot be argued that our IEF fractionation resulted in perfect clonal fractionation. Nevertheless, considering the antigenic simplicity of each immunogen we would expect only a rather small number of phage specific antibody species, and we would thus expect considerable segregation of these antibody species within the set of IEF fractions.

Conclusions

1. Fidelity of Clonal Pattern Recognition. From results 1 and 2 we conclude that, again, hypothesis A seems to be correct. Not only do the clonal antibody species react best with the immunizing antigen, but each species shows analogous cross-reaction patterns. Thus, the recognition of the epitope pattern is conserved with high fidelity at the clonal level.

2. Difference between Primary and Secondary Recognition Mechanisms. Result 4 shows that at the level of memory response, cross-stimulation of specific lymphocytes is readily obtained. Why, then, are there no antibody species observed which react, for example, better with a cross-reacting antigen than with the immunizing antigen, or with a different cross-reaction pattern, i.e. one corresponding to another phage as immunogen? That each animal could have in principle responded in this fashion is made plausible by result 3. That antigen is not limiting is shown by the fact that at least 60% of the injected bacteriophage circulates freely in the blood until day 3, i.e. shortly before antibody becomes detectable. Also, the cross reaction patterns are independent of the dose of priming antigen, over a 100-fold dose range. We therefore conclude that the primary recognition apparatus is distinctly different from the secondary. The primary is obviously much more discriminating than the secondary. That this should be due to differences in repertoire would imply that the primary repertoire is greater than the secondary. This is difficult to accept. The difference in discrimination could also be due to different thresholds of stimulation. However, in view of the excess free circulating antigen this explanation is also implausible.

Some Speculations. On the other hand, it is possible that an adaptive mechanism resides in the system of primary antigen-reactive lymphocytes, which permits the evolution of an antiserum with the striking specificity that we are accustomed to in immunology. The memory response could then be characterized by the absence of such plasticity. This adaptive mechanism would throw up a number of new problems of which some may be quite familiar in current research on synaptic mechanisms. It might require, in the same sense as for the nervous system [3], the introduction of receptor types as units of analysis of the immune system, next to the analysis in terms of cells and clones. In any case, however, it would not solve the completeness problem because we still must assume the existence of a complete primary recognition system. However, it would represent some economy on the part of diversity of this primary receptor system necessary to cover the universe of antigens.

Our experimental observations suggest that we sharpen the question: 'Why are immune reactions so specific?' to: 'Why are they specific in the particular fashion that we have described?' This question is understood as one addressed to the immune system and addressed to molecules and genes only in second order.

References

1 Abeles, M.; Goldstein, M.H., Jr.: J. Neurophysiol. *33:* 172–187 (1970).
2 Little, J.R.; Eisen, H.N.: J. exp. Med. *129:* 247–265 (1969).
3 Shepherd, G.M.: Yale J. Biol. Med. *45:* 584–599 (1972).

Dr. Helmut Pohlit, UNICET/U.R.D., Faculté de Médecine, 45, rue des Saints-Pères, F–75006 Paris (France)

The Immune System, vol. 1, pp. 155–168 (Karger, Basel 1981)

The Joint Evolution of Antigens and Antibodies

S. Fazekas de St.Groth[1]

Basel Institute for Immunology, Basel, Switzerland

An immune response is complex because many kinds of cells are involved in it; it is also heterogeneous because different cells make different products. Thus even the simplest antigen may trigger hundreds of clones, each making its own kind of antibody. Variety of this order is unusual, possibly unique, among natural products. This, by itself, is no cause for alarm. What is worrying about this variety is that the immune system does not operate in a vacuum: its products are part and parcel of the molecular ecology of an organism, together with all sorts of antigens. And it would be a disaster if an imprecise response, aimed at some foreign intruder, would also react with any of these self-antigens, let alone if self-antigens themselves would stimulate a response. We should like to find out therefore what made Nature choose such an inelegant solution and how she gets away with it.

This we do, of course, not by asking philosophical questions nor by giving our version of what an Omniscient Immunologist should have done, but by finding out what's going on and then seeing how this fits into the scheme of things. What we are trying to do then is to answer the simplest question in immunology: 'Who's combining with whom?' Or, to put it into a more scientific form, what is the structure of the combining site, how many different kinds of such structures can an individual or a species make, and what part of the information is heritable? Because of the nature of the question we cannot make use of the fashionable tools of immunology, such

[1] *S. Fazekas de St.Groth* has been a member of the Basel Institute for Immunology intermittently since its founding. His contribution to this volume is based on a lecture prepared at the request of *Niels Jerne.*

Table I. Cross-reactions among influenza A3 strains

Epidemic strain of	Antibody against strain			
	1968	1970	1971	1972
1968	100	92	114	87
1970	18	100	105	92
1971	7	9	100	100
1972	10	9	22	100

as isolated cells, homogeneous myeloma proteins or even clonally restricted responses to any antigen. What we need are methods by which antibody populations can be dissected. We further need a series of antigens, preferably a series of single-step mutants of the same antigen, for the analysis of cross-reactions.

As far as the antigen is concerned, we have no major problem: influenza viruses meet all our requirements and, by Special Providence, an unending series of further antigenic mutants is given unto us in the shape of yearly epidemics, each caused by a different but cross-reacting virus. Let us start then with a short detour into virology and see what evidence we have for cross-reactivity and how good this evidence is.

Antigenic Evolution

We take viruses from consecutive outbreaks of the present era, starting with the Hong Kong virus which caused the pandemic of 1968–69. Against each of these viruses antibodies are prepared; it makes no difference whether we use convalescent human sera or antibodies prepared in rabbits, the results are the same. With these reagents all possible reactions are performed, giving a matrix of neutralizing titres. Since we are not interested here in absolute values, the titres have been normalized, calling the homologous reaction 100%. Hence the values of 100 in the main diagonal of table I.

What is remarkable about this matrix is the asymmetry of cross-reactions. All high titres lie above the main diagonal and all low ones below it. These antisera seem retrospectively effective but not prospectively, and if we wished to make a facile observation, we might say that this is *why* a new virus could establish itself in a population which was immune to its predecessors. But we are not here to make facile observations, so we rather ask *how* this could come about. The first intelligent answer assumed that these viruses were antigenic mosaics, the initial member of the series sharing one antigen with the rest, the last all four. After a chequered career of some 20 years, this view became untenable on more recent chemical and immunological evidence: we know that the antigen involved is a homogeneous repeating unit, and any one particle carries and can transmit only one kind of it.

The viral antigen, being homogeneous, cannot be blamed for lopsided cross-reactions. So we are left with the only other component in the reaction, antibody. And here we know that the response to any antigen is heterogeneous. If we represent an antigenic determinant (i.e. a small area on the surface of a protein molecule, made up of perhaps 10–12 amino acid side chains) as a fork with prongs of different sizes and shapes, then an antibody population could complement this in many different ways. There may be one species of antibody that fits the determinant perfectly, and there will be many species which fit it imperfectly or only in part. What happens then if such a mixture of antibodies is brought together with an antigen that differs minimally from the homologous, say, by having leucine instead of valine at a particular locus? The situation is like trying to fit a glove to a hand with a swollen finger: all species complementary to valine will be sterically hindered by the bulkier leucine, the rest combining equally with both antigens. In the converse situation – fitting a large glove to a small hand – all antibodies are permitted to combine: a cavity fit for leucine will take in sundry smaller amino acids such as valine. To give it a name, we call the antigen with the bulkier amino acid *senior* to the one carrying the smaller amino acid. In these terms the asymmetry of cross-reactions rests on the exclusion of part of a junior antibody population by a senior antigen. The converse reaction is permissive, as there is no steric hindrance between antigen and antibody. This is a testable proposition, and could be confirmed by several independent methods. Symmetric cross-reactions cannot be close since they must rest on at least two point-mutations, involving different loci in the pair of antigens.

And this, precisely, is what we do not find in our matrix. If these epidemic strains were survivors of random antigenic mutations, we should expect all to be senior to the original parent, but not to each other. The observed picture is compatible with sequential changes at the same locus, establishing a hierarchic order of increasingly senior mutants. Such an ordered set of mutants we call a subtype. Since the first isolation of influenza virus some 50 years ago we have had four subtypes, each era beginning with a severe pandemic. In each case a junior virus caused the first pandemic, to be replaced by more senior forms over the following decade. This process is readily imitated in the laboratory: there is abundant evidence in the literature that some of the viruses isolated from immune hosts or from culture systems in the presence of antibody carried a different antigen, were genetically stable, and were senior to their parent. Sequential selection starting with a junior antigen yielded a series of increasingly senior forms, and each sequence came to a point where the same techniques failed to select further mutants, without passing from one subtype to the next. When starting from the senior member of a subtype, failure was immediate.

Model Experiments in Antigenic Evolution

We have performed a large number of such selection tests, not so much to imitate Nature but rather to find out what kind of mutants arise and with what frequency. An answer to this question would, of course, also tell us what scope an adequate antibody response should have. To our great surprise it turned out that most of the over a thousand mutants we selected were not antigenic mutants, but what we call *adsorptive mutants.* The virus escaped neutralization not by changing its antigenic determinant but by inserting a positive charge in the area reserved for attachment to susceptible cells. This little trick tips the scales in favour of infection but is of no relevance to immunology. Among the true antigenic mutants there were two subsets, differing by several criteria. The first, which we call *hierarchic mutants,* are senior to their parent and, like it, give only distant and usually symmetric cross-reactions outside their subtype. Under pressure of homologous antibody they yield further, even more senior mutants. The second kind we call *bridging mutants.* These, too, are senior to their parent but, unlike it, are also senior to the early members of another subtype – they form bridges between two subtypes. They have also other distinguishing features: they arise with a frequency about two orders of magnitude below that of hierarchic mutants and behave like terminal forms, i.e. the same methods that led to their selection yield no further mutants.

We have carried such selection tests through several generations, starting with the junior members of two subtypes. These experiments allow some general conclusions. First, the series is hierarchic: consecutive mutants within a line stand in junior-senior relation to each other. Second, the series is bounded: after three or four steps of selection no further antigenic mutant can be obtained by the standard method of selection. Third, the series is degenerate: the same antigens are occasionally obtained in one, two or three steps, or through different intermediates. Fourth, the series is convergent: while in the first and second generation there are several distinguishable new antigens, they tend to give rise to the same terminal forms on further selection. Fifth, adsorptive mutants may arise at any point in the series, including the terminal stage.

A Molecular Model

All these observations may be summed up in a molecular model. We start with an antigenic determinant made up of a handful of amino acids. In the case of influenza we have another bit of information: the virus-antibody union is entirely entropy-driven, its latent heat being indistinguishable from zero, sometimes even positive. This indicates hydrophobic bonding as the mechanism, and we should expect both the antigenic determinant and the combining region of antibodies to be made up largely, if not entirely, of hydrophobic amino acids. This is the reason why the hypothetical antigenic site looks like it does (fig. 1).

We have then our junior antigen, made up of small to middling amino acids, and if it is to escape neutralization it has to offer steric hindrance to antibodies. This it can do in two ways. The usual option is to replace its largest component by an even larger one, in the example Val → Leu at position 3. This is the first step on the ladder of seniority, and the mutant will have all the properties of the class we call hierarchic mutants. Alternatively, the parent may opt for a bulky substitutent at some other spot, say, Thr → Ile at locus 5. This form will have the properties ascribed to bridging mutants, except that as yet we do not see any reason why such a pattern should be terminal, i.e. throw no further forward mutants.

Continuing along the hierarchic line, locus 3 passes through the series Val → Leu → Phe → Tyr and can, at any stage, hold antibodies at arm's length, whether they were directed against the immediate parent or against any of its predecessors. Tyrosine represents an obvious terminal stage: it could turn only into smaller amino acids or terminating signals. Thus subtype 3 is ripe for extinction.

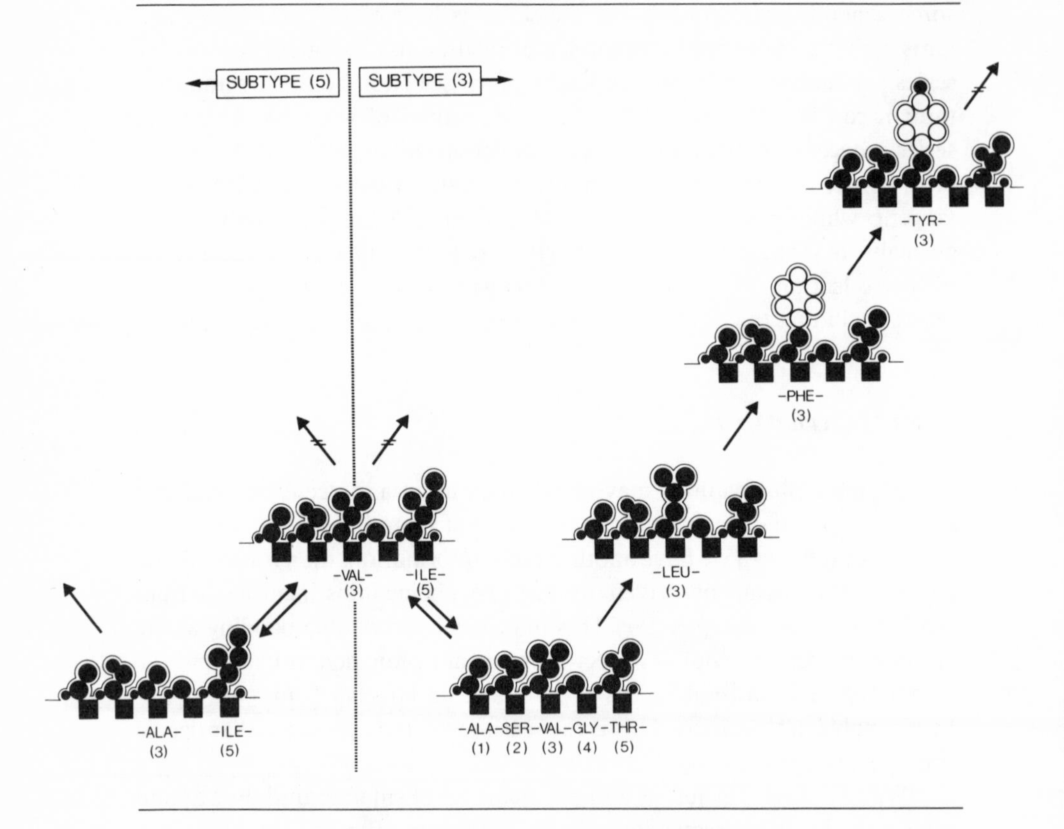

Fig. 1. Evolution of a junior antigen. Hypothetical antigenic determinants are represented by the van der Waals contours of five hydrophobic amino acids.

The alternative to the hierarchic line is a set of bridging strains. These may be regarded as double mutants of a protoantigen since they combine the junior loci of two subtypes. Such were found to behave as terminal forms under the pressure of antibody. That means they have no viable senior mutants, as homologous antibody would allow the survival of senior forms only. Double mutants, however, have also another escape route, as one of their back mutants represents the first member of another subtype. Such mutations are as likely as any other, the survival of a particular pattern depending only on the environment, in the case of influenza on herd immunity.

The model, then, employs the simplest mechanism of natural selection: mutants arise at random and those senior to the parent have marked survival advantage. The only special assumption is that the sterically hindering amino acids are not placed at random. Each subtype has its private locus at which sequential substitutions occur, and the subtypes differ in the position of this locus. Such a constraint was originally postulated simply to make the model consistent with observation. From it followed the hierarchic evolution within subtypes, their roughly equal life span, their limited number, as well as the transition between subtypes and their periodic recurrence.

In the light of recent developments in the energetics of protein conformation, the special postulate appears as an instance of a general phenomenon, in fact, follows from the nature of the antigenic determinant. When a peptide chain folds, the first rule is that all charged groups should be on the outside, and the second that all hydrophobic side chains should be hidden. The actual solution is always a compromise on the second rule: most proteins have 20–30% of their hydrophobic side chains exposed to the aqueous medium. Since these side chains are surrounded by structured water, this arrangement adds a negative entropy term of the conformational energy of the molecule, i.e. an element of instability. Still, single substitutions by amino acids further up the hydrophobicity scale are acceptable as long as there is conformational energy to spare. The situation is different for hydrophobic areas (sets of 4–8 contiguous hydrophobic side chains). If these carry more than one large side chain and the side chains interact (i.e. eliminate structured water), the result is a 'hydrophobic flip', amounting to a fundamental conformational change.

The antigenic determinant of influenza A viruses represents just such a hydrophobic area, as has been evident from thermodynamic measurements for the past 20 years. In these terms, then, only hierarchic substitutions are permitted, while bridging strains represent the maximal hydrophobic bulk compatible with the native conformation of the molecule. A hydrophobic flip is, of course, lethal for the virus, even though it might count among the most advanced methods of breaking emulsions.

This brings us to the end of our journey into virology. Since travel is supposed to broaden the mind, there are two lessons we should take home from this journey. First, a positive one, that cross-reactions are asymmetric. This means that the average antibody will react with many antigens, to wit, with all that are homologous or junior to it. Conversely, the average antigen has a chance of stimulating a host of antibody-producing cells, to wit, all that can make antibodies homologous or senior to it. This excuses Nature for settling for a sloppy solution: with cross-reactions being what they are there is no way of avoiding redundancy. And the second lesson, a negative one, is that the evolution of influenza viruses cannot represent the general case. It may be a useful didactic example, with viable mutants restricted to sequential changes at a single locus, and it is also convenient in practice, for the same reason. But this type of evolution or, better, revolution in the literal sense, is restricted to hydrophobic areas. A less exceptional antigenic determinant is bound to contain also charged groups and, indeed, we know that most antigen-antibody unions are characterized by enthalpy changes of 5 kcal/mol or more. Thus, what influenza does at one spot at a time, the average antigen should be able to do simultaneously at several loci. This does not upset the principle of cross-reactions nor the rules of antigenic evolution, just makes their study considerably more difficult. So we shall retain influenza viruses when we are turning now to the immunological side, but do this with our eyes open, not forgetting that they are a paradigm, not the general rule.

The Immune Response

Let us see then for a start what it takes to trigger an immune response. We vaccinate groups of mice with different doses of virus. After a fortnight the sera are tested, establishing the primary dose-response relationship (fig. 2a).

We observe that: (1) it takes less junior virus to induce a response than it takes senior virus; (2) after high doses of virus the response asymptotes, with the same ceiling for junior and senior antibodies; (3) the average quality of anti-junior antibodies is consistently better than that of anti-senior antibodies.

6 weeks later we boost these groups with a uniform dose of 1 μg antigen, and test their sera a week after that, establishing the secondary dose-response relationship (fig. 2b).

We observe that: (1) even some of the groups which produced no measurable primary antibody give a good secondary response; (2) the ceiling reached by anti-junior and anti-senior antibodies is the same *and* also the same as was the primary ceiling; (3) the average quality of antibodies improves in both groups, but the difference between junior and senior becomes even more marked.

The secondary response can be tested also in another way. We take large groups of animals, all of which had received a primary dose of antigen

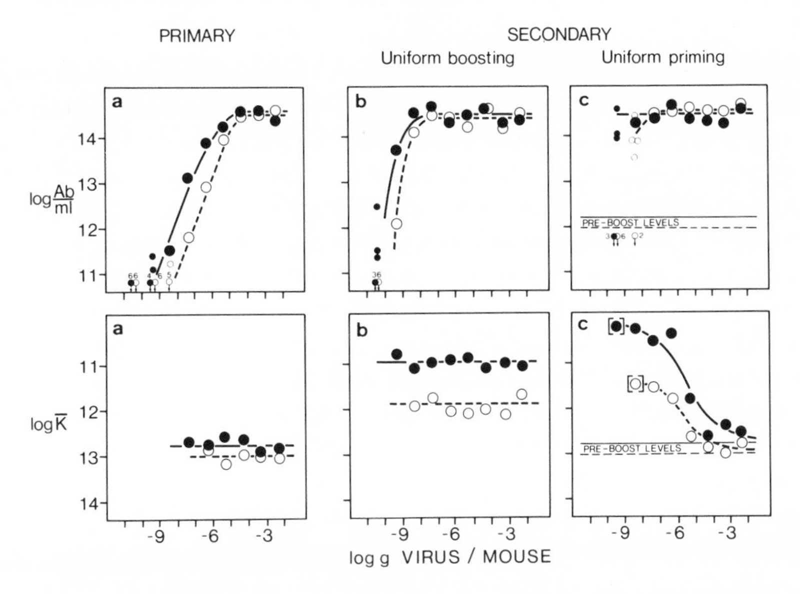

Fig. 2. The antibody response to junior (•) and senior (○) antigens. The number of antibody molecules at the peak of the response is shown in the upper panels, their quality (average equilibrium constant in cgs units) in the lower panels.

such that they responded with 5×10^{13} antibody molecules/ml at 14 days. These groups are then subdivided and boosted with varying doses of the same antigen secondarily (fig. 2c).

We observe that: (1) the quantity of antibody molecules is independent of the boosting dose over a wide range; (2) the ceiling is, once again, the same for junior and senior antibodies; (3) the quality of antibodies improves after boosting with small doses of antigen but stays at the primary level after large doses, the difference in favour of junior antibodies being maintained throughout.

These results are quite orthodox as long as we consider only one type of antigen at a time and, especially, if we look only at the central portions of the curves. This, in fact, is the region usually studied by immunologists since, generally, an antigen induces tolerance rather than immunity at both ends of the dose range. With influenza viruses neither low-zone nor high-zone tolerance could ever be demonstrated, in spite of repeated and serious efforts. So we are in the unfortunate position of having to stretch orthodoxy over immunogenic doses that differ by eight orders of magnitude, and over pairs of very similar antigens which trigger quite dissimilar responses.

Well, orthodoxy does not stand up too well to such Procrustean treatment. The conventional view of an affinity controlled immune response is compatible with what we see on boosting primary immunity: smaller secondary doses select for better average quality. This, actually, was the experimental basis for the postulate. But such a thing does not happen in the primary response, and even in the secondary the relation of quantity to quality is different for junior and senior antigens. What is worse, the maximum number of antibody molecules is much the same in primary and seconary responses, while the corresponding equilibrium constants may differ by two orders of magnitude.

Such conclusions are disquieting, even though they come from small groups of animals and some of the points are widely scattered about their means. So we repeat the critical parts, this time on large groups of mice and using inbred strains for greater homogeneity. What we find, and can state this time firmly, is that each of the strains responds well to a junior antigen, producing about 10^{15} antibody molecules/ml, and the average quality is equally high in each of the groups. When using a senior antigen, the quantity of antibodies is once again about 10^{15} molecules/ml, but their quality is poor and highly variable, even among members of the same inbred strain.

Analysis of Antibody Populations

It looks as if our test animals would find it easy to respond to junior antigens and rather difficult to respond to senior ones. This is a trivial conclusion which actually follows from the fact that there are more ways of complementing a junior than a senior antigen. What is not so trivial is the greatly increased variation in the response to senior antigens. To look into this problem somewhat more closely we have to dissect the antibody response. This we can do in several ways, none of them entirely satisfactory or complete.

If we separate antibodies according to their isoelectric points, we find that an anti-junior response is much more heterogeneous than an anti-senior response. This is a crude qualitative statement, as the technique is unsuited to resolving heterogeneities of the order encountered in standard antibody responses.

We can also fractionate antibody populations according to their affinity, by chromatographing them on antigenic columns. Even though the resolution here is worse than in isoelectric focussing, the technique allows one firm conclusion: the lower average affinity of anti-senior antibodies does not rest on peculiarities in the distribution of subclasses but in the complete lack of antibodies of the highest affinity. This finding accounts for the poorer average quality of a response to senior antigens. It also presents us with a paradox: we should expect that a wider scatter of subclasses would also be reflected by a larger variance about the mean. What we find is the opposite. But the paradox is resolved as soon as we look at the performance of individual animals. They vary within narrow bounds in response to a junior antigen, the chromatographic profiles being very similar within any one strain and, indeed, even between strains. After vaccination with a senior antigen there is a wide scatter in the response, and this scatter is largely due to the presence or absence of the more firmly binding subclasses of antibody.

This variation among members of the same inbred strain is even more evident by the third method of analysis. Using a series of related antigens – and such are readily available among influenza viruses – for sequential absorption of an antiserum, we can quantitate the various cross-reacting subclasses making up that serum. It turns out that there are significant differences among the three inbred strains we have tested: each can be characterized by the particular subset of antibodies they *fail* to produce. Within this strain – let us not shy away from the word, genetic – restriction, members of the same inbred line still vary greatly in terms of the best-fitting antibodies they can make.

Taken together, these observations teach us something about the immune response in general. First, the number of antibody molecules tends to an asymptote after adequate stimulation. This asymptote does not depend on the quality of antibodies, as we have seen by comparing primary and secondary responses, or by comparing responses to junior and senior antigens. We shall therefore need some mechanism that can set and maintain such a ceiling. Second, junior antigens (i.e. antigens of low profile) can be complemented in many ways. The response to such antigens is highly heterogeneous, with antibodies ranging from the lowest affinity to the highest. All animals, irrespective of species or genetic background, seem to be capable of making most, if not all, varieties of anti-junior antibodies. Third, senior antigens (i.e. antigens of marked profile) can be complemented in fewer ways, yet the individual response is even less heterogeneous than would be expected since antibodies of the highest affinity are usually absent. These individual responses vary within an inbred strain of animals, and inbred strains differ from one another by the gaps in the spectrum of anti-senior antibodies they can make. We need then some mechanism which limits the expression of antibodies closely fitted to senior antigens, does this variably within sets of genetically uniform animals, applies different rules to different genetic constitutions, and places no visible restrictions on the response to junior antigens.

Speculations on the Evolution of the Immune Apparatus

If we start from the germline theory, i.e. believe that each individual carries the information for all antibodies the species can make, we have to admit that at least in the response to a senior antigen the organism displays only part of its repertoire, with many of the potential clones remaining silent. Since this partial display is also variable within a genetically uniform population of responders, we have an unpalatable element of randomness added to a doctrine designed to avoid such heretic notions as somatic development of diversity.

We may also choose a less committed starting position, saying that an organism always does as well as it can, and some clones are not induced simply because there is nothing there to be induced. Such a simple-minded approach begs the question of what is there to be induced by the time an organism has to face up to a foreign antigen. Being simple-minded, we have no pat answer to this question; all we can say is that an organism is a large assortment of antigens, and what should *not* be there is antibodies inducible against these self-antigens. Thus, not unmindful of the theory of games and the more reputable military authors, we first minimize our losses by making self-antigens as senior as possible. This, we now know, is the only safeguard against unwanted cross-reactions. The number of maximally senior antigenic patterns is small: an informed guess (based on the size of the antigenic area and on the minimum change in free energy to make an antigen-antibody complex stable) would make this number 210. Thus, if an individual would be characterized by only four such antigens, we could distinguish, and distinguish perfectly, 78 million individuals; if we took ten self-antigens, the number would come to 3×10^{16}.

What this amounts to is that a system of strictest surveillance would need only 210 senior antibodies. It goes without saying that the subset directed against the individual will have to be eliminated from this set of 210 maximally senior antibodies. When trying to do this, we note that the choice which minimized our losses also maximized our gains. It follows from the nature of these unwanted patterns that *any* change at any of their anti-senior loci will render them junior relative to self. A very small number of mutations is thus sufficient to disarm the offspring of these cells and make them acceptable to the organism. While this still leaves us with the host of unreformed parents, the model has features to take care of that, too. The cells we want to get rid of form the only subset of the population which can practise *auto-surveillance:* they are making both self-antigens and their complementing structure at the same time. This provides the means to commit suicide; they are also faced with the *Russell-Frege* antinomy, which might provide the motive.

It is altogether surprising that such a naive construct should have the properties we expect of a system of surveillance. It has also the stability such a system should have. Any drift on the antigenic side is counter-selective as it raises the danger of unwanted cross-reactions, and that without any compensating advantage. A drift on the antibody side looks harmless, i.e. selectively neutral and thus unlikely to last long in evolutionary time. When trying to find some use for these harmless by-products, we notice that our scheme of surveillance is a modest immune system as well. The small set of inherited senior antibodies will recognize and react with a very large number of junior antigens, in fact, is an ideal first-line defence capable of taking on all comers. And the harmless by-products are harmless insofar as they cannot react with senior self-antigens; they are patently useful in dealing with a still large sector of junior antigens and are the first stage in the development of diversity.

At this point we also notice that any further speculation along these lines would be paraphrasing *Jerne's* theory and that, surely, is out of place in a volume like this. What is left for us is to see whether our model bears any relation to reality. There are many possible check-points – I shall pick out a few of the more critical ones.

First, we should like to see whether the immune system has a compartment characterized exclusively by anti-senior receptors. We should expect this to be in the most ancient sector of the system, the thymus-derived lymphocytes or T cells. The experiments have been done by *Roelants,* by *Mäkelä,* by *Möller* and they find that T cells recognize the same universe of antigens as do B cells, but they do this in a homogeneous fashion and at low affinity and the response does not mature. This is precisely what should happen if T cells had only very senior paratopes as receptors, while B cells displayed the whole gamut of possible complementary structures.

Second, since anti-self clones have been deleted by the suicidal mechanism built into the system, the diversity of antibodies should have different starting points, depending on the genetic constitution of the organism. Further development from these starting points is a stochastic process, and hence members of an inbred strain should be similar in antibodies readily derived from the deleted clones, but may be dissimilar in antibodies that can be reached only by a long random walk over the codon table. In immunological terms, inbred lines should be indistinguishable by the junior antibodies they can make and distinguishable by the senior antibodies they can make. Also, individual variation should be marked in senior antibodies, negligible in junior ones. This, essentially, is the summary of our findings.

And, third, a more sophisticated – and hence more critical – corollary. The assignment of most senior antigens as self-markers has an interesting implication. Since such antigens carry side chains of uneven length, their native conformation is less rigid than that of more compact determinants – they are liable to distortion by molecules placed next to them in the cell membrane. These distortions, however, are not random: changing the spatial relationships of one or the other long side chain can result only in a shift from one senior antigenic pattern to another. Such changes will be readily spotted and acted upon by the T cell compartment, and the model implies several surprising side effects. Thus, some foreign molecule, say, a viral antigen, may stimulate killer cells restricted to one or the other end of the major histocompatibility region, some to both, some to neither. Furthermore, these 'specific' killer cells may ignore gross antigenic differences (as is the case with influenza viruses) or become alloreactive against unrelated haplotypes marked by a senior pattern into which the self-marker has been distorted. Each of these implications is also an experimental fact, baffling all who hold with double recognition by T cells.

It is not my aim here to test a particular hypothesis, but rather to demonstrate what consequences the simple rules of cross-reactivity may have. And if I am leaving the impression that their effect is all pervasive and the main reason why the immune system is not simpler than it is, I have left the right impression. To maintain such a system will need a complex set of checks and balances, worthy of the attention of an Omniscient Immunologist. As we are privileged of having not one but several contributing to these volumes, I may safely stop here and leave the final answers to them.

Prof. S. Fazekas de St.Groth, Basel Institute for Immunology, Postfach, CH–4005 Basel (Switzerland)

The Immune System, vol. 1, pp. 169–175 (Karger, Basel 1981)

Neighborly and Unneighborly Determinants of Site-Specific Mutation Rates

John W. Drake[1]

Laboratory of Molecular Genetics, National Institute of Environmental Health Sciences, Research Triangle Park, N.C., USA

Large differences are observed among mutation rates of similar genes in different organisms, among diverse genes in a single organism and among particular sites within a given gene. Many aspects of mutation rates are established on a genome-wide basis, for instance by the fidelity characteristics of the enzymes of DNA replication and repair. From time to time, however, it may be advantageous for an organism to establish unusually high or low mutation rates at specific regions of select genes. The somatic diversity demanded of the immune response is a striking candidate for such mutational fine tuning. *Niels Jerne's* enthusiasm for the problem of immune diversity [8] helped to highlight the potential role of somatic mutation in this system. The recently discovered contributions of multiple genes and specialized recombination mechanisms may have reduced the need to invoke mutagenesis as the prime source of antibody diversity, but a role for somatic mutation somewhere within the system is far from excluded. This brief paper will therefore summarize our current understanding of how an immense range of site-specific mutation rates may be determined.

While geneticists had frequently noted that individual-gene mutation rates varied markedly even within the same species, there were few insights into such variation until the first attempts at saturation fine-scale mapping were carried out in the late 1950s in the T4*rII* system [2]. It was then

[1] *John Drake* was an associate of *Niels Jerne* at Caltech in 1954–55.

revealed that enormous variations in mutation rates can be observed at individual sites. It is now clear that the many point mutations studied by *Benzer* and his associates are comprised of both base pair substitutions and base pair additions and deletions (many of which are frameshift mutations). The mechanisms generating these two broad classes of mutations are distinctly different, and will be discussed separately.

First, however, it is important to note that the efficiency of detection of individual mutations may be very different depending upon their molecular nature, so that it is necessary to inquire whether apparent rate differences reflect differences in the underlying mutational processes or merely in the probabilities that the mutations will be scored. Because of their drastic effects upon protein primary sequence, frameshift mutations are generally believed to be detected with very high efficiency. On the other hand, in the T4r*II* locus some 85–95% of potential base pair substitutions are simply not observed [5], and those that are detected are very likely to involve chain-termination codons whose effects are as drastic as those of frameshift mutations. In the first place, at least a fourth of all base pair substitutions are likely to generate nothing more than synonymous codons. Moreover, one expects that many more of the 'missing' mutants are just too 'leaky' (insufficiently phenotypically expressed) to be readily observed. The possible existence of many mutants having no easily determined phenotypic alteration (cryptic mutants) has been examined experimentally in the T4*rII* system [10]. When nitrous acid-induced *rII* mutants were collected in a temperature-sensitive-*rII*-mutant background at a permissive temperature, many mutants were recovered which expressed so little mutant phenotype when separated from the 'sensitizing' *ts* mutation that they virtually always escape detection as *r* mutants under laboratory screening conditions. (They might not, however, have escaped natural selection, since most could be distinguished from the wild type when directly compared under special conditions.) Mapping experiments revealed two interesting characteristics of these cryptic mutations. First, the sites they defined were typical of previously defined base pair substitution sites in that both single occurrences and distinct hot spots (highly mutable sites) appeared. Second, the cryptic-mutant sites tended to cluster around the sensitizing *ts* sites, suggesting that a large set of *ts* sensitizers would have revealed a very large set of cryptic sites. Many of the 'missing' T4*rII* sites therefore appear to result from the failure to phenotypically detect altered DNA sequences rather than from actual immutability. Immutable sites would in any case be thermodynamically difficult to justify.

The true differences in mutation rates that have been observed are particularly well rationalized by theory in the case of frameshift mutations. The base pair addition and deletion mechanism proposed by *Streisinger* and his associates predicts a crucial role for local base sequence redundancy in the formation and stability of strand-slippage intermediates in this mutational pathway. This prediction was dramatically confirmed by the demonstration that a T4*e* frameshift mutational hot spot consisted (in the language of mRNA) of a sequence of five adenine residues [11]. Furthermore, mutants arising at this hot spot by the addition of one adenine reverted back to the wild type (6 → 5 adenines) at a frequency some 150 times greater than the site's forward mutant frequency. It is remarkable that this is even faster than the entire gene's detected forward mutant frequency.

This high susceptibility of redundant sequences to frameshift mutation has since been amply confirmed in bacterial systems. Among *Ames' Salmonella* tester strains now so familiar to genetic toxicologists, for instance, are two bearing highly revertible frameshift mutations. The reverting sequences (written for one strand only) are GGGG for strain TA1537 [1] and CGCGCGCG for TA1538 [7]. In the *E. coli lacI* gene, a frameshift-prone sequence consists of CTGG repeated as a trimer [6]. Mutants reduced to the dimeric sequence revert infrequently, while mutants increased to the tetrameric sequence revert very rapidly. Thus, the rate and even the specific composition of frameshift mutations can in principle be finely adjusted at specific sites within a gene.

Base pair substitution as well as frameshift mutation rates can differ greatly at different sites; for instance, the mutability of amber and ochre codons at diverse T4*rII* sites can vary over several orders of magnitude [12]. The first advance in understanding this phenomenon was the demonstration that such variation could in fact depend upon neighboring base pairs, rather than upon more distant, perhaps even extragenic influences. This was illustrated by an analysis of mutation rates at the first two positions of a pair of allelic amber (UAG) and ochre (UAA) mutations at a particular T4*rII* codon [9]. Using special selective devices, A:T → G:C transitions at these two sites were measured and compared as a function of the base pair, A:T or G:C, at the third position. An approximately 15-fold change was seen in the transition mutation rate at the adjacent (second-position) base pair, and 4-fold at the penultimate (first-position) base pair, as the base pair in the third position of the codon changed. These results suggested two generalizations, both chemically plausible but neither as yet confirmed: first, that neighboring-base-pair effects on site-specific mutation rates might extend at least two base pairs from affected sites (thus permitting a large number of potentially different rates); and second, that such effects might decrease with increasing distance.

While further evidence for great specificity in base pair substitution mutation rates has continued to accumulate in the ensuing decade, insights into its mechanistic basis have been rare indeed. Recent progress in this area has followed upon the development of efficient methods for nucleic acid sequencing that circumvent the ambiguities of sequences deduced solely from amino acid analyses. In addition, certain sequences associated with unusual mutation rates have actually helped to explain the causes of those rates. Two good examples are now available, both based upon extensive studies carried out with the *E. coli lacI* system.

In the first example, prominent G:C → A:T spontaneous hot spots were observed at the sequence ...CCAGG..., specifically at the cytosine adjacent to adenine [4]. These hot spots are the result of two circumstances. The first is the intrinsic tendency of the cytosines in any sequence to deaminate to produce uracil, which upon replication would pair like thymine. Instead, however, the uracil is rapidly removed by an efficient excision repair system. The second circumstance is the enzymatic modification of this particular cytosine: the sequence ...CCAGG... signals a specific DNA methylation enzyme which acts upon the C5 position of cytosine. When the resulting 5-methylcytosine deaminates, thymine is produced rather than uracil. The thymine residue is not subject to excision; it therefore persists as a mutational intermediate and contributes to the observed G:C → A:T hot spot.

In the second example, prominent G:C → A:T ultraviolet-induced hot spots were observed to reside at cytosine residues adjacent to another pyrimidine, thus constituting pairs potentially susceptible to the UV-induced formation of cyclobutane-type pyrimidine dimers; while these pyrimidine pairs in turn reside in very special longer sequences [13]. More specifically, each hot spot resides in a region composed of at least partly inversely repeated sequences capable of forming hairpin structures. Such hairpin structures, if they actually did form in vivo, would be composed of stems in which the bases were mostly hydrogen-bonded, and of terminal loops in which the bases were unpaired. The cytosine UV mutational hot spots are in each case located within the potential terminal loop. While the mechanistic reasons for the high mutational lability of these sites is not yet clear, it may be reasonable to anticipate that the stability of the stem and the configuration of the loop are both important. Thus, both the presence of the hot spot and possibly also its specific 'temperature' could depend upon contributions from DNA sequences at a considerable distance. From present observations, bases as far away as a dozen sites might be involved.

An example of site-specific mutability modified by probably even more distant (unneighborly) base pairs has been described by *Conkling* et al. [3] in the T4*rII* system. In this case transition mutation rates at each of the three base pairs of a chain-terminating codon were examined as a function of a single base pair substitution (recognized as a temperature-sensitive mutation) located between 20 and 80 base pairs away. Striking changes in mutation rates were seen for transitions induced by nitrous acid and by 2-aminopurine, as well as for spontaneous mutations. For example, 2AP-induced mutagenesis at the three measured sites showed a 2-fold change at the third position of the codon, an 8-fold change at the second and no change at the first position. Unfortunately, the relevant DNA sequences are not yet available, so that it remains unclear what clues they may provide; both elements of DNA secondary structure and protein bridging effects are attractive hypotheses. Even without understanding its mechanism, however, the observation clearly suggests that the number of base pairs which may affect the mutation rate at a given site is potentially very large. The resulting range of possible mutation rates may thus be great and may be virtually continuously populated.

To date, the extremes of point mutation frequencies (below 10^{-11} to above 10^{-4}) have appeared to be those generating frameshift mutations. There exist hints, however, of non-frameshifting mutations which may occur at extremely high rates. Unpublished and still preliminary work in this laboratory has revealed a class of T4*rII* mutants, reasonably frequent in occurrence, which are so unstable that their stocks may contain 1% or more of revertants instead of the more typical 0.0001% or less. Some of these mutants appear to consist neither of tandem duplications (since they can form stable recombinational heteroduplex heterozygotes, a characteristic of point mutations but not of physically large mutations) nor of frameshift mutations (since they fail to inactivate the function of a distal gene segment). While their further characterization is incomplete, the occurrence of such site-specific hypermutability suggests that further mechanisms may be available to modulate mutation rates and might therefore be available to organisms for generating great diversity in protein structure under special circumstances.

It would seem from the above examples that the cell could readily evolve very localized hypermutability in specific genes if this were adaptively useful. Clearly such sequences would tend to be strongly excluded from the vast majority of genes, and would only be detected either as unstable point mutations or within adaptively hypervariable genes. What kinds of genes, then, would be candidates to contain such sequences? They would most probably consist of genes involved in aspects of development which require a large number of different specificities, demanding more informational diversity than could be carried by a purely germ-line coding mechanism. Some obvious examples are the spatial orientation of individual cells in complex and highly differentiated tissues such as the eye and the optic tectum; memory; and antibody synthesis.

Although the potential role of somatic mutation in generating antibody diversity remains almost as obscure now as it was a decade ago, it is clear that the required mutation rates *could* be designed into antibody genes, and specifically into those very restricted portions of antibody genes that determine the hypervariable regions. In his classic 1971 paper, *Jerne* [8] estimated that a mutation rate of 10^{-7} per base pair per cell generation would suffice. However, the average mutation rate per base pair is not at all well measured in mammalian somatic cells. Extrapolations from lower organisms, mostly microbes, suggested a rate in the neighborhood of 10^{-10} or less [5], which would presumably be too low by some orders of magnitude for present purposes. This figure, however, is subject to at least two major uncertainties: the lack of reliable estimates of mutation detection efficiencies in mammalian cells, and confusion about the contributions of extensive untranslated sequences to mutational target sizes. More direct tests are therefore needed of the mutabilities of these particular sequences.

References

1 Ames, B.N.; Lee, F.D.; Durston, W.E.: Proc. natn. Acad. Sci. USA *70:* 782–786 (1973).
2 Benzer, S.: Proc. natn. Acad. Sci. USA *47:* 403–415 (1961).
3 Conkling, M.A.; Koch, R.E.; Drake, J.W.: J. molec. Biol. *143:* 303–315 (1980).
4 Coulondre, C.; Miller, J.H.; Farabaugh, P.J.; Gilbert, W.: Nature, Lond. *274:* 775–780 (1978).
5 Drake, J.W.: The molecular basis of mutation (Holden-Day, San Francisco 1970).
6 Farabaugh, P.J.; Schmeissner, U.; Hofer, M.; Miller, J.H.: J. molec. Biol. *126:* 847–863 (1978).
7 Isono, K.; Yourno, J.: Proc. natn. Acad. Sci. USA *71:* 1612–1617 (1974).
8 Jerne, N.K.: Eur. J. Immunol. *1:* 1–9 (1971).

9 Koch, R.E.: Proc. natn. Acad. Sci. USA *68:* 773–776 (1971).
10 Koch, R.E.; Drake, J.W.: Genetics *65:* 379–390 (1970).
11 Okada, Y.; Streisinger, G.; Owen, J.E.; Newton, J.; Tsugita, A.; Inouye, M.: Nature, Lond. *236:* 338–341 (1972).
12 Ronen, A.; Rahat, A.; Halevy, C.: Genetics *84:* 423–436 (1976).
13 Todd, P.A.; Glickman, B.W.: Proc. natn. Acad. Sci. USA *78* (in press, 1981).

Dr. John W. Drake, Laboratory of Molecular Genetics, National Institute of Environmental Health Sciences, Research Triangle Park, NC 27709 (USA)

The Immune System, vol. 1, pp. 176–181 (Karger, Basel 1981)

Facts and Speculations on T Cell Activation

Eva-Lotta Larsson[1]

Department of Immunology, University of Umeå, Umeå, Sweden

The study of lymphocyte activation constitutes a fundamental part of immunology, because the majority of the cells that compose the immune system are resting throughout the life of the animal. The ability to mount an immune response depends, therefore, not only on the ability of a given set of lymphocytes to recognize a particular antigen, but also on their reactivity, i.e. their ability to be activated by the available concentrations of that antigen.

Since antigens are so extremely diverse and growth receptors in a given cell type are likely to be structurally identical, direct interactions of antigens with growth receptors appear impossible. Therefore, the control of lymphocyte growth most likely involves two sets of specific recognition reactions, one providing the immunological specificity of the reaction, and another providing the cell biological specificity of the response given by a given cell type. To study lymphocyte activation is to describe the details in the successive steps leading a resting cell to clonally expand and to mature and to perform effector functions. Although these steps are interconnected in a normal immune response, it is likely that they can be analyzed independently and that each of them is controlled by a variety of mechanisms. Such dissection of the pathways of lymphocyte activation should lead to a precise understanding of the strategies used by lymphocytes to couple those two sets

[1] *Eva-Lotta Larsson* was a student at the Basel Institute for Immunology in 1978 and 1979.

of specific recognition reactions, and this understanding will be fundamental for clarifying the basis of self-nonself discrimination. Thus, the resting lymphocytes in a normal immune system ignore antigenic self structures while they readily respond to nonself substances. Furthermore, since such discrimination of self versus nonself appears, at least in the case of T lymphocytes, to be learned ontogenically and to involve extensive cell proliferation, it could be expected that the understanding of the rules controlling T lymphocyte activation and proliferation will give us insight into the ontogenic development of T cell repertoires. I will try here to confront our current knowledge on the activation and growth of killer T lymphocytes with *Jerne's* hypothesis for the generation of T cell diversity. I will not consider helper T lymphocytes, since the mechanisms of activation of growth of these are far from well understood at the present time.

Work with mitogenic lectins which polyclonally activate large numbers of T cells is largely responsible for the rapid progress in this area. The recent discovery of T cell specific growth factors induced by such polyclonally mitogenic lectins has provided the analytic tools for the investigation of each one of the steps involved in the response. It is now clear that cytotoxic T cell growth is strictly dependent on the availability of such growth factors and that no cytotoxic T cell proliferation or activation can take place in their absence. A corollary, which has now been experimentally confirmed, is that all cytotoxic T cell responses depend on the simultaneous activity of helper cells that produce growth factors. This strict helper cell dependence of all killer cell responses makes the latter secondarily dependent on Ia-positive accessory cells which are fundamental for the activation of helper cells. A primary requirement for an antigen-dependent cytotoxic T cell response is, therefore, the macrophage-dependent activation of helper cells for the production of growth factors with specificity for cytotoxic precursor cells.

Since growth factors are not antigen-specific, the process of T-killer cell activation cannot only consist of growth factor induced activation and proliferation of cytotoxic cell precursors. Killer T cell activation *must* necessarily involve a discriminatory, specific step of immunological recognition that ensures the clonal nature of an immune response. The discovery that such growth factors have no effect on normal, resting cytotoxic T cell precursors demonstrated that the specific ligand not only induces the production of growth factors, but it also renders resting T killer cells sensitive to the mitogenic activity of these growth factors.

Four important points have now been demonstrated which concern this acquisition of reactivity of cytotoxic T cell precursors to growth factors:

(a) induction of responsiveness is a property of antigens or other mitogenic ligands, the recognition of which by specific cellular receptor sites leads to the expression of functional receptors for growth factors which are not present on the surface of resting cells; (b) this first reaction per se does not lead to mitosis and no further activation to proliferation or effector functions will take place in the absence of growth factors; (c) this reaction appears to be autonomously performed by cytotoxic T cell precursors upon direct recognition of the specific ligand, i.e. it appears to be completely independent of other accessory or immunocompetent cells; (d) finally, once the specific ligand has been successfully recognized and the cell has expressed receptors for growth factors there is no more requirement for the initial specific interaction and the 'poised' cytotoxic cell will now grow independently of the initiating ligand under the sole control of the available concentrations of growth factors. Actually, even the cells in the clonal progeny that revert to a noncycling stage of a small lymphocyte will maintain functional expression of surface growth receptors and will dispense with specific ligand for a secondary restimulation that will depend exclusively on growth factors. From these results we conclude that antigen recognition in the induction of cytotoxic T cell responses is limited to the initial phase of this response and that the requirements for persistence of antigen are exclusively due to its role in the generation of growth factors. It follows that cytotoxic T lymphocytes only use their clonally distributed specific receptors at two points in a response: for a very short period at the induction phase and then for an even shorter period when cytolytic activities are expressed. This means that proliferating cytotoxic T cells are blind in terms of direct regulation by specific antigen and they do not use their specific receptors in clonal expansion.

These conclusions are based on experiments studying mature, post-thymic T cells, and I will try now to use them as a basis for understanding the process of acquisition of T cell repertoires. I will assume that the same principles of activation and expansion of cytotoxic T cell precursors control differentiation, expansion, and acquisition of immunocompetence by precursor cells within the thymus. As postulated by *Jerne* in 1971, precursor cells express a complete set of clonally distributed receptors with specificity for all MHC antigens within the species, including self. Taking into account the modifications of the theory presented in 1978, precursors for cytotoxic T lymphocytes express receptors with specificity for K/D products of the species. Precursor cells are postulated to acquire immunocompetence by intrathymic differentiation. But while allo-reactive cells do not alter their germ-line repertoire and leave the thymus with the same specificities, precursor cells with receptors for self K or D products will be confronted with the appropriate antigen and stimulated by this interaction to proliferation; only then will they be rescued by reverting to a resting state whenever their specific receptors are altered by mutation to such an extent that self K or D products are no longer recognized. We can now hypothesize that the mechanisms inducing proliferation in precursor cells, within the thymus, are basically the same as in the mature cell pool; i.e. these cells are dependent on the availability of cytotoxic T cell specific growth factors which act selectively on those cells that have been exposed to and have recognized specific antigen. The first point to stress is that prethymic precursor cells found in thymus-less (nude) mice already express clonally distributed receptors, at least those with specificity for allo-antigens. As recently shown, such cells, when exposed to specific allo-antigen and an exogenous source of growth factors, will be activated into proliferation and effector functions, which bear the same clonal specificities. It must be concluded, therefore, that prethymic precursors not only express clonally distributed receptors before entering the thymus but, in addition, they are fully competent in their ability to recognize specific antigen and respond by expression of functional growth receptors, and they are also competent to engage in growth factor dependent proliferation and maturation to effector functions. This implies that at least allo-reactive cytotoxic T cells do not require intra-thymic processing to reach a fully mature and competent stage. The role of intrathymic differentiation in the acquisition of T cell competence appears to be limited to the differentiation of growth factor producing helper cells, which definitely do not develop in the absence of a thymus.

Since the same experiments demonstrate that restricted T killer cells, to non-MHC antigens, are not included in this set of prethymic T cells that are fully reactive to antigen and growth factors, *Jerne's* postulates remain untouched or rather receive further support. The essence of the problem, however, remains the acquisition of immunocompetence of restricted cells. If intrathymic proliferation is dependent on the same mechanisms that apply to peripheral T cells, two incongruities can easily be found. The first arises from the observation that mature T cells, once activated to the expression of surface receptors for growth factors, will retain such expression and functional reactivity to mitogenic growth factors even after they have reverted to a noncycling small lymphocyte. If the same rules apply to precursor cells proliferating in the thymus and the mitogenic influence of growth factors and if the arrest of proliferation is due to a decrease in active concentrations of growth factors, the mature cells leaving the thymus would in any case retain the ability to react with mitogenic growth factors independent of antigen recognition. This, we know, is not correct. It must be postulated, therefore, that either the process of intrathymic proliferation is not determined by TCGF-dependent mechanisms or else that precursor cells, in contrast to peripheral mature lymphocytes, do not retain expression of functional growth receptors when they revert to small lymphocytes.

The other incongruity is more serious, as it relates to the heart of the problem – to the acquisition of a mature repertoire by somatic diversification of germ-line-specific receptors for self K/D antigens. As pointed out above, mature cytotoxic T cell precursors require recognition of antigen for a very short period of exposure until functional receptors for growth factors are expressed. Once this has been achieved, there is no longer a requirement for the specific ligand, and consequently for specific receptors, in the process of clonal expansion. It follows that if intrathymic precursor cell proliferation is determined by the same rules as mature T cell proliferation, mutations in the genes coding for specific receptors, or any other modification of these receptors after a cell has engaged in expansion, would in no way result in limitation of growth and, consequently, in the selection for cells that have diversified those receptors away from anti-self specificity. Therefore, either the process of arrest of proliferation within the thymus is determined by other mechanisms than TCGF-reactivity or else precursor cell activation and growth are determined by rules which are completely different from those governing peripheral activation and growth.

Another very unlikely alternative is that the postulates in *Jerne's* hypothesis are not fully correct. Since *Niels Jerne,* however, has always been right, this alternative is a priori disregarded. Thus the conclusion must be that intrathymic activation and expansion of cytotoxic T cell precursors follow quite distinct rules from those derived from the study of peripheral mature T cells. This conclusion, on the other hand, indicates that the study of mature T cell activation, while perhaps important for understanding the regulation of immune responses, will not contribute to our understanding of self-nonself discrimination, as it does not give us insight into the ontogenic learning of this discrimination and the generation of the mature T cell repertoire.

Acknowledgement. I thank *Antonio Coutinho,* who helped to put these thoughts in writing, and *Carin Häggström,* who typed the manuscript.

Selected Reading

Boehmer, H. von; Haas, W.; Jerne, N.K.: Proc. natn. Acad. Sci. USA *75:* 2439–2442 (1978).
Hunig, T.; Bevan, M.J: J. exp. Med. *152:* 688–702 (1980).
Jerne, N.K.: Eur. J. Immunol. *1:* 1–9 (1971).
Larsson, E.-L.; Coutinho, A.: Nature, Lond. *280:* 239–241 (1979).
Larsson, E.-L.; Coutinho, A.; Martinez-A, C.: Immunol. Rev. *51:* 61–91 (1980).
Morgan, D.A.; Ruscetti, F.W.; Gallo, R.C.: Science *193:* 1007–1008 (1976).

Dr. Eva-Lotta Larsson, Department of Immunology, University of Umeå, S–901 87 Umeå (Sweden)

The Immune System, vol. 1, pp. 182–186 (Karger, Basel 1981)

The Thy-1-Positive Cells of Nude Mice

Berenice Kindred[1]

Max-Planck-Institut für Biologie, Abteilung Immungenetik, Tübingen, FRG

The nude mouse is widely used as a T cell-deficient system for studies on T cell function, as a source of B cells and for determining whether antigens are T-dependent [13]. However, as well as a large number of 'weak' Thy-1-positive cells [20], nude mice do have some lymphocytes which are, with regard to the Thy-1 marker, comparable to the T cells of normal mice.

The frequency of Thy-1-positive cells found in nude mice varies somewhat. Thus *Raff and Wortis* [27], using cytotoxicity, found 5–6% while *Raff* [26] detected only 0.5–1.6% using fluorescein conjugated allogeneic anti-Thy-1 serum. Put this way it sounds insignificant but if one estimates that a mouse has 10^9 lymphocytes, 1% is 10^7, which is appreciable.

What Is Their Origin?

One possibility is that some T cells differentiate in the early embryonic thymus before it degenerates. Light microscopic surveys of the thymus rudiment from 12 days' gestation to birth have revealed no maturing lymphocytes [8, 24] and electron microscopic investigation of the 12- to 13-day nude thymus showed no lymphoblasts within the thymic lobes [7]. Nevertheless, although it seems improbable, it cannot be excluded that a few T cells do mature in the early thymus and divide to produce the majority of the Thy-1-positive cells.

[1] *Berenice Kindred* was a member of the Basel Institute for Immunology from 1972 to 1977.

Since most nu/nu mice are bred from +/nu mothers some of the Thy-1-positive cells are probably of maternal origin, particularly as nude offspring of homozygous nude parents have been reported to have fewer Thy-1-positive cells than those of heterozygous parents [19]. However, offspring of homozygous parents did have Thy-1-positive cells which could hardly have come from the mother.

The most likely explanation is that some Thy-1-positive cells can develop without passing through the thymus. A wide variety of substances including thymus extracts cause a dramatic increase in the number of Thy-1-positive cells both in vitro and in vivo [13]. The finding of *Scheid* et al. [30] that nudes infected with hepatitis virus may have up to 40% Thy-1-positive lymph node cells suggests strongly that a thymus is not necessary for the development of Thy-1-positive cells. Further, anti-SRBC responses have been reported in sick nudes [2, 15].

Recent work comparing young and old nudes also supports this conclusion. *MacDonald* et al. [21] claim that the frequency of cells bearing the marker Thy-1, LyT-1 and LyT-2 is considerably higher in the older animals although *Gershwin* et al. [4] using a different strain were unable to find evidence for such an increase. Further, *Maryanski* et al. [22] were able to demonstrate the presence of alloreactive cytotoxic T cell precursors in old, but not in young, nudes.

Are They Functional T Cells?

It is important that, as shown by *Hoffmann-Fezer* et al. [9], the majority of these cells are found in the T-dependent areas of secondary lymphoid organs and, although *Sprent* [31] found no recirculating T cells in nudes, *Roelants* et al. [29] did find a low number of Thy-1-positive lymphocytes in the thoracic duct lymph of nudes. Therefore, there does not appear to be a mechanical reason why these cells should fail to function. Yet 10^7 normal, congenic thymocytes are sufficient for reconstitution of immune function in nudes [12] so the Thy-1-positive cells of nudes are not equivalent to normal thymus cells. Possibly only a subset of T cells are able to mature in the absence of an actual thymus or possibly the Thy-1 marker does not necessarily denote the presence of functional T cells.

The first T cell function reported in nudes was the anti-Thy-1-sensitive response to monomeric flagellin reported by *Kirov* [17]. Most other functions described involve the recognition of alloantigens. Thus, cells from nude mice respond in MLR [3], generate CTL if the nudes are older [22] and recognize alloantigens in the PAR test [28]. They also reject skin grafts if injected with rabbit complement [18] and produce cytotoxic antibody against alloantigens [16, 25]. Apart from the CTL which could be removed by anti-Thy-1 treatment, it is not clear whether these functions depend on a low number of T cells or are T-independent.

Nude mice are also able to resist the growth of some xenogeneic tumours [13] and virus infected cells [23]. This resistance has been attributed to 'natural killer' (NK) cells. However, as some NK cells are now described as Thy-1-positive [23] it seems to be a semantic question whether or not these are T cells. In any case the presence of Thy-1-positive NK cells has been reported in nude mice and these therefore account for at least part of the Thy-1-positive population.

Significance of Nude Thy-1-Positive Cells

If T cells can differentiate in the absence of a thymus, the specificities of these cells can provide much information about the basic T cell repertoire and the role of the thymus in its development. Until now the most clearly defined specificities both in untreated nudes [22] or after administration of T cell growth promoters [5, 33] are for alloantigen-specific cytotoxic cells.

According to *Jerne's* [11] theory on the acquisition of the T cell repertoire, alloantigens are the only antigens which should be recognized in the absence of a thymus. It is therefore important to examine the other responses which have been induced in nude mice or nude mouse cells. Primary responses to various antigens have been reported following administration of activated T cell supernatants in vivo [15] or thymus extract in vitro [1]. However, actual T cell specificity was not demonstrated. What was shown was increase in response to a specific antigen which could have been the result of stimulation of specific B cells which had bound antigen by non-specific T cells. *Stötter* et al. [32] have presented a more careful analysis in which T cells from treated nudes were titrated with previously unstimulated B cells. However, the antigen used was SRBC and the T cells stimulated responses to both SRBC and HRBC and at present it is not clear whether this shows lack of specificity or a cross-reaction at the T cell level between SRBC and HRBC.

The interpretation of the response to SRBC is further complicated by the finding of *Gisler* et al. [6] that peritoneal cells from nude mice, after culture without antigen, could produce PFC response against SRBC but only SRBC. *Ishikawa and Saito* [10] have also shown anti-SRBC activity in nude peritoneal cells. Until these findings are clarified an anti-SRBC response must be considred a special case.

Lack of Secondary Responses

In most of the work on induction of helper T cells in nudes, secondary responses have not been tested. However, an anti-SRBC response induced in vivo [14] was not followed by a secondary or even a second primary response after boosting in vivo, although an in vitro response could still be detected indicating that the in vivo unresponsiveness could be reversed. *Wagner* et al. [33] also showed a secondary in vitro response to alloantigens after treatment with interleukin II.

Thus, augmented primary responses, particularly to partially thymus-independent antigens, both in vivo and in vitro can be demonstrated although secondary, in vivo, responses have yet to be shown. *Kirov's* [17] results possibly fall into the category of partially thymus-independent antigens, since, although monomeric flagellin is T cell dependent, the polymerized form of the antigen is T cell-independent and, also, she reports that 2 years previously nudes from the colony had not responded to monomeric flagellin so the response clearly depended on the condition of the mice.

In conclusion, the Thy-1-positive cells of nude mice are not the equivalent of T cells of normal mice since they fail to support functions which can be induced by injection of normal, congenic thymocytes [12] but may be a particular T cell sub-population. It seems likely that the increased numbers of such cells which appear during virus infection or after treatment with activated T cell supernatants are confined to the same functions since they appear to serve only to amplify activities which were present in healthy or untreated nudes and new functions, e.g. the ability to mount a secondary response after boosting, do not appear.

Allo-cytotoxic T cells and possibly T cells with other specificities can develop slowly in the absence of a thymus but the precursors of allospecific T cells are also present in young nudes and can be demonstrated after appropriate treatment [5, 33], thus emphasizing the special status of alloantigens. It still remains to be shown whether specificity for other antigens can arise in the absence of a thymus.

References

1 Armerding, D.; Katz, D.H.: J. Immun. *114:* 1248 (1975).
2 Beattie, G.; Lannom, R.; Osler, A.G.; Kaplan, N.O.: Proc. 3rd Int. Workshop on Nude Mice (in press).
3 Croy, B.A.; Osoba, D.: J. Immun. *113:* 1626 (1974).
4 Gershwin, M.E.; Merchant, B.; Steinberg, A.D.: Immunology *32:* 327 (1977).
5 Gillis, S.; Union, N.A.; Baker, P.E.; Smith, K.A.: J. exp. Med. *149:* 1460 (1979).
6 Gisler, R.H.; Pagés, J.M.; Bussard, A.E.: Annls Immunol. *126C:* 231 (1975).
7 Groscurth, P.; Kistler, G.: Beitr. Path. *156:* 359 (1975).
8 Hair, J.: Proc. 1st Int. Workshop on Nude Mice, p. 23 (Fischer, Stuttgart 1974).
9 Hoffmann-Fezer, G.; Rodt, H.; Thierfelder, S.: Beitr. Path. *161:* 17 (1977).
10 Ishikawa, H.; Saito, K.: J. exp. Med. *151:* 965 (1980).
11 Jerne, N.K.: Eur. J. Immunol. *1:* 1 (1971).
12 Kindred, B.: Cell. Immunol. *28:* 174 (1977).
13 Kindred, B.: Prog. Allergy, vol. 26, p. 137 (Karger, Basel 1979).
14 Kindred, B.; Bösing-Schneider, R.: Submitted for publication.
15 Kindred, B.; Bösing-Schneider, R.; Corley, R.B.: J. Immun. *122:* 350 (1979).
16 Kindred, B.; Schirrmacher, V.: Cell. Immunol. *52:* 266 (1980).
17 Kirov, S.M.: Eur. J. Immunol. *4:* 739 (1974).
18 Koene, R.A.P.; Gerlag, P.G.G.; Jansen, J.J.; Hagemann, J.F.H.; Wijdeveld, P.G.A.B.: Nature, Lond. *251:* 69 (1974).
19 Kramer, M.H.; Gershwin, M.E.: Transplantation *22:* 529 (1976).
20 Loor, F.; Roelants, G.E.: Ann. N.Y. Acad. Sci. *254:* 226 (1975).
21 MacDonald, H.R.; Lees, R.K.; Sordat, B.; Zaech, P.; Maryanski, J.L.; Bron, C.: J. Immun. *126:* 865 (1981).
22 Maryanski, J.L.; MacDonald, H.R.; Sordat, B.; Cerottini, J.-C.: J. Immun. *126:* 871 (1981).
23 Minato, N.; Bloom, B.R.; Jones, C.; Holland, J.; Reid, L.M.: J. exp. Med. *149:* 1117 (1979).
24 Owen, J.J.T.; Jordan, R.F.; Raff, M.C.: Eur. J. Immunol. *5:* 653 (1975).
25 Piguet, P.F.; Vassalli, P.: J. Immun. *120:* 79 (1978).
26 Raff, M.C.: Nature, Lond. *246:* 350 (1973).
27 Raff, M.C.; Wortis, H.H.: Immunology *18:* 931 (1970).
28 Ramseier, H.: Immunogenetics *1:* 507 (1975).
29 Roelants, G.E.; Loor, F.; von Boehmer, H.; Sprent, J.; Hägg, L.B.; Mayor, K.S.; Rydén, A.: Eur. J. Immunol. *5:* 127 (1975).
30 Scheid, M.P.; Goldstein, G.; Boyse, E.A.: Science *190:* 2111 (1975).
31 Sprent, J.: Cell. Immunol. *7:* 10 (1973).
32 Stötter, H.; Rüde, E.; Wagner, H.: Submitted for publication.
33 Wagner, H.; Hardt, C.; Heeg, K.; Röllinghof, M.; Pfizenmaier, K.: Nature, Lond. *284:* 278 (1980).

Dr. Berenice Kindred, Max-Planck-Institut für Biologie, Abteilung Immungenetik, Correnstrasse 42, D–7400 Tübingen (FRG)

Immune System, vol. 1, pp. 187–194 (Karger, Basel 1981)

The MHC: Still a Few Loose Ends

N.A. Mitchison[1]

Imperial Cancer Research Fund Tumour Immunology Unit, Department of Zoology, University College London, London, England

These are years of triumph for the Basel Institute of Immunology.This decade has seen a solution to the two great problems of immunogenetics. For one of them, the nature of the immunoglobulin genes, the Institute has undoubtedly led the way. For the other, the function of the major histocompatibility complex, the Institute has made notable contribution. This essay surveys the solution for the MHC, and looks at some favourite loose ends.

The Central Theory: the MHC as a Group of T Cell Guides

Briefly summarizing our understanding of the functions of the MHC, we can identify three important steps forward. The first was to recognize that all of the apparently diverse functions controlled by single 'regions' within the complex are in fact carried out by the same molecule. Thus, a single molecule, for example the 44K membrane glycoprotein encoded by H-2K, functions as a restriction element, a serologically defined allo-antigen, a T cell-defined allo-antigen, and an immune response gene. The second was to recognize that the apparently different functions of 44K molecules and 28/32K molecules (respectively class I and class II in *Klein's* nomenclature) are misleading, and that both types of molecule function in the same way but in relation to different T lymphocyte sets. And the third

[1] *Avrion Mitchison* has been a member of the Board of Advisors of the Basel Institute for Immunology since 1977.

was to recognize that the natural function of these molecules, as distinct from their artificial functions as allo-antigens, it to guide T lymphocyte parallel sets: 44K molecules guide cytotoxic T cells, 28/32K molecules guide helper T cells, and (but the evidence for this is weaker) I-J molecules, of as yet undetermined size, guide suppressor T cells.

Up to a point these parallel sets resemble the parallel sets of B lymphocytes specialized for the production of immunoglobulin classes: in both cases each set contains essentially a full repertoire of receptors, subject presumably only to perturbations resulting from antigenic selection. Consequently a given antigen is potentially able to stimulate a subset of cytotoxic, helper, and suppressor T cells, and which it chooses is determined by the MHC molecule which the T cells see it to be associated with. Thus, the guidance of appropriate T cells is at least partially understood at the molecular level, something which is still lacking for B cells.

Dual recognition of antigen and MHC molecule kills two birds with one stone. The requirement for the MHC molecule ensures that the appropriate T cell parallel set is chosen for activation. At the same time it ensures that T cells are not activated prematurely, by encounter with antigen free in body fluids. This is obviously more important for a cell whose effector function is essentially local, than for a B cell which releases antibody to act at a distance.

Thus far the ground is largely uncontested. We turn now to the loose ends.

Immune Response Genes: Presentation versus Repertoire Still Unresolved

It is now generally agreed that allelic substitution of the MHC guidance molecules alters the range of immune responsiveness. This is a powerful mechanism of genetic control, and can on occasion be a major determinant of disease susceptibility. It is almost certainly the driving force behind the polymorphism of the MHC. But one should keep a sense of proportion: other genes also affect responsiveness, and indeed I find it hard to imagine a gene substitution which would have no impact on anything as complex as the immune system. At any rate, a great deal of effort has gone into trying to work out just how the MHC genes exercise this control. It is disappointing that the problem is still unsolved, and even more so to find that cellular immunology as practised at present has no way of solving it.

There are two competing theories about how MHC genes control the response. One postulates direct binding of foreign antigen by MHC products on the surface of antigen presenting cells. Only if binding takes place, it is assumed, can T cells recognize the antigen. MHC allelic products vary, it is further assumed, in their ability to bind particular antigens, and it is the strength of binding which determines whether or not a response occurs. Thus, MHC immune response genes act at the level of antigen presentation. The alternative theory holds that MHC products influence the repertoire of T cell receptors, presumably as *Jerne* first proposed via selection within the thymus. It is easy to see how variation in the repertoire could control immune responsiveness, but less easy to specify in detail what goes on in the thymus.

Much of the controversy has concerned the antibody response of mice to protein antigens, but this is of course only one facet of the problem. Responses of cytotoxic T cells to viral antigens are no doubt of greater evolutionary relevance. An interesting instance is the competitive interaction involved in the generation of cytotoxic T cells specific for viruses and for the cell surface allo-antigen H-Y. Both viruses and H-Y can be recognized by murine lymphocytes only in the context of K, D or L molecules, and in the normal way there is a choice of two in inbred mice (either K or D) and of four in hybrid mice. Both types of antigen tend to make sharply exclusive choices, in the sense of choosing to associate with either K^a or D^a but not both. Furthermore, when K^a is chosen in preference to D^a in an inbred mouse, it may lose out in preference to K^b in a hybrid. Thus, one can construct a hierarchy of associative MHC molecules, and the ordering of the hierarchy for one antigen will be different for another. These hierarchies can be equally well explained in terms of (a) competition between MHC products for limiting numbers of antigenic molecules (i.e. at the level of presentation), and (b) competition between clones of T cells recognizing different MHC molecules as guides, for limiting amounts of surface-bound antigen (i.e. at the level of repertoire).

This controversy has been notable for the unscholarly way in which the opponents tend to omit reference to the opposing theories, and even more so for the failure to realize just how ambiguous are the experimental designs. This applies particularly to chimaera experiments, which for a while convinced many of us (myself included) that the controversy had been settled in favour of thymic influence on the repertoire. It now seems that no cellular approach is likely to resolve the matter. We shall probably have to await chemical studies on the interaction between MHC molecules and foreign antigens, something for which there is at present no suitable experimental system. In the meanwhile perhaps the best advice for the aspiring student of selection in the thymus is: desist.

Allo-Specific T Cells Do Not Match Their MHC Components Accurately

The concept of MHC molecules as guides for T cell parallel sets rests primarily on 'restriction' experiments. These involve the guidance of helper cells by self-28/32K molecules, and of cytotoxic cells by self-44K molecules. However, the contribution made by studies with allo-MHC molecules should not be forgotten. If neither *Katz* nor *Zinkernagel* had started, perhaps *Eijsvogel* and *Bach* would have got there in the end. True, studies with allo-MHC molecules got off to a false start with emphasis on a misleading distinction between 'serologically defined' and other components. True also, this type of study yields information which on its own is hard to interpret. Nevertheless, very early in the game students of the mixed lymphocyte reaction had grasped the central ideal that helper and cytotoxic T cells recognize different components of the MHC, and had managed correctly to identify those components as 28/32K and 44K molecules, respectively.

Even now, alas, we still do not understand quite why a given T cell parallel set recognizes the same type of MHC components in allo-specific as in conventional responses. It seems extremely likely that some kind of cross-reaction is concerned, the precise nature of which is debatable. Yet a theoretician as skillful as *Cohn* can still suppose that the allo-specific T cell receptor actually belongs to the anti-X class (in his nomenclature; i.e. the class which recognizes foreign antigens such as viruses, as distinct from the class which recognizes self-MHC). As argued by him in person the reasons seem cogent, but in his absence common sense prevails.

In considering how allo-specific T cells have fared lately, it is worth making a distinction between two kinds of functional assays. In one, as practised originally for T-to-T and later for T-to-B collaboration, the regulatory cell and its effector partner are syngeneic. They interact with one another in much the same way as they would in a normal immune response, except that the antigen happens to be an allogeneic cell. In a second kind of assay, the effector cell is allogeneic, usually a B cell bearing an allo-antigen with which the regulatory T cell interacts. Assays of the two types may be termed respectively 'allo-specific, syn-regulatory', and 'allo-specific, allo-regulatory'. Only the first is properly physiological. This terminology is not meant as prejudicially as it sounds. Allo-regulatory assays using B cells are sensitive and precise, and very useful for cloning experiments.

Allo-specific T cells seem to match MHC guides imprecisely, far more so than might have been hoped from the original *Eijsvogel-Bach* data. To take the most difficult example, nearly all the clones which *Dupont* and his colleagues have isolated from HLA-D-stimulated mixed lymphocyte reactions turn out to be cytotoxic. Well, they have not been tested for regulatory activity yet, and may possibly have the expected helper activity as well (although that would still make things complicated); and the cloning procedure which depends on growth factor may have been highly selective. But it's hard to fit with the central theory.

Equally in work on syn-regulation of B cells we have regularly been able to generate helper T cells specific for H-2K and D. This type of activity develops so slowly under our schedules of immunization that we term it latent help; it certainly develops far slower than H-2I-specific help. Nevertheless, it seems to be real, and as such again to transgress the central theory.

Exceptions of much the same type have been described in allo-regulatory systems by *Swain, Dutton, Dennert* and their collaborators. Fortunately, they begin to offer a way out of the puzzle. Cells with abnormal specificity (cytotoxic cells responding to 28/32K, helper cells responding to 44K molecules) consistently have abnormal Ly phenotypes. This does not explain what is going on, but at least it reassures us to find that cells which are odd in one way are also odd in others.

I-J

The I-J gene(s) and their products have a short but dramatic history. First described 5 years ago by *Murphy* and by *Tada,* the I-J allo-antigen occurs on suppressor T cells and a factor TsF which they produce. Other suppressor T cells (distinguishable from the TsF-releasing cells by surface markers and reactivity) recognize antigen only in association with TsF. Thus, I-J seems to play much the same role for suppressor cells as other I-region encoded molecules do for helper cells. In keeping with this view, anti-I-J antiserum has an up-regulating effect when used as an agent of intervention in the immune response.

Yet there are problems. I-J molecules have not been isolated by radioimmune precipitation from the cell surface, and what little we know of their chemistry rests solely on *Taniguchi's* work on TsF. Even the biological work on I-J as a membrane molecule is controversial. *Niederhuber* finds it on the surface of conventional antigen presenting cells, while *Tada* does not (he entertains the interesting hypothesis that the suppressor cell pathway starts with an MHC-guided lymphocyte which assumes the function of antigen presentation).

Work on the regulatory activity of allo-specific I-J responses offers another approach. *Streilein* has found circumstances in which a skin graft can be protected against rejection by containing a minority of cells bearing allo-I-J antigens. Admittedly, his claim for a special role allo-I-J in immunological tolerance has not been accepted by *Holan* and the Prague group. My colleague *Czitrom* has confirmed and extended his claim in an in vitro system, where allo-I-J suppresses the proliferative response to other MHC components. However, the system is proving hard to work with, and we are quite uncertain where it will go. It seems to us important to understand what is going on, not only from a theoretical point of view, but also because allo-I-J may have a future as an agent of intervention. For us the immediate task is to establish a workable system for I-J serology.

Splitters and Lumpers

Whatever is one to make of all these T cells? A recent review by *Tada* has 8, and that does not include *Gershon's* contrasuppressors or feedback suppressor-inducers. The questions are: (1) how many of these represent true parallel sets, in the sense that they constitute a group of stably differentiated cells rather than a particular phase in an activation cycle, and (2) does each true parallel set have a unique MHC component as a guide? Immunologists divide into the splitters and the lumpers, those who enthusiastically continue to subdivide the regulatory network, and those who hope that some of the subdivisions will evaporate.

The I-J subspecificities are a case in point. Up to four subspecificities are now on offer, based on absorption of mixed antisera with hybridoma cells bearing I-J antigens. These have no known genetic counterpart, nor, as mentioned above, do they have a counterpart among molecules isolated from cell membranes by conventional immunochemical methods. *Janeway's* separation of idiotype-specific regulatory cells presents a difficulty of another type. If he is correct, idiotypes and antigens do not belong as *Jerne* proposes to one universe, but to two. The original economy and simplicity of the immunological network would disappear. It is at this point that cell cloning assumes such importance. One looks forward to a time when the validity of any hypothetical T cell set depends on its presence in the great future Basel library of clones.

MHC Molecules on Inappropriate Cells

The doctrine of T cell guidance makes no provision for 28/32K molecules on haemopoietic stem cells or on glandular and endothelial cells, locations where they have been observed by *Greaves* and *Thorsbe.* There is still some doubt whether these molecules have been actively synthesized or passively acquired. However, this does not help much, for the molecules on endothelium, whether passively acquired or not, are clearly able to function in antigen presentation. A possibility has been canvassed that these molecules serve a residual function in cell-cell interactions of a non-immunological character, and that their immunological activities with which we are familiar have evolved from this primitive function.

Alien Specificities on Malignant and Virus-Infected Cells

The *Bodmer-Festenstein* hypothesis holds that MHC molecules are encoded by clusters of pseudo-alleles and (at least in the *Festenstein* version) that alien specificities may be expressed on occasion particularly in cancer and virus infection. Some but not all studies of cancer cells made by others have now confirmed the hypothesis; the strongest support has come from an immunochemical study which has identified in some detail the peptides of an alien specificity on one malignant cell line. Yet many questions remain. Why are the alien specificities so unpredictable? Why do they not cause tumour rejection? Has the possibility of contamination been rigorously excluded?

These, like other questions raised in this essay, can best be answered by the recombinant DNA approach to the MHC. This has become a Worldwide industry. Our own contribution, in collaboration with *Flavell,* is to transform cells with cosmid-linked alien DNA and search for MHC expression. We hope thus to count MHC genes.

Class III Molecules: the Intruders

Complement components, the class III of *Klein's* classification, seem to be encoded ubiquitously within the MHC. In terms of function this makes no sense. However, their genetics begin to offer a clue. In man and mouse they share the hallmarks of the class I and II genes; tandem replication, polymorphism, and high linkage disequilibrium. Is it possible that they have joined the MHC for the linkage disequilibrium ride? And if so, who got onto this piece of chromosome first?

Prof. N.A. Mitchison, Department of Zoology, University College London, Gower Street, London WC1E 6BT (England)

The Immune System, vol. 1, pp. 195–201 (Karger, Basel 1981)

Influence of the H-2 Complex on the Response of T Cells to Sheep Erythrocytes in vivo

J. Sprent[1]

Division of Research Immunology, School of Medicine, University of Pennsylvania, Philadelphia, Pa., USA

Largely due to the development of the Jerne hemolytic plaque assay, heterologous erythrocytes such as sheep red blood cells (SRC) have proved an invaluable tool for dissecting the immune response. Although many workers in recent years have turned to more refined antigens, SRC continue to be a useful reagent for studying many aspects of basic immunology, particularly the mechanism by which T lymphocytes recognize antigen.

Unlike B cells, T cells do not appear to recognize free antigen but antigen associated with gene products of the major histocompatibility complex (MHC) [4, 13]. In mice, T cells involved in T-B collaboration, delayed-type hypersensitivity and proliferative responses to antigen are restricted by determinants coded by the *I*-region of the *H-2* complex; cytotoxic T cells, by contrast, are restricted by K, D and L H-2 determinants. For each of these four T cell functions it has been shown that the capacity of T cells to recognize antigen in association with a given set of H-2 determinants depends upon the T cells encountering these determinants during differentiation in the thymus.

Despite intensive speculation, the precise mechanism by which T cells recognize the association of antigen plus H-2 determinants is still unclear. In particular it remains to be established whether T cell specificity is mediated by two receptors – one for antigen and the other for H-2 determinants – or via one receptor with specificity for neoantigenic determinants created by the association of the two molecules.

[1] *Jonathan Sprent* was a member of the Basel Institute for Immunology from 1972 to 1974.

Much of our current knowledge of the phenomenon of H-2 restriction has come from studies conducted in vitro. Although the information from such studies has proved invaluable, it is obviously of considerable importance to determine to what extent the data are applicable to the normal immune response. In this respect I will summarize certain aspects of my own findings on the influence of H-2 gene products on T helper cell function in vivo.

T Cell Selection to Antigen in Irradiated Mice

When T cells encounter antigens such as sheep erythrocytes (SRC) in vivo, the antigen-specific cells leave the circulation, e.g. thoracic duct lymph (TDL), for 1–2 days and become selectively sequestered in the spleen, the main site of antigen localization [8]. During this stage of *negative selection* the sequestered cells proliferate extensively. Thereafter, the activated progeny of these cells reenter the circulation in large numbers. This stage of *positive selection* is maximal at day 5–6 after antigen injection.A similar sequence of events occurs if T cells are transferred with antigen into heavily irradiated mice and then harvested from TDL of the recipients; T cell function is monitored by adoptively transferring the TDL together with B cells and antigen into further irradiated mice to measure antibody formation [9]. The lymph-borne cells collected at day 1–2 after antigen injection show no response to the injected antigen but respond normally to third party antigen. By day 5–6 post-transfer, however, the lymph-borne cells show marked hyperreactivity to the injected antigen.

In the case of normal homozygous strain *a* T cells, selection to antigen requires that the donor T cells and the selection hosts share H-2 determinants [9]. Thus, in contrast to H-2-identical or H-2-semiallogeneic hosts, purified T cells fail to undergo either negative or positive selection in totally H-2-different hosts. In this situation, even with massive doses of antigen, the T cells ignore antigen and recirculate in a resting state. These findings apply to T-B collaboration with syngeneic strain *a* or $(a \times b)F_1$ hybrid B cells; it is important to point out that in our hands normal strain *a* T cells do *not* collaborate with H-2-different strain *b* B cells [9].

Selection of heterozygous $(a \times b)F_1$ T cells is more complicated [9]. When these cells are exposed to antigen in irradiated F_1 mice, the donor cells collected from the lymph at day 1–2 fail to collaborate with B cells from strain *a*, *b* or F_1 mice. By day 5 the donor cells give hyperreactive responses with each of these three B cell populations. With selection of F_1 T cells in irradiated strain *a* mice, by contrast, selection is apparent only when T helper function is monitored with homozygous strain *a* B cells. Thus, at day 1–2 post-transfer the lymph-borne donor F_1 T cells fail to respond with strain *a* B cells but give unimpaired responses with strain *b* and F_1 B cells. At day 5, the donor cells give very high (activated) responses with strain *a* and F_1 B cells and unchanged (unprimed) responses with strain *b* B cells. Reciprocal findings occur with selecion in strain *b* mice.

The major conclusion from these studies is that, as in vitro, T cells do not respond to free antigen. Antigen is recognized in association with H-2 determinants of the radioresistant cells of the host used for selection.[2] With this in mind, the above findings are interpreted as follows. With normal *homozygous* strain *a* T cells, the helper cells are restricted entirely by strain *a* H-2 determinants, i.e. antigen is recognized only in association with strain *a,* not strain *b,* H-2 determinants. This restriction of the 'anti-a' T cells applies both during the activation (selection) of the T cells to antigen and during the effector phase, i.e. during T-B collaboration. Anti-*b* T cells are not detectable in normal strain *a* mice. Normal *heterozygous* T cells, by contrast, consist of a mixture of anti-*a* and anti-*b* cells. Only the anti-*a* cells are activated with exposure to antigen in irradiated strain *a* mice, whereas both T cell subgroups undergo selection in F_1 mice.

H-2 Restriction Controlled by Both I-A and I-A/E Hybrid Molecules

Genetic mapping studies have invariably led to the conclusion that the determinants which control the function of I-region restricted T cells map in the *I-A* subregion [4, 13]. Based on biochemical evidence, it is known that there are two major classes of Ia antigens present on most B cells and some macrophages [3, 12]. Both are two chain structures with a small (29K) β chain and a larger (33K) α chain. In the case of classic I-A molecules, both the α and β chains (A_α and A_β) are coded by genes situated in the *I-A* subregion. For the second class of Ia molecules, which are termed I-A/E hybrid molecules, the α chain (E_α) is coded in the *I-E* subregion whereas the β chain (E_β) is coded in the *I-A* subregion.

[2] Adoptive transfer studies have shown that the antigen-presenting cells in the irradiated host are a class of macrophage-like, non-T, non-B accessory cells which express both I-A and I-E Ia antigens [10].

At face value, the fact that the restricting elements for T cell function appear to map only in the *I-A* subregion might be taken to imply that restriction is controlled only by I-A molecules and not by I-A/E hybrid molecules. A priori, however, it could be argued that certain T cells are restricted by I-A/E hybrid molecules, the restriction being controlled either by (1) the conformational determinants created by the association of the α and β subunits, or (2) the β subunit alone. It should be noted that if the α subunit alone mediated restriction, the restriction would map to the *I-E* subregion. This is not the case, as shown by the following example. If T cells from F_1 hybrids between B10 *(I-A^b^, I-E^b^) (b b)* and B10.A *(k k)* are positively selected to SRC in irradiated B10.A *(k k)* mice, the activated T cells collaborate well with B10.A B cells but, significantly, do not interact with B10.A(5R) *(b k)* B cells, despite the fact that B10.A and B10.A(5R) are compatible in the *I-E* subregion [11].

According to the above reasoning, the function of T cells restricted by a particular set of I-A/E hybrid molecules, e.g. $I\text{-}A^k/E^k$, would depend upon the antigen being presented by cells which were of the *k* haplotype in both the *I-A* and *I-E* subregions. This requirement would apply both during positive selection to antigen (T-macrophage interaction) and during T-B collaboration. Thus, in the above example, T cells activated to antigen plus the $I\text{-}A^k/E^k$ hybrid molecules of B10.A would interact only with B cells which expressed identical hybrid molecules, i.e. B10.A B cells but not B10.A(5R) B cells. Function of the I-A-restricted T cell subgroup, by contrast, would require only that the antigen presenting cells or B cells be matched at the *I-A* subregion, and not necessarily at the *I-E* region. Two approaches have now provided evidence for the existence of I-A/E-restricted T cells. Both approaches rely on the selective removal of I-A-restricted T cells.

Effects of Blocking Positive Selection with Anti-Ia Antibodies

If T cells from F_1 hybrids between CBA *($I\text{-}A^k$ $I\text{-}E^k$ (k k))* and B6 *(b b)* are positively selected to SRC in irradiated F_1 mice, the activated T cells collaborate well with B10.BR *(k k)*, B10.A(4R) *(k b)* and B10 *(b b)* B cells. Addition of large doses of monoclonal anti-$I\text{-}A^k$ (anti-Ia.17) antibody during selection prevents the activation of cells able to collaborate with 4R *(k b)* B cells but fails to impede positive selection of cells which help either B10.BR *(k k)* or B10 B cells [11]. The interpretation of this finding is that the anti-$I\text{-}A^k$ antibody binds to $I\text{-}A^k$ determinants on the host F_1 macrophages and thereby prevents the activation of the I-A^k-restricted T cell subgroup, i.e. cells which can interact with either B10.BR or 4R B cells but not with B10 B cells. Activation of a subgroup of I-A^k/E^k-restricted T cells is not impeded by anti-I-A^k antibody. Once activated, these latter cells give high responses with B10.BR *(k k)* B cells but do not interact with 4R *(k b)* B cells. Likewise selection of the I-A^b-restricted T cells is not inhibited; these cells interact only with B10 *(b b)* B cells and not with either 4R *(k b)* or B10.BR *(k k)* B cells. Different findings occur with positive selection of F_1 T cells in F_1 mice in the presence of A.TH anti-A.TL antiserum, i.e. a reagent which contains a mixture of anti-I-A^k and anti-I-E^k antibodies (the latter specificity is directed to the E_α chain of the I-A/E^k molecule). Here one generates high levels of help for B10.B cells but not for either B10.BR or 4R B cells. In this situation inhibition of positive selection thus applies to both the I-A^k- and I-A/E^k-restricted subgroups of cells but not to the I-A^b-restricted cells.

According to the above interpretation, selection of (CBA × B6)F_1 T cells to SRC in irradiated 4R mice would activate only a single subgroup of T cells, i.e. the I-A^k-restricted cells.[3] If so, in contrast to selection in irradiated F_1 mice, adding specific anti-I-A^k antibody during selection in 4R mice should suppress help for *both* B10.BR *(k k)* and 4R *(k b)* B cells. This is indeed the case [11].

Collectively, these data provide strong if indirect evidence for the existence of T cells restricted to I-A/E^k molecules. To prove the existence of these T cells will require studying the inhibitory effects of a mixture of monoclonal anti-I-A^k and anti-I-E^k antibodies, i.e. rather than the ill-defined assortment of antibodies in A.TH anti-A.TL antiserum. Such studies are in progress.

Negative Selection of Homozygous T Cells in Irradiated H-2 Recombinant Mice

The second approach for demonstrating the existence of I-A/E^k-restricted T cells has come from studies on negative selection of homozygous T cells [*Sprent, Alpert,* unpublished]. If CBA *(k k)* T cells are negatively selected to SRC in irradiated CBA, B10.BR (also *k k*) or (CBA × B6)F_1 mice, the lymph-borne CBA cells collected at 1–2 days post-transfer are

[3] To avoid confusion it should be pointed out that *I-E^b* is a null allele; the E_α chain is not expressed and for this reason the E_β chain remains in the cytoplasm.

greatly depleted (>90%) of T cells able to stimulate anti-SRC responses by CBA B cells. Similar results apply with selection in B10.AQR mice, i.e. mice which are H-2-compatible with CBA only in the I-region (*I-A* through *I-E*). Selection in 4R *(k b)* mice, by contrast, leads to only a partial (20–50%) reduction in T cell help for B10.BR B cells; with 4R B cells, however, selection is near-complete.

These findings imply that, in addition to I-A^k-restricted T cells, CBA mice contain an extra subgroup of cells which is restricted by determinants controlled by genes situated between *I-A* and *I-C*. These cells are not restricted by E^k determinants per se since selection of CBA T cells through B10.A(3R) mice $(I\text{-}A^b\ I\text{-}B^b\ I\text{-}J^b\ I\text{-}E^k\ I\text{-}C^d)$ *(b k)* causes little if any depletion of the response with CBA B cells. Likewise double negative selection of CBA T cells through 4R and then 3R mice reduces the response with CBA B cells by <50%. Selection is near complete, however, if CBA cells are negatively selected through (4R × 3R)F_1 mice, i.e. a situation in which A/E^k hybrid molecules can be formed as the result of *trans* chain association ($E_\beta{}^k$ from 4R and $E_\alpha{}^k$ from 3R).

Concluding Comments

The evidence that A/E hybrid molecules can act as T cell restriction sites (and also serve as alloantigens [2]) provides a convenient explanation for the phenomenon of β-α *Ir* gene complementation. In this respect it has been known for several years that responses to certain antigens, e.g. GLΦ, are controlled by two genes, β and α, situated in the *I-A* and *I-E* subregions, respectively [1]. The studies of *Schwartz* et al. [7] have shown that T cell proliferation to GLΦ in vitro requires that the antigen-presenting cells express the products of both responder alleles. From this and other evidence these workers postulate that GLΦ-specific T cells recognize antigen only in association with particular Ia molecules, namely 'responder' A/E hybrid molecules created by the association of the β and α gene products of the two complementing strains. This explanation obviously hinges on the assumption that *Ir* gene products and Ia antigens are synonymous. Although this thesis has yet to be proved, the fact that monoclonal antibodies specific for A/E hybrid molecules can block the proliferative response to GLΦ [6] is clearly consistent with this notion.

Whereas homozygous mice presumably express a single set of A/E-restricted T cells, in theory four different subgroups of these T cells could exist in F_1 hybrids made between strains which differed in both the *I-A* and *I-E* subregions. Two subgroups would be restricted to the parental *(cis)* molecules whereas two more subgroups would be restricted to *trans*-associated molecules. In practice there are probably only two subgroups of A/E-restricted T cells since, in contrast to the E_β chain, the E_α chain (which is missing in many haplotypes) is relatively nonpolymorphic [12].

The notion that *trans* chain association in F_1 hybrids can create additional A/E hybrid molecules (plus complementary T cells restricted to these molecules) raises the question of whether analogous hybrid molecules can be formed as the result of *trans* association of the A_α and A_β subunits of the I-A molecules. Recent studies of *Kimoto and Fathman* [5] with cloned lines of antigen-specific T cells suggest that this is indeed the case. The existence of A/A hybrid molecules could account for β-β *Ir* gene complementation, i.e. where the two complementing *Ir* genes both map in the *I-A* subregion [1].

Acknowledgement. This work was supported by USPHS Grants CA-15022, AI-15822, AI-10961, AI-15393, and CA-09140.

References

1 Benacerraf, B.; Dorf, M.E.: Cold Spring Harb. Symp. quant. Biol. *41:* 465–475 (1976).
2 Fathman, C.G.; Hengartner, H.: Proc. natn. Acad. Sci. USA *76:* 5863–5866 (1979).
3 Jones, P.P.; Murphy, D.B.; McDevitt, H.O.: J. exp. Med. *148:* 925–939 (1978).
4 Katz, D.H.; Benacerraf, B.: Transplant. Rev. *22:* 175–205 (1975).
5 Kimoto, M.; Fathman, C.G.: J. exp. Med. *152:* 759 (1980).
6 Lerner, E.A.; Matis, L.A.; Janeway, C.A.; Jones, P.P.; Schwartz, R.H.; Murphy, D.B.: J. exp. Med. *152:* 1085 (1980).
7 Schwartz, R.H.; Yano, A.; Stimpfling, J.H.; Paul, W.E.: J. exp. Med. *149:* 40–57 (1979).
8 Sprent, J.: in Marchalonis, The lymphocyte: structure and function, pp. 43–112 (Dekker, New York 1977).
9 Sprent, J.: Immunol. Rev. *42:* 108–137 (1978).
10 Sprent, J.: J. Immun. *126:* 2089 (1980).
11 Sprent, J.: J. exp. Med. *152:* 996 (1980).
12 Uhr, J.W.; Capra, J.D.; Vitetta, E.S.; Cook, R.G.: Science *206:* 292–297 (1979).
13 Zinkernagel, R.M.; Doherty, P.C.: Adv. Immunol. *27:* 51–177 (1979).

Dr. Jonathan Sprent, The University of Pennsylvania, School of Medicine,
Division of Research Immunology, Philadelphia, PA 19104 (USA)

The Immune System, vol. 1, pp. 202–206 (Karger, Basel 1981)

Some Immunological Mysteries

Susan L. Swain[1], *Richard W. Dutton*[1]

Department of Biology, University of California, San Diego, La Jolla, Calif., USA

The most entertaining questions in research are those for which one's ingenuity is unable to provide a plausible and convincing answer. The danger (of course) is that one can be so intrigued by the mystery that one can ignore or reject the mundane explanation.

We discuss here three questions in an area in which *Niels Jerne* has already made a most striking proposal [3]. Our questions concern the activities of T cells that seem to us to present a 'mystery' and for which we at least cannot see any convincing explicit explanation.

The first is the question: What *is* a restricting determinant? The second is: What is the significance of a class 1 (K, D) restricting determinant as opposed to a class 2 (I) restricting determinant? Thirdly: What is the significance of the Lyt1 and 23 gene products as markers of T cell subsets? These questions are clearly related to one another. We will consider these in turn, and we will choose our examples from the mouse. In these discussions we will use a dual recognition model of the T cell receptor but the argument is the same for an altered self model if the necessary transpositions are made.

First, the restricting determinant. One might take the view that *Jerne* [3] has already provided us with the most plausible answer to this mystery in his 1971 paper and its sequel [1]. In the T cell response to a virus the cytotoxic T cell that develops in the primed response will kill only the target that carries the virus antigen and the appropriate K or D gene product. The K or D gene product is a molecule that has many serologically detectable

[1] *Susan Swain* and *Richard Dutton* were members of the Basel Institute for Immunology in 1978.

antigenic determinants and can be shown (in the allogeneic response) to have a number of determinants that can be recognized by T cells [7, 8]. It is generally believed, however, that the virus-specific T cell must recognize the *allele-specific* determinant of self K and D as the restricting determinant. That is to say that T cells that recognize virus and a determinant that is shared by or cross-reacts with a different haplotype do not develop, or if they do develop, do not function.

The evidence for this belief, however, has not been obtained in experiments specifically designed to establish this assertion. It has come instead from experiments which show, on the one hand, a remarkable lack of cross-reaction of the K or D restricted anti viral killing [12] and, on the other hand, experiments which show quite extensive cross-reaction in the response to allogeneic K and D [6]. This point will be returned to below.

There would seem to be only two classes of explanation for this apparent 'fact'. Either there must be something 'special' about the restricting determinant itself so that it can serve some purpose that the other determinants cannot or only T cells recognizing this determinant are present in the anti self plus X repertoire.

The consequences of adopting either explanation seem to us quite profound. In the first case, one is lead to see a *functional* role of a particular *part* of the K or D molecule. Recognition of the restricting determinant produces a (conformational?) change or some other effect that does not result when another part of the molecule is recognized. We have some difficulty in accepting such a model since allo-killing can be directed against the I gene product and this is not K- or D-restricted [4]. The difficulty is not insurmountable and it is possible to construct ad hoc models that get around this problem.

In the second explanation, in which the restricting determinant is *defined* in the T cell repertoire, one is driven to a somewhat extreme germ line model of the T cell repertoire. In this model the germ line repertoire must consist, at least predominantly, of anti-'restriction determinants' for the recognition of MHC products. The somatic generation of diversity from this repertoire then provides the simplest explanation of antigen specific Ir gene control. This leads to a rather stark uncompromising model of the type most explicitly stated by *Langman* [5] and by *Cohn* et al. [2].

We are not quite comfortable with either explanation for the difference in the cross-reaction of the two types of MHC determinant, the self MHC restriction determinant and the self MHC determinants as viewed by allogeneic T cells. It is thus, by our criteria, a mystery.

There is a mundane explanation, however, that must be seriously considered. It is simply that the original premise that there is a difference in the two types of cross-reactivity is unfounded. This possibility seems worthy of some further discussion.

The demonstration of the lack of cross-reactivity of the restriction determinant is best seen in a secondary anti-viral response. In this case the initial stimulation may have selected a relatively rare cell or group of cells and expanded them into a sizable population of restricted heterogeneity. The response has a tendency to be 'monoclonal' and cross-reactions on other alleles would be rarely seen. In the primary response to allo K or D a *large number* of T clones, displaying considerable heterogeneity, may respond and cross-reactions on other alleles are to be expected.

We suggest that the distinction may disappear if cross-reactions are compared under conditions in which the two responses are more comparable with respect to the number of T cells that respond. One should compare the secondary response to a whole panel of minor histocompatibility antigens with the *secondary* response to allogeneic K or D using a very small pool of precursors as responders in the primary stimulation. Experiments currently in progress suggest that when this is done the difference in cross-reactivity may indeed disappear.

The second and third questions concern the association of helper function, the Lyt1 phenotype and the I region restriction determinant on the one hand and the cytotoxic function, Lyt2, 3 phenotype and the K- or D-restricting determinant on the other. This correlation holds good in most cases and is clearly a sound generalization.

In our studies of the past 2 years, however, we have shown that when the triple correlation between Lyt phenotype function and the MHC subregion does break down, the Lyt phenotype is better correlated with the MHC subregion recognized rather than with T cell function.

The evidence for this assertion comes from studies of responses to allogeneic MHC rather than the self MHC-restricted response to a non-MHC antigen. Thus, the helper cell induced by allogeneic K or D has the Ly12 phenotype rather than Ly1 [10]. Furthermore, the secondary allo-helper response to K or D is not I-restricted and is blocked by anti-Lyt2 [9]. The cytotoxic response to allogeneic I of the long-term alloreactive line C.C3.11.75 has the phenotype Lyt1^{+}2^{-} [11] and, unlike normal CTL, is not blocked by the same anti-Lyt2 that blocked the allohelp to K or D [9].

Nonetheless, in the normal self plus X response K, or D, do serve as the restricting determinants for CTL while I serves as the restricting determinant for help. We do not understand why this is so. One obvious explanation would be that killing is effected through the K or D molecule and help is delivered through the I molecule. An alternative explanation is that killers are derived only or predominantly from cells that recognize K or D and helpers only from those that recognize I. The connection is thus at the level of the T cell repertoire, not at the function of the K or D versus the I molecule. Neither explanation seems satisfactory (at least in the allogeneic response) since there are examples in which the correlations are reversed (see above). We feel this data is inconsistent with different roles for the two subclasses of MHC molecules. (It is possible, however, that one should consider models in which the alloreactive repertoire is derived in a manner different from the self plus X repertoire.) Again we are left with a mystery.

The Lyt phenotype has been seen as a marker of T cell function and the simplest model would suggest that the different Lyt molecules were in some way involved in the delivery of that function. Lyt1 does something in the delivery of help, Lyt2 does something in the delivery of the killing signal. Our recent results suggest that this is untenable: there is a killer cell to I which *does not* express Lyt2 and is not blocked by Lyt2 antibodies; there are helper cells to K and D which express Lyt2 and the activity of which is blocked by Lyt2 antibodies [9–11]. The logical alternative would be to propose that the Lyt molecule had something to do with the recognition itself or the consequences of that recognition. Thus, for example, recognition of K or D on the stimulator or target cell would initiate a chain of events which requires the subsequent participation of the Lyt2 molecule on the effector cell. Conversely, the recognition of I initiates a different chain which does not require the participation of Lyt2 but (perhaps) requires the participation of Lyt1. This is clearly speculation and in our view the basis for the correlations between Lyt phenotypes, T cell function, and MHC subregion remain unresolved.

Acknowledgements. This work was supported by grants from the National Institutes of Health (AI 08795) and the American Cancer Society (IM-1N).

References

1 von Boehmer, H.; Haas, W.; Jerne, N.K.: Proc. natn. Acad. Sci. USA *75:* 2439 (1978).

2 Cohn, M.; Langman, R.; Geckler, W.: Diversity 1980; in Progress in Immunology, vol. 4 (in press).

3 Jerne, N.K.: Eur. J. Immunol. *1:* 1 (1971).
4 Klein, J.; Chiang, C.L.; Hauptfield, V.: J. exp. Med. *145:* 450 (1977).
5 Langman, R.E.: in Review of physiology, biochemistry and pharmacology, vol. 81, p. 1 (Springer, Heidelberg 1978).
6 Lindahl, K.F.; Peck, A.B.; Bach, F.H.: Scand. J. Immunol. *4:* 541 (1975).
7 Melief, C.J.M.; van der Meulen, M.Y.; Postma, P.: Immunogenetics *5:* 43 (1977).
8 Melief, C.J.M.; deWaal, L.P.; van der Meulen, M.Y.; Melvold, R.W.; Kohn, H.I.: J. exp. Med. *151:* 993 (1980).
9 Swain, S.L.: Submitted.
10 Swain, S.L.; Bakke, A.; English, M.; Dutton, R.W.: J. Immun. *123:* 2716 (1979).
11 Swain, S.L.; Dennert, G.; Wormsley, S.; Dutton, R.W.: Eur. J. Immunol. (in press).
12 Zinkernagel, R.M.: J. exp. Med. *143:* 437 (1976).

Dr. Susan L. Swain, Department of Biology, University of California, San Diego, La Jolla, CA 92093 (USA)

The Immune System, vol. 1, pp. 207–218 (Karger, Basel 1981)

On the Influence of the Major Histocompatibility Gene Complex on the Specificity Repertoire of T Lymphocytes

Wulf Droege[1]

Institut für Immunologie und Genetik, Deutsches Krebsforschungszentrum, Heidelberg, FRG

Niels Jerne [9] proposed more than 10 years ago that the specificity repertoire of the immune system is driven by self antigens and in particular by the individual's own major histocompatibility antigens (MHA). In view of the available experimental evidence about the specificity repertoire of T lymphocytes, this concept is still appealing and likely to be correct.

For the sake of clarity, I would like to begin by defining some terms: 'conventional antigens' are antigens that are not encoded in the MHC; 'T lineage induction' is the process that commits a cell clone to a given T cell function such as helper or cytotoxic function. The term 'T cell repertoire' shall be defined in this paper as the qualitative and quantitative composition of the T cell pool at a given time. This term refers essentially to the different T cell functions and specificities, but also to the relative frequency of T cells with a given function and specificity.

This paper deals with the hypothesis that the T cell repertoire at any given time equals the T cell memory which has accumulated at that time and which has been generated by practically all (cell surface) antigens in the organism, including relatively many self antigens and somewhat fewer foreign antigens. The mechanism of T cell memory involves, according to this hypothesis, not only clonal expansion of the T lineage cells but also the commitment to a given T cell function and to the corresponding restriction pattern. The overrepresentation of alloreactive cells in the T cell population can be explained by this hypothesis on the basis of a theoretical principle that has been recently described [4, 5].

[1] *Wulf Droege* was a member of the Basel Institute for Immunology from 1973 to 1976.

The immune system of humans and higher animals contains two major cellular components: B lymphocytes, which produce antibodies that defend the organism against a variety of structures including soluble materials; and T lymphocytes, which contribute to the defense by killing virus infected cells (and possibly tumor cells) and by regulating the activity of other immunologically relevant cells in the system such as B lymphocytes, monocytes and macrophages. Because of these functions, it makes sense that T lymphocytes recognize antigens often, and possibly always, on cell surfaces and not in solution [17, 18]. The cytotoxic and the regulatory T lymphocytes distinguish their respective groups of target cells from each other and from soluble antigens through special cell surface markers, namely the MHA. Cytotoxic T lymphocytes (CTL) recognize conventional antigens (i.e. non-MHA) mainly in association with cell surface molecules that are encoded in the K or D region of the major histocompatibility gene complex (MHC), while helper cells (and cells mediating delayed hypersensitivity) recognize conventional antigens mainly in association with I region controlled cell surface molecules [10, 15, 17, 18, 21, 24]. This phenomenon is known as MHC restriction. The tissue distribution of these MHC-encoded cell surface components is correlated with the respective target cells of the T cell classes: I region determinants are present on immunologically relevant cells, and K or D region determinants are present on practically all cells in the organism. Recent experimental evidence suggests that the recognition of the MHC-encoded cell surface markers is 'learned' through selection during ontogenetic development rather than determined by the genetic program of the T cell itself [1, 2, 11, 18, 25]. The evolutionary and physiological significance of this phenomenon is obviously related to the extraordinarily high degree of polymorphism of the MHA and to the requirement for complementary recognition structures on T cells. The observation that cells of different MHC haplotypes interact in parent $\rightarrow F_1$ radiation chimeras where lymphocytes of both parents mature in an F_1 hybrid environment) [1, 2, 18] indicates that complementarity is not mediated by dimerization (like-like interaction) as proposed earlier [12, 13], nor by a combination of genes which code for complementary cell surface structures, but rather by selection of complementary receptors from a yet unknown original repertoire. The widely observed restriction of T cell responses to products of self MHC [18] together with the observation that $A \times B \rightarrow A$ radiation chimeras generate T cell responses against conventional antigens in the context of MHA from parent A but not parent B [18, 23, 25] suggests that individual T cells recognize preferentially those allelic variants of MHA which they have experienced in their environment during maturation. These observations suggest also that self-restrictedness exists in the T cell population before intentional immunization. The present paper is based on the assumption that this is essentially correct. The available data, however, do not rigorously exclude the alternative possibility that *all* the individual T cells have anti-MHA receptors with high affinity for allogeneic MHA and only intermediate affinity for self-MHA; and this possibility is the central assumption in the theory of *Janeway* et al. [7].

The available experimental evidence [18] supports the assumption that both conventional antigens and MHA are recognized by a single T cell receptor. T cell receptor may recognize newly formed neoantigenic determinants which are formed by the interaction of conventional antigen and MHA ('altered self model'), or may recognize unmodified determinants on both types of antigens ('dual recognition') [23].

The Role of Self Antigens for the Generation of the T Cell Repertoire and for the Development of Alloreactivity

This paragraph describes a theory of the generation of the T cell repertoire which explains the phenomenon of self-restriction and also the extraordinarily high proportion of alloreactive cells in the T cell repertoire. The theory is based on the following set of *assumptions:* (1) T cells recognize conventional antigens together with MHC encoded determinants through one receptor with one or more distinct V regions. (2) The T cell receptor usually recognizes determinants of conventional antigens *plus* MHC encoded determinants ('dual recognition'), and only occasionally neoantigenic determinants that are formed by interaction between conventional antigen and MHA ('altered self'). (3) Conventional antigens and MHA determinants contribute *approximately* additively to the total binding energy and, therefore, in a multiplicative way to the total binding affinity of the receptor, irrespective of what proportion each determinant contributes to the binding ($A_T = A_{CA} \cdot A_{MHA}$). (4) The mature T cell repertoire is mainly the result of a selection through two independent types of stochastic processes, namely 'T lineage induction' (i.e. the positive selection which commits a cell to a given T cell function) and self-tolerance induction (negative selection). The effect of these processes can be described by the probability function: $P_m = P_o \cdot P_{ind} \cdot (1-P_{tol})$, indicating that the frequency of a given receptor in the mature repertoire P_m is derived from its relative frequency in the original (i.e. germ line) repertoire P_o by multiplication with the corresponding probability for T lineage induction P_{ind} and by subtraction of the probability for tolerance induction P_{tol}. (5) The probability for these selection processes is a function of the affinity to corresponding antigens on inducing or tolerogenic cells, respectively, in the environment.[2] These antigens are mostly conventional self antigens in combination with self-MHA. It is possible but not obligatory that both processes use the same affinity threshold: $P_{ind/tol} = f(A_T)$. (6) Receptors with high affinity for alogeneic cells have low or intermediate affinity for self antigens due to cross-reactivity between the various allogeneic MHA [20]; and the special assumption is that these receptors fall approximately into the range of affinities, which is highly enriched by the two selection processes to be described later.

No assumptions are made with respect to several points: (1) Tolerance induction may occur either before, after, or in parallel to the 'T lineage induction'. (2) While it is tempting to assume two distinct V regions in analogy to the B cell receptor, the theory would also work on the basis of one V region only. (3) The theory is compatible with the assumption that the T cell V gene repertoire is essentially random, but it is also compatible with the alternative assumption that the germ line repertoire codes primarily for receptors with medium or high affinity for MHA-*like* structures as suggested by *Jerne* [personal communication].

Our six assumptions imply that the T cell repertoire can be described as a probability function of the affinity for self antigens. The frequency of a given specificity in the mature T cell repertoire is determined by its affinity to self antigens: $P_m = f(A_T)$. Mathematical equations of this kind are at present of only theoretical interest, but their validity may soon be tested experimentally.

The six assumptions explain the generation of alloreactivity as a side effect during the ontogenetic development of the T cell repertoire. This is best illustrated by plotting all possible receptors in a two-dimensional distribution profile according to their affinity for conventional self antigens on the one hand and self MHA on the other hand (fig. 1). One expects in a random repertoire a high frequency of receptors with a low affinity for a

[2] This assumption probably does not apply to receptors with both very high affinity for conventional self antigens and very low affinity for self MHA. In this case, the interaction with conventional antigen in the absence of self MHA would be expected to competitively inhibit the interaction with the antigen in complex with MHA and would therefore prevent the MHA-dependent induction as described in a subsequent paragraph.

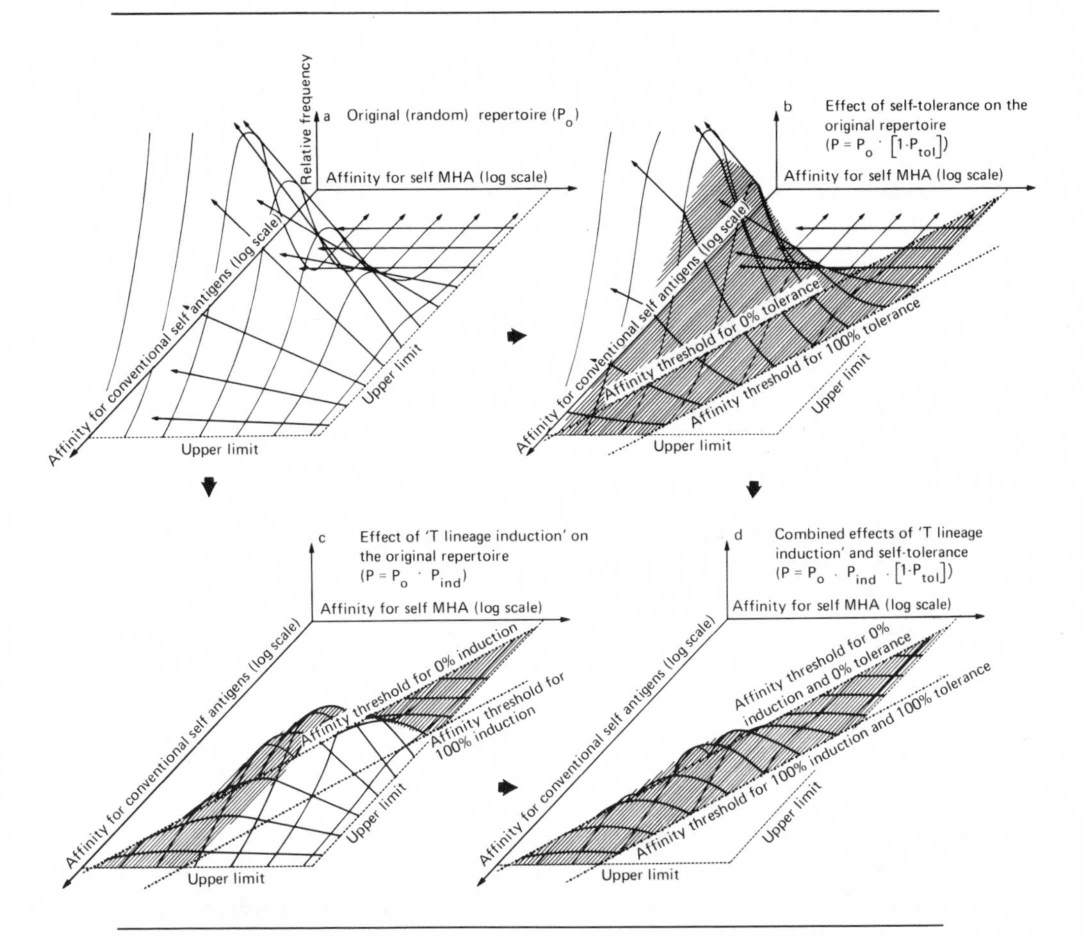

Fig. 1. Hypothetical probability distribution profile for the T cell receptor repertoire before and after 'T lineage induction' and self-tolerance induction. The original repertoire (a) is assumed to be random and distributed in a bell-shaped profile. Self-tolerance eliminates selectively the high affinity portion (b) and 'T lineage induction' eliminates the low affinity portion below the indicated affinity thresholds (dotted lines) (c). The combined processes select for intermediate affinities that were located between the two threshold lines (d). Some receptors in the mature repertoire (d) have high affinity for self MHA and low affinity for conventional self antigens implying specificity for foreign conventional antigens. Most receptors, however, are alloreactive, and recognize the allogeneic cells with two binding sites: one site has intermediate or high affinity for conventional self antigens including nonpolymorphic antigens, and the other site has low or intermediate affinity for self MHA and according to the model by chance high affinity to allogeneic MHA (for details see text).

given self MHA, and a decreasing frequency for receptors with increasing but finite affinity for this antigen (fig. 1a). The many conventional self antigens, on the other hand, express in comparison to the MHA more antigenic determinants, and the majority of receptors in a random repertoire is therefore expected to bind with intermediate affinity to one of the many conventional self antigens while receptors with very low or very high affinity for the best fitting self antigen are expected to be rare. We therefore obtain a bell-shaped distribution profile along the affinity axis for conventional antigens (fig. 1a), but we also expect an upper limit for the affinity. The multiplication of this frequency distribution with a probability function for 'T lineage induction' preserves the entire high-affinity portion and reduces selectively the low-affinity portion as illustrated by the shaded area in figure 1c. The threshold for this reduction is characterized by a given total affinity and therefore by a constant product of the individual affinities: $A_T = A_{CA} \cdot A_{MHA} = \text{constant}$. This threshold is a straight line on a logarithmic plot for the affinities (dotted line in fig. 1c). The subtraction of the probability function for self-tolerance from the original frequency distribution leaves the low-affinity portion of the original repertoire intact and reduces only the high-affinity portion (shaded area in fig. 1b). The affinity thresholds for self-tolerance induction (fig. 1b) were assumed to be the same as for 'T lineage induction' (fig. 1c). Figure 1d illustrates, finally, that the combined selection processes result in a population of intermediate affinities, regardless of whether self-tolerance occurs before or after 'T lineage induction', as was expected from assumption number 4.

A selection for intermediate affinities was also postulated in the theory of *Janeway* et al. [7] although only for the MHA-binding sites. The Janeway theory concluded therefore that *all* T cells have low or intermediate affinity for self MHA and by chance high affinity against one of the many allogeneic MHA implying that all T cells are essentially allorestricted. The present theory leads, in contrast, to the conclusion that the receptors with specificity for foreign conventional antigens and low affinity for conventional self antigens in the mature repertoire (fig. 1d) are more likely to have higher affinity for self MHA than for allogeneic MHA, implying self-restriction at the level of the individual T cell. The majority of cells in the mature repertoire, on the other hand, is expected to be highly reactive against allogeneic cells, since they have low or intermediate affinity for self MHA and by chance (this was assumption number 6) high affinity towards one or the other foreign MHA. These same cells have intermediate or high affinity for conventional self antigens including nonpolymorphic structures that are also present on allogeneic cells (fig. 1d). This concept states, in contrast to the *Janeway* [7] theory, that the allogeneic cells are recognized through MHA *plus* non-polymorphic conventional self antigens. Figure 1 gives, indeed, support to the notion of *Matzinger and Bevan* [16] that it is mainly the *multiplicity* of the conventional nonpolymorphic *self antigens* which is responsible for the overrepresentation of alloreactive cells.

Prediction of a Novel Type of Ir Gene Phenomenon

Since different allelic variants of a given class of MHA differ in amino acid composition and sequence, it is expected that they differ also with respect to the quality of binding to the best fitting receptor and also with respect to the average binding energy with a random repertoire of receptors. The same receptor repertoire is therefore expected to show different affinity distribution profiles if tested with two different MHA (fig. 2, original repertoire A and B); and the mature repertoires after completing the selection procedures towards the two MHA would also be different (fig. 2). This effect applies essentially to receptors for all kinds of conventional antigens, although receptors with specificity for conventional self antigen are expected to be less affected than receptors for foreign antigen. MHA with particularly good binding properties are expected to determine a higher degree of self-restrictedness especially in the immune responses towards foreign antigens with little cross-reactivity for self antigens (fig. 2). This effect is expected to produce a novel type of MHC-controlled immune response pattern (Ir gene phenomenon) which determines, in contrast to the classical MHC-linked Ir gene phenomena, a particularly low or high responsiveness not only to one or a few but to many unrelated antigens.

The Hypothesis that the Process of 'T Lineage Induction' and the Antigen-Driven Generation of T Cell Memory Are Identical

Helper cells are preferentially restricted towards I region determinants and cytotoxic T cells towards K or D region determinants. The process that commits a lymphocyte to one or the other T cell function also commits it to a corresponding restriction pattern; this process has been called 'T lineage induction' for the purpose of this paper. It is widely assumed that this process is a single event in the early history of a T cell clone and likely to occur

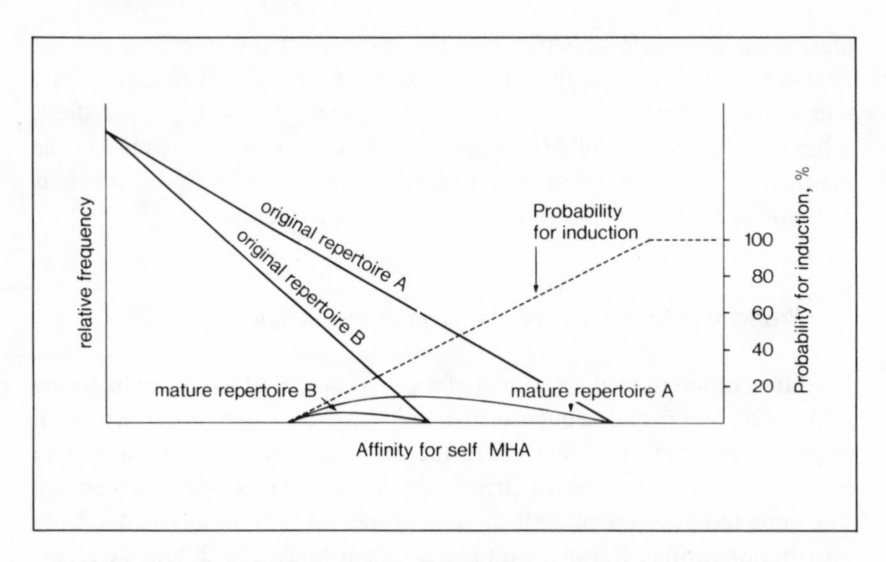

Fig. 2. The effect of different MHC alleles on the generation of the T cell repertoire. Two allelic variants A and B of a given class of MHA are assumed to express different average and maximum affinities to receptors of a given random repertoire. The same original receptor repertoire shows therefore different affinity distribution profiles with respect to MHA of type A and type B. If this original receptor repertoire is subjected to 'T lineage induction' and self-tolerance according to the indicated probability function (dotted line), it will result in the corresponding mature repertoires A and B. The mature repertoire A contains a higher proportion of receptors with high affinity for self MHA and would, therefore, mediate a higher degree of self-restrictedness. This effect is an MHC-linked Ir gene phenomenon, which applies to all conventional antigens, although more strongly to foreign antigens than to conventional self antigens.

in the thymus [25]; but there is the possibility that the thymus primarily controls other maturation steps and that 'T lineage induction' occurs subsequently and at various anatomical sites. The mechanism that links a given T cell function to the corresponding restriction pattern is not known. One possibility, which was essentially suggested by the *Jerne* [9] theory, is that T lineage commitment for one or the other function leads to the activation of a corresponding V gene pool which codes for receptors with preferential recognition of I region determinants *or* K/D region-like structures, respectively. Another possibility is that the restriction for products of a given MHC subregion is achieved by a selection processs during the induction of the corresponding T cell function. The V gene pool may be random and it may be used for I region-restricted *and* K/D region-restricted T cells.

The mechanism that determines the restriction of a T cell towards a given allelic variant of a K/D or I region determinant may be related to the mechanism of subregion restriction and 'T lineage induction'. There is suggestive although not conclusive evidence that restriction for self MHA results from a maturation process that selects for receptors with preferential recognition of MHA alleles on cells in the selecting environment. This selection process occurs largely [25] but not exclusively [14] in the thymus.

The simplest explanation for this set of observations is that one essentially random V gene pool is used by both helper cells and cytotoxic cells, and that a single selection process determines: (1) the preferential recognition of a given allelic variant of MHA; (2) the recognition of MHA of a given subregion, and (3) the corresponding T cell function. This implies, however, that the original V gene repertoire does not recognize the MHA as exceptional antigen, and MHA must, therefore, play an active role during the selection process in order to exclude selection towards cell surface determinants other than MHA and to ensure that MHA of a given subregion determines the corresponding function. This active role may occur in several ways: (1) The MHA molecule might be secreted as a factor and may contain a constant region which delivers the inductive signal and determines the function of the T lineage cell (fig. 3, left side). So far, soluble factors with K or D region-encoded determinants have not been described while I region-controlled determinants have been observed on soluble factors in several instances. But this model would require that the I region determinants on the factors are the same as those on B cells and macrophages, which has not yet been shown. (2) The more attractive possibility is that the MHA molecule on a specialized inducer cell recognizes the interaction with the antigen receptor of a T lineage cell and causes the inducer cell to deliver the inductive signal. The resulting inductive signal must be different for I region versus K/D region encoded MHA (fig. 3, middle and right side).

Recent experimental evidence revealed that functionally mature helper T cells and cytotoxic T lymphocyte precursors require different factors for their activation and proliferation: cytotoxic T lymphocyte precursors require interleukin 2 which is produced by helper T cells, and the helper T cells require interleukin 1 which is produced by macrophage-like cells [19]. *Wagner* et al. [22] reported moreover that Ly $1^{+}2^{-}3^{-}$ cells precede the appearance of Ly 123^{+} cytotoxic T lymphocyte precursors in the

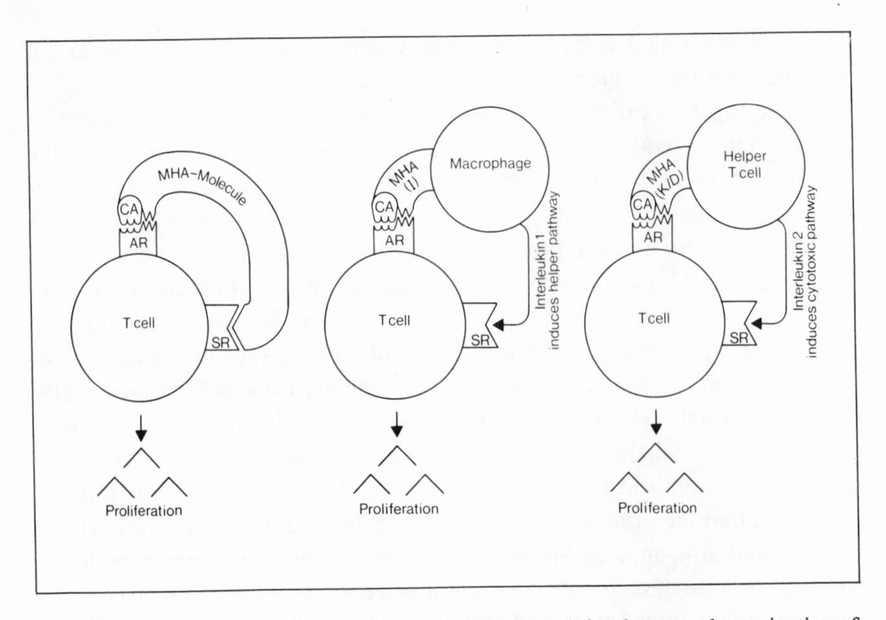

Fig. 3. Possible models for the induction of T cells and simultaneous determination of MHC restriction. The T cells are assumed to express originally a random repertoire of antigen receptors (AR) in addition to a set of signal receptors (SR) for interleukin 1 and interleukin 2 or for another yet unknown type of signal. Induction occurs in one of the following ways: (1) The antigen receptor binds a given conventional antigen (CA) in association with a MHA molecule. This presents an active site to the corresponding signal receptor SR and induces a given function (left side). (2) The antigen receptor of a T cell binds to conventional antigen plus MHA on the surface of an inducing cell (macrophage or helper T cell) and stimulates this cell to deliver interleukin 1 or interleukin 2, respectively. Interleukin 1 commits the T cell to the helper function (middle), interleukin 2 commits the cell to cytotoxic function (right side).

thymus and hypothesized that interleukin 2 is indeed not only involved in the activation of mature cytotoxic T lymphocyte precursors but also in the early thymic maturation of this T cell lineage. I propose that the process of 'T lineage induction' is equivalent to the early events of the antigen-driven and interleukin-dependent activation and proliferation of mature T cells. Of course, this assumption is only meaningful if 'T lineage induction' by itself does not generate active effector cells but rather a state of 'T cell memory' which is characterized by the antigen-driven expansion of specific sets of clones. This hypothesis states that the T cell repertoire is T cell memory.

The hypothesis has one important and testable implication: The interaction at the surface of the helper T cells which triggers the helper cells to release the inducing signal for the cytotoxic cells (i.e. interleukin 2), must occur at the K- or D-encoded cell surface antigens of the helper cells rather than at their I region-restricted antigen receptors; and it must be the I region-encoded determinants on the macrophages which recognize the interaction with the I region-restricted receptor of the helper cell and trigger the macrophages to release the inducing signal for the helper T cells (i.e. interleukin 1) (fig. 3). Such a result would be in concordance with the postulate of *Cohn and Epstein* [3] that MHA function not only as antigens but also as signal-channeling receptors.

Self and Non-Self

The theory on alloreactivity in this paper required the assumption that 'T lineage induction' is (mainly) driven by conventional self antigens in combination with self MHA. The hypothesis that 'T lineage induction' equals the generation of T cell memory implies, therefore, that memory is not only generated against foreign conventional antigens but also and predominantly against self antigens. The mechanism of self-tolerance prevents, according to the theory, only the generation of T memory cells with high affinity but not intermediate affinity for self antigen (fig. 1d), but it must completely prevent the generation of active effector cells against self antigens. There is a possibility that either the activation of mature effector cells or the lytic phase requires higher affinity (i.e. longer interaction times) than the generation of T cell memory. A suppressor cell system with a very stringent effect on the maturation of active cytotoxic effector cells and considerably less stringent effects on the generation of T cell memory [6] provides an additional fail-safe mechanism to prevent auto-aggression.

It should be emphasized, however, that the theory makes no distinction between self antigens and foreign antigens, provided they persist in the organism in sufficient concentration to drive the selection processes. Only the sudden introduction of a previously inexperienced foreign antigen into the organism is expected to result temporarily in the coexistence of antigen and large numbers of T cells with high affinity for this antigen; this temporary disequilibrium in the selection processes may be a prerequisite for acute immune responses. The discussion of these events is, however, beyond the scope of this paper.

References

1 von Boehmer, H.; Sprent, J.: Transplant. Rev. *29:* 3 (1976).
2 von Boehmer, H.; Hudson, L.; Sprent, J.: J. exp. Med. *142:* 989 (1975).
3 Cohn, M.; Epstein, R.: Cell. Immunol. *39:* 125 (1978).
4 Droege, W.: Immunobiology *156:* 2–12 (1979).
5 Droege, W.: Cell. Immunol. *57:* 251 (1981).
6 Droege, W.; Süssmuth, W.; Franze, R.; Müller, P.; Franze, N.; Balcarova, J.: in Quastel, Cell biology and immunology of leukocyte function pp. 467–483 (Academic Press, New York 1979).
7 Janeway, C.A.; Wigzell, H.; Binz, H.: Scand. J. Immunol. *5:* 993 (1976).
8 Janeway, C.; Jones, B.; Binz, H.; Frischknecht, H.; Wigzell, H.: Scand. J. Immunol. *12:* 83–92 (1980).
9 Jerne, N.: Eur. J. Immunol. *1:* 1–9 (1971).
10 Reviewed in Katz, Lymphocyte differentiation, recognition, and regulation (Academic Press, New York 1977).
11 Katz, D.H.: Cold Spring Harb. Symp. quant. Biol. *41:* 611–624 (1976).
12 Katz, D.H.; Benaceraff, B.: Transplant. Rev. *22:* 175 (1975).
13 Katz, D.H.; Hamaoka, T.; Dorf, M.E.; Benaceraff, B.: J. exp. Med. *137:* 1405 (1973).
14 Katz, D.H.; Katz L.R.; Bogowitz, C.A.; Bargatze, R.F.: J. Immun. *124:* 1750–1757 (1980).
15 Kindred, B.; Shreffler, D.C.: J. Immun. *109:* 940 (1972).
16 Matzinger, P.; Bevan M.: Cell. Immunol. *29:* 1 (1977).
17 Reviewed in Göran Möller (ed.): Immunol. Rev. *35* (1977).
18 Reviewed in Göran Möller (ed.): Immunol. Rev. *42* (1978).
19 Reviewed in Göran Möller (ed.): Immunol. Rev. *51* (1980).
20 Teh, H.S.; Phillips, R.A.; Miller, R.G.: J. Immun. *120:* 425 (1978).
21 Vadas, M.A.; Miller, J.F.A.P.; Whitelaw, A.; Gamble, J.R.: Immunogenetics *4:* 137 (1977).
22 Wagner, H.; Hardt, C.; Heeg, K.; Pfizenmaier, K.; Solbach, W.; Bartlett, R.; Stockinger, H.; Röllinghoff, M.: Immunol. Rev. *51:* 215 (1980).
23 Zinkernagel, R.M.: Immunol. Rev. *42:* 224–270 (1978).
24 Zinkernagel, R.M.; Doherty, P.C.: Contemp. Top. Immunobiol. *7:* 179 (1977).
25 Zinkernagel, R.M.; Callahan, G.N.; Althage, A.; Cooper, S.; Klein, P.A.; Klein, J.: J. exp. Med. *147:* 882 (1978).

Dr. Wulf Droege, Institut für Immunologie und Genetik, Deutsches Krebsforschungszentrum, D–6900 Heidelberg (FRG)

The Immune System, vol. 1, pp. 219–225 (Karger, Basel 1981)

The HLA: the Point of View of the Mouse

M. Trucco[1], *G. Garotta*[1]

Basel Institute for Immunology, Basel, Switzerland

When a man meets a pretty woman for the first time, he is generally most attracted by her charming eyes, her upturned nose, her delightful smile, and warm voice. But when a woman is foreign, the attraction lies in her exotic qualities – her height, her strange complexion, the unusual color of her hair. Difference, not beauty, is the compelling factor and details go, for the moment, unnoticed.

So it is when the mouse views man immunologically; the mouse will recognize immediately (more frequently) the gross differences, and only later (less frequently) the nuances. But the 'myopia' of the mouse confers the advantage of a more complete outlook of human antigens. With its antibodies, the mouse can also recognize molecules or part of molecules that are non-immunogenic in an allogeneic system. Consequently, the library of monoclonal antibodies to HLA structures produced so far in the mouse includes many γ-globulins that recognize non-polymorphic determinants but only a few that are directed to allogeneic epitopes (table I).

The recent publication of the complete sequences of the A2 and B7 HLA antigens by *Strominger's* group [8] has allowed speculations as to *what* the mouse can actually recognize as 'non-self' on the HLA molecules, *where* the different epitopes lie within these structures, and *why* mouse and man 'see' the HLA products from different perspectives. However, it is difficult to correlate antigenic configurations with the amino sequence because of the folding of the polypeptides.

[1] *Massimo Trucco* was a member of the Basel Institute for Immunology from 1977 to 1980, and *Gianni Garotta* was a member of the Institute from 1978 to 1980.

Table I. Monoclonal antibodies already characterized

Clone	Specificity	Cdl[1]	Correlation coefficient
S1.16/3, S1.24/23, S1.26/114	$\alpha\beta_2$m	+	
S1.34/24, F.10.17	αHLA heavy chain common part	–	
F8.11, F8.24	αHLA heavy chain common part	+	
S1.5/1, S1.6/1	αHLA DR dimer common part	–	
S1.19/9, E3.18, E2.17	αHLA DR dimer common part	+	
S1.10/3	αB5 + B7 + B14	+	0.72
F18.7	αB7 + B27 + Bw39	+	0.55
F18.103	αB7 + B27	+	0.53
F18.20	αA28	+	0.52
E2.2	αA2 + A28	+	0.65
F10.13/19	αB8 + B7 + Aw19	+	0.74
E3.15/4 (8w1247)	αMT2	+	0.79
F9.45	αDR 1 + 2 + 3 + 7	+	0.75
E1.21	α8w13 + DRw6	+	0.70
E1.3	αMB3	+	0.40
E3.13/9	αMTI + DR3 + DR5	+	0.55
E3.21	αALL DR but MB2	+	–071 with MB2

[1] Complement-dependent lysis.

From *Strominger's* work, we can say that the HLA-A,B,C products are intrinsic molecules that cross the entire plasma membrane. Most of these molecules are located outside the cell and can be solubilized by papain cleavage. 25 residues span the membrane, with the carboxy-terminal hydrophilic piece extending into the cell. Papain-solubilized HLA-B7 heavy chain consists of 270 residues, and because of a peculiar chemical cleavage, it can be further split into three fragments: the α1, α2 and α3 domains (fig. 1). The HLA heavy chains (43,000 dalton) are non-covalently associated with β_2-microglobulin (β_2m), a non-glycosylated polypeptide of molecular weight 12,000. The β_2m primary structure is almost identical with that of the HLA-α3 domain (residues 181–270) and homologous with the domains of the constant region of the immunoglobulin chains. In addition, the sequence within the disulfide loop of α3 is totally conserved in different HLA alleles (e.g. HLA-A2, HLA-B7, HLA-B40). Similar results were obtained when a mixture containing different HLA specificities was used for sequencing [16].

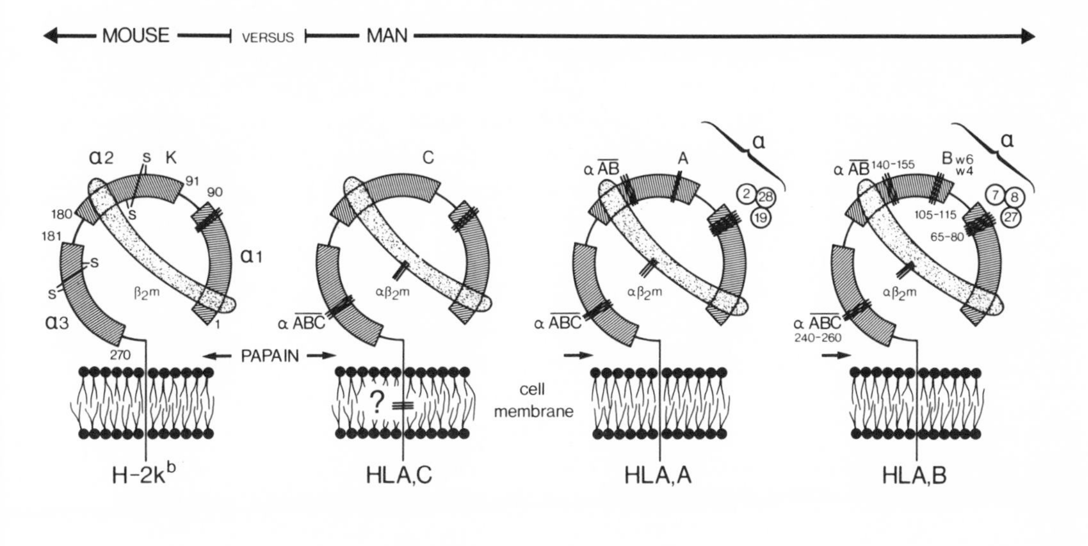

Fig. 1. Schematic comparison of the HLA and H-2 structures. The hypothetical position of the HLA epitopes recognized by the murine monoclonal antibodies studied thus far is indicated.

However, if one compares the sequence of the human α3 domain with that of the mouse (H-2K^b), some differences are found within the 240–260 residue stretch. We can tentatively assign the epitope recognized by our monoclonal anti-HLA common part (anti-HLA-$\overline{\text{ABC}}$) to this location, based on the following points. First, antibodies with similar specificity were never obtained in the allogeneic system. Second, the antibodies react in co-capping experiments with the HLA-A and with B and C molecules [18], but are unable to inhibit the binding of allogeneic specific antisera normally employed in tissue typing; thus, the epitope must lie far from the allotypic site. Finally, the Scatchard analysis is consistent with homogeneous binding characterized by a single value of the equlibrium constant [19]; therefore, the epitope must be the same on the A, B and C chains. Fowl non-cytotoxic anti-β_2m antibodies [2] and murine monoclonal anti-β_2m are able to inhibit the binding of the anti-HLA-$\overline{\text{ABC}}$ as well as all the other allogeneic anti-HLA-A,B,C antisera. The special role of β_2m in conferring new conformational properties and stability on the HLA heavy chains must be involved here. It has been shown that the anti-HLA-$\overline{\text{ABC}}$ antibodies are not able to bind the HLA target structure once the structure becomes separated from β_2m: our S1.34/24 anti-HLA-$\overline{\text{ABC}}$ monoclonal antibody, for instance, failed to react with the cytoplasmic HLA heavy chains present on Daudi cells. As no β_2m is produced by these cells, the heavy chains of the HLA structures, which require β_2m to drive them through the plasma membrane [11], cannot reach the cell surface. Both modification of the tertiary structure of the HLA dimer after reacting with the antibodies with β_2m and simple steric hindrance are possible explanations for this inhibition phenomenon.

The sequence of β_2m does not seem to differ very much among species [5], but the prominent position of this molecule, once bound to the HLA heavy chain [15], could explain its high immunogenicity in the mouse [17]. Presumably, the portions of the α1 and α2 domains that bind β_2m have also changed little during evolution. The 'masking' by β_2m of these sites probably prevented their immunogenicity, even in a xenogenic system. Thus, we can explain why the beginning of the α1 domain and the end of the α2 domain do not seriously differ in the sequences of HLA-A and HLA-B, and why probably no monoclonal antibody recognizes these domains.

However, a monoclonal antibody that reacts with all HLA-A and HLA-B structures, the anti-HLA-$\overline{\text{AB}}$ antibody, has been described; it shows different affinity for different alleles at the HLA-A and B loci [14]. Once diluted, it is significantly correlated with HLA-Aw19 as well as HLA-B8 and B7 specificities. Absorption-elution and co-capping experiments have shown this monoclonal antibody to be similar to the CYNAP (cytotoxic-negative, absorption-positive) alloantisera [3]. Since the affinity of anti-HLA-$\overline{\text{AB}}$ is influenced by the HLA specificity, the antigenic determinant must be closer to the allotypic site than the anti-HLA-$\overline{\text{ABC}}$ monoclonal antibody (fig. 1). The mouse might recognize as non-self the stretch between residues 145 and 155 of the α2 domain, or alternatively, the initial stretch of the α1 domain, which are not immunogenic in the allogeneic system because they are homologous between A and B.

Structures common to the allele at both the A and B loci must, in any case, be considered. Mouse monoclonal [14], human allogeneic [1, 7] and rabbit heteroantisera [13] have been described that are able to recognize a determinant shared by supertypic HLA-A (Aw19, A9) and HLA-B (Bw4, Bw6) alleles. We propose the existence of a particular sequence between residues 105 and 115, a region of little homology between HLA-B7 and

HLA-A2, that determines (perhaps by tertiary structure modification) whether the actual allotypic site (residues 65–80) will be either A *or* B *or* C. It is, of course, also possible that the stretch of residues 105–115 is the allotypic site influenced by the 65–80 residues sector. Though the sequences of two alleles of the same locus are not available, we can assume that the 'determining' sequences of antigens of the same locus are absolutely homologous (e.g. HLA-A2 and HLA-A28, or HLA-B7 and HLA-B8). It follows then that *the same* mutation in the allotypic site can be recognized as different alleles characteristic of the HLA-A *or* B *or* C locus, whereas *different* mutations at the allotypic site can result in a determinant shared by different loci (A9 ~ Bw4 or Aw19 ~ B8 and B7) because of the different determining sequences.

We may also predict that in the case of the B locus, differences at the position 105–115 (or 65–80) will be found by comparing products of the Bw4 with products of the Bw6 diallelic antigen system (fig. 1). Because the sequences determining A and C are not polymorphic in the allogeneic system, it is impossible to find antisera that react with all the alleles at the A locus or at the C locus. However, we can occasionally find sera directed against the Bw6 or Bw4 determining sequences in donors homozygous for Bw4 or Bw6 alleles, respectively. Interestingly, there are twice as many alleles at the B locus as there are at the A locus and most of the subtypic antigens at the B locus are associated with Bw4 or Bw6 [7]. The hypervariable stretches of the α2 (105–115) and of the α1 (65–80) domains are then immunologically separate determinants [12]. However, the recognition of α1 65–80 by the antibodies is dependent on α2 105–115 (or vice versa). It will be no surprise if a mouse monoclonal antibody is obtained that specifically reacts with all the alleles at the HLA-A locus. Perhaps such an antibody has already been obtained but discarded as uninteresting.

Monoclonal antibodies that recognize the HLA polymorphic determinants defined by allogeneic antisera have also been obtained. The monoclonal 'polymorphic' antibodies available [18] are generally directed against well-known cross-reacting specificities like HLA-A2 × A28 or HLA-B7 × B27 (table I), suggesting that the mouse may also recognize the hypervariable part of the HLA structure in position 65–80 (or 105–115), but less frequently than the intraspecies non-polymorphic HLA structures.

The HLA-C locus seems to be only slightly immunogenic in both the mouse and in humans and, to date, a satisfactory anti-HLA-C antibody, either from hybridoma products or from sera obtained by planned immunizations in human volunteers [4], has not been described. It is possible that the C locus did not mutate in the 105–115 (or 65–80) determining sector once duplicated from the ancestor common to the A and B loci and perhaps to the H-2 loci. A more convincing explanation, however, that is supported by some experimental data [9], is based on the assumption that a change occurred in one of the more constant parts of the HLA-C molecule which resulted in an instability of the C dimer at the cell surface (fig. 1); desensitization, shedding and exhaustion experiments clearly suggest that the biological properties of the HLA-C products are different from those of the HLA-A or B locus. A short stay at the cell surface may then be the cause of its low antigenicity in humans and in the mouse.

It is certainly premature to apply this reasoning to the HLA-DR locus products since we know so little about their biochemical and serological characteristics.

Today the study of histocompatibility antigens with the help of monoclonal antibodies seems to be mandatory for a better understanding of the problems concerning tissue transplantation, as well as those related to cell-to-cell recognition. *Jerne* [6], in fact, in 1971 proposed that receptors, coded for by germ line genes of lymphocytes and directed against major histocompatibility complex gene products of the individual itself, became able to recognize non-self antigens as a result of mutational events.

The point of view of the mouse will certainly inspire a second look at the HLA system, and it may be some time before we understand the common features of the two (allogeneic versus xenogeneic) perspectives. The production of human monoclonal antibodies à la *Olsson and Kaplan* [10] directed against the HLA structures may be the best approach to this problem.

References

1 Belvedere, M.; Mattiuz, P.; Curtoni, E.S.: Immunogenetics *1:* 538–548 (1975).

2 Bernoco, D.; Bernoco, M.; Ceppellini, R.; Poulik, M.D.; van Leevwen, A.; van Rood, J.J.: Tissue Antigens *8:* 253–261 (1976).

3 Ceppellini, R.: in Amos, Progress in immunology, p. 973 (Academic Press, New York 1971).

4 Ferrara, G.; Tosi, R.; Longo, A.; Castellani, A.; Viviani, C.; Carminati, G.: J. Immun. *121:* 731–735 (1978).

5 Gates, F.T., III; Coligan, J.E.; Kindt, T.J.: Biochemistry, N.Y. *18:* 2267–2272 (1979).

6 Jerne, N.K.: Eur. J. Immunol. *1:* 1–9 (1971).

7 Kostyu, D.D.; Cresswell, P.; Amos, D.B.: Immunogenetics *10:* 433–442 (1980).

8 Krangel, M.S.; Orr, H.T.; Strominger, J.L.: Scand. J. Immunol. *11:* 561–571 (1980).
9 Mayr, W.R.; Bernoco, D.; De Marchi, M.; Ceppellini, R.: Transplant. Proc. *5:* 1581–1593 (1973).
10 Olsson, L.; Kaplan, H.: Proc. natn. Acad Sci. USA (in press, 1980).
11 Ploegh, H.L.; Cannon, L.E.; Strominger, J.: Proc. natn. Acad. Sci. USA *76:* 2273–2277 (1979).
12 Richiardi, P.; Carbonara, A.O.; Mattiuz, P.L.; Ceppellini, R.: in Dausset, Colombani, Histocompatibility testing 1972, pp. 455–464 (Munksgaard, Copenhagen 1973).
13 Richiardi, P.; Pellegrino, M.A.; Ferrone, S.: Transplantation *28:* 333–338 (1979).
14 Richiardi, P.; Amoroso, A.; Crepaldi, T.; Ceppellini, R.; Trucco, M.: Immunogenetics (in press, 1980).
15 Tanigaki, N.; Pressman, D.: Transplant. Rev. *21:* 15–34 (1976).
16 Tragardh, L.; Rask, L.; Wiman, K.; Fohlman, J.; Peterson, P.A.: Proc. natn. Acad. Sci. USA *76:* 5839–5842 (1979).
17 Trucco, M.M.; Stocker, J.W.; Ceppellini, R.: Nature, Lond. *273:* 666–667 (1978).
18 Trucco, M.M.; Garotta, G.; Stocker, J.W.; Ceppellini, R.: Immunol. Rev. *47:* 219–252 (1979).
19 Trucco, M.M.; De Petris, S.; Garotta, G.; Ceppellini, R.: Human Immunol. *3:* 233–243 (1980).

Dr. Massimo Trucco, The Wistar Institute, Thirty-Sixth Street at Spruce, Philadelphia, PA 19104 (USA)

The Immune System, vol. 1, pp. 226–233 (Karger, Basel 1981)

Polyclonal and Monoclonal Xenoantibodies to Polymorphic Determinants of Human Histocompatibility Antigens

M. Belvedere[1], *M.A. Pellegrino, S. Ferrone*

Institute of Medical Genetics, University of Torino, Torino, Italy; Department of Molecular Immunology, Scripps Clinic and Research Foundation, La Jolla, Calif., USA

In humans, like in other animal species, gene products of the major histocompatibility complex display a high serological polymorphism which has been characterized with operationally specific alloantisera. Although these reagents are available in limited quantities and contain low titer and affinity antibodies, there have been limited attempts to produce xenoantisera to allotypic specificities of HLA antigens [for review see 8]; furthermore, the resulting xenoantisera have been regarded with a certain suspicion and distrust in the HLA field, are not part of the library of sera used for HLA typing and with few exceptions have not been included among the sera studies during national and international histocompatibility workshops. The skepticism towards HLA xenoantisera stems from the 'dogma' that the major antibody population in sera elicited by immunization across a species barrier are directed to common determinants of HLA antigens and therefore the sera are of no value for HLA typing. However, the development of the hybridoma technique to prepare monoclonal antibodies has stimulated interest in xenoimmunization as a viable approach to prepare reagents to allospecificities of HLA antigens; this is best illustrated by the fact that the next international histocompatibility workshop will address itself to the characterization of monoclonal xenoantibodies to HLA antigens.

[1] *Mariaclara Belvedere* has, on several occasions, worked as a visiting scientist at the Basel Institute for Immunology.

In this paper we will summarize the experimental evidence proving that xenoantisera to HLA alloantigens have been prepared, discuss the variables which appear to influence the production of xenoantibodies to allotypic determinants of HLA antigens and present some of the information provided by HLA xenoantibodies.

Polyclonal and Monoclonal Antibodies to Allotypic Specificities of the HLA-A, B and Ia-Like Antigenic System

Following early unsuccessful attempts, operationally specific antisera to HLA-A, B allotypic specificities have been elicited in monkeys, goats, rabbits, and mice [for review see 8]. Some xenoantisera (i.e. 1828 and 167) are specific for subtypic specificities indicating that the discriminatory power of xenoantisera can be as high as that of alloantisera. Representative examples are shown in table I. It is of interest that most of the false positive reactions occur with lymphocytes carrying cross-reacting allospecificities: for instance anti-HLA-A9 xenoantisera lyse HLA-A2-bearing lymphocytes and anti HLA-B15 xenoantisera lyse HLA-B17-bearing lymphocytes. This phenomenon may reflect a higher affinity of xenoantibodies than of alloantibodies for the cross-reactive alloantigen and/or the increased sensitivity of the lymphocytotoxic test performed without human serum components that bind natural antihuman lymphocyte antibodies present in rabbit complement and thus decrease its lytic efficiency [2]. In view of the skepticism about the specificity of HLA xenoantisera, it is noteworthy that most of our xenoantisera have been tested by several investigators who have confirmed the specificity of our reagents; in particular our goat anti-HLA-B15 antiserum has been included among the sera tested during the VIII International Histocompatibility Workshop and the results of the assays performed by the participating laboratories against a large panel of donors (table I) have confirmed the specificity of this xenoantiserum [unpublished results].

The data on xenoantisera to human Ia-like antigens are limited since this antigenic system was identified only a few years ago; and the development of the hybridoma technique has changed the emphasis to preparation of monoclonal antibodies. However, xenoantisera to allotypic specificities of this system have been described [for review see 8]; three of our xenoantisera have been tested during the II American Histocompatibility Workshop and were shown to display a reaction pattern similar to that of DR alloantisera.

Table I. Specificity of sera from animals immunized with human × murine somatic hybrid cells and with serum HLA antigens

Xenoantiserum					Xeno-/alloantisera reactivity				r	p	HLA specificity[g]
code	animal	immunogen	adsorption	dilution	+/+	+/–	–/+	–/–			
0745	rabbit	Raji × IT22[a]	RPMI 4098	1:10	18	3	1	240	0.89	$<10^{-10}$	B27
0746	rabbit	Raji × IT22[b]	RPMI 1788	1:5	47	4	5	206	0.89	$<10^{-10}$	A3
0806	rabbit	M21 × IT22[c]	none	1:1	37	4	9	166	0.81	$<10^{-10}$	A3
1595	rabbit	M10 × CL1D[d]	none	1:10	108	3	3	202	0.96	10^{-10}	A2
1828	rabbit	serum[e]	none	1:10	81	6	6	209	0.90	$<10^{-10}$	Aw24
2958	rabbit	serum[e]	none	1:10	53	3	2	182	0.94	$<10^{-10}$	A9
167	goat	serum[f]	human leukocytes	1:1	564	138	59	3,977	0.82	$3\cdot10^{-10}$	Bw62

[a–d] Hybrid cells (code No. respectively 60-2, 60-3, 74-15, 4DJM 17s12) were obtained by fusion of murine fibroblasts IT22 with cultured human lymphoid cells Raji (HLA-A3) ([a, b]) and with cultured human melanoma cells M21 (HLA-A3, A9, B5) ([c]) and of murine fibroblasts CL1D with cultured human melanoma cells M10 (HLA-A9, A10, B5, B12) ([d]).

[e, f] These immunogens were serum fraction obtained by ion exchange chromatography on Sephadex-QAE 50 ([e]) and serum high-density lipoproteins obtained by KBr flotation ([f]) from normal donors No. 5 (A9, A28, B12, Bw15) and S.G. (A2, A9, B5), respectively.

[g] The HLA specificity was established in our laboratory and in the laboratories of Drs. *F. Bach* and *P.I. Terasaki* (USA) (HLA-A3 and B27), *E.S. Curtoni* (Italy), *F. Bach* and *P.I. Terasaki* (HLA-A2), *D.B. Amos, D. Cross, R. Duquesnoy, G. Rodey, J.J. van Rood, R. Walford* and *E. Yunis* (USA) (HLA-Aw24), and *P. Richiardi* (Italy) and 8th International Histocompatibility Workshop (HLA-Bw62).

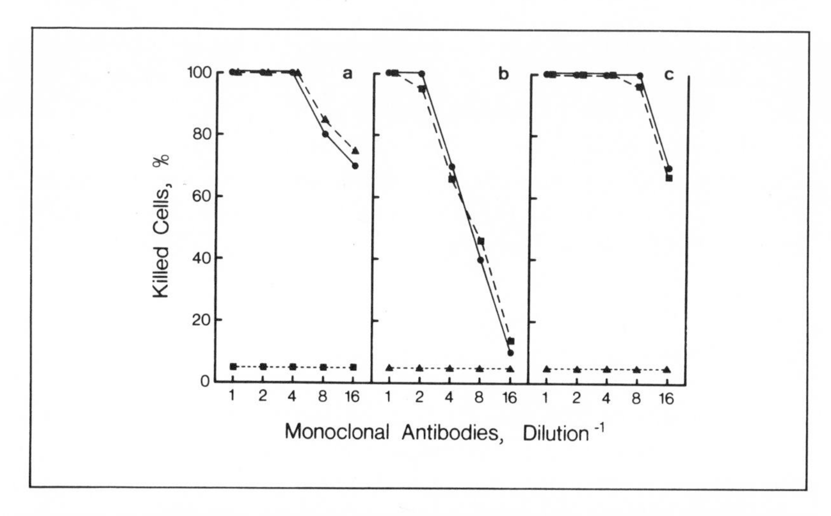

Fig. 1. Susceptibility to complement-dependent lysis of human B lymphoid cells treated with Fab_2 fragments from rabbit antisera to framework determinants of HLA antigens. Cultured human B lymphoid cells Victor were incubated with Fab_2 fragments from normal rabbit serum (●), rabbit antiserum 8492 (anti HLA-A, B, C framework determinant(s) (▲), and rabbit antiserum 8802 (anti Ia-like framework determinant(s) (■). Cells were then used in the standard microcytotoxicity test as targets of monoclonal antibodies anti rat histocompatibility antigens HPA (A), K.2.3.17 (B) and K.2.7.2 (C).

More recently, monoclonal antibodies to HLA antigens have been developed [3, 4]. The majority recognize determinants which had already been identified with polyclonal allo- and xenoantibodies, i.e. framework, common to all specificities coded for by one locus, public and private determinants. Furthermore, a certain number of monoclonal antibodies react with determinants expressed on subpopulations of molecules carrying allotypic specificities and defining a heterogeneity which had not been identified with conventional allo- and xenoantisera because of the complexity and heterogeneity of their antibody populations. Thus, for instance, our anti HLA-A,B monoclonal antibody Q1/28 binds about 70% of the molecules carrying the allospecificities HLA-A9 or B5 [5] and our anti-Ia-like antigen monoclonal antibody Q2/70 binds about 40% of the molecules carrying the allospecificities DR1, 2, 3, 4, 7, DRw6 or DR4 × 7 [6]. This type of heterogeneity does not appear to reflect a post-translational event since the monoclonal antibodies which define it react with histocompatibility antigens isolated from lymphoid cells treated with tunicamycin [unpublished results], an inhibitor of N-asparagine-linked glycosylation. Finally monoclonal antibodies elicited with cells from other animal species can recognize determinants expressed on HLA molecules, as indicated by the susceptibility to specific blocking by Fab_2 fragments from xenoantisera to human β_2-μ or to human Ia-like antigens (fig. 1).

Discussion

The available data indicate that antisera to allotypic specificities of HLA antigens can be prepared by immunization across a species barrier. The majority of the xenoantisera have become specific to allotypic determinants of HLA antigens following absorptions with cells with an appropriate HLA phenotype; however, some xenoantisera did not require any absorption or acquired specificity to HLA-A,B alloantigens following absorption with cells lacking this type of antigen (i.e. red blood cells, Daudi cells) or to Ia-like alloantigens following absorptions with cells lacking this type of antigen (i.e. human red blood cells, T lymphoid cells in long-term culture) [for review see 8]. The latter findings suggest that certain xenoantisera do not contain significant amounts of antibodies to framework determinants of human histocompatibility antigens and that the major antibody populations react with allotypic determinants. How can one explain these results? Histocompatibility antigens are highly conserved through evolution, as indicated by the high degree of structural homology and by immunological cross-reactivity among histocompatibility antigens from phylogenetically distant animal species. Therefore, it is our contention that in certain xenoimmunizations the antigenic disparity between the HLA antigens used as immunogens and the histocompatibility antigenic profile of the animal being immunized may mimic the degree of incompatibility which occurs in allogeneic combinations. Xenoimmunizations under these conditions may result in antisera directed to allotypic specificities of human histocompatibility antigens. Another important variable influencing the type of antibody response is the characteristics of the immunogen: membrane-bound antigens and soluble antigens concentrated in a high density lipoprotein milieu are efficient in eliciting xenoantibodies to allotypic specificities of the HLA system [for review see 8]. On the other hand, HLA antigens purified by conventional biochemical procedures or by affinity chromatography on

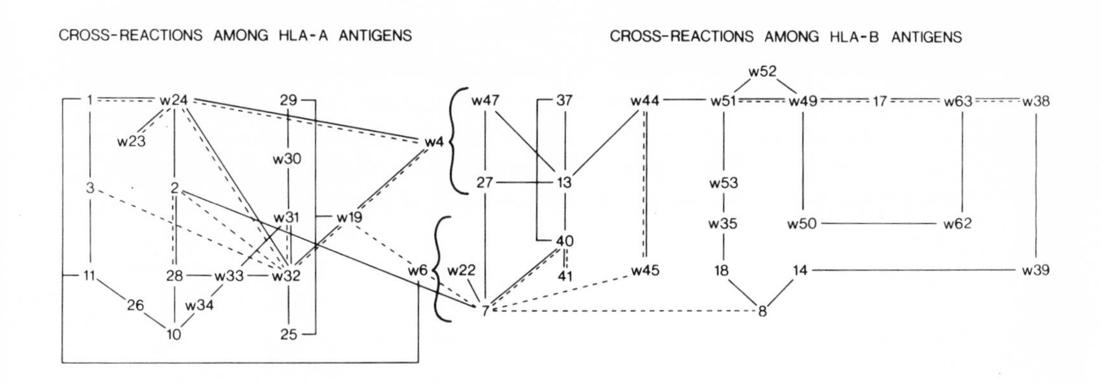

Fig. 2. Patterns of cross-reactivities of HLA-A, B antigens as detected by alloantibodies (–) and by xenoantibodies (––).

xenoantibodies bound to immunoadsorbents are efficient in eliciting xenoantibodies to framework determinants of HLA antigens [for review see 8]. HLA antigens during purification may acquire a molecular configuration which favors the immunogenicity of framework determinants over that of allotypic determinants. The degree of purity of the antigen preparations used as immunogens does not appear to play a major role in the successful development of xenoantisera to allotypic specificities of HLA antigens, since such reagents have been developed in rabbits with the high density lipoprotein fraction isolated from 3 *M* KCl extracts of cultured B lymphoid cells and from serum which contains several components besides HLA antigens and with human-mouse hybrids which still contain several human chromosomes besides the one carrying the human MHC.

Have HLA xenoantibodies provided any useful information which could not be obtained with alloantibodies and which justifies the efforts required to develop them? Because of their high affinity, HLA xenoantibodies have facilitated the use of immunochemical techniques to define the structural profile of these antigens and their relationship to other components of the immune system. Thus, an anti HLA-A9 xenoantiserum has been successfully used to purify HLA-A9 antigens which have been analyzed for their N-terminal amino acid sequence [1]. Monoclonal antibodies have been used to purify HLA-DR antigens from the pool of human Ia-like antigens [7]; peptide mapping analysis of the α and β chain of HLA-DR antigens and of Ia-like antigens encoded by a locus other than HLA-DR has revealed significant differences in both subunits of human Ia-like antigens [unpublished results], therefore indicating that the structural organization of Ia-like antigens is different from that of HLA-A, B, C antigens in which the nonpolymorphic subunit, i.e. β_2-microglobulin, is common to the alloantigens coded for by the three loci of the HLA-A, B, C system. Xeno-immunization has provided antibodies reacting with determinants which are poorly immunogenic or occur rarely or not at all as incompatibilities in allogeneic combinations. These xenoantibodies have detected cross-reactivities between HLA allospecificities which had not been previously identified with alloantisera; the results are summarized in figure 2. Furthermore, xenoantibodies, especially monoclonal antibodies, have identified cross-reactions of human histocompatibility antigens with their counterparts in other animal species, thus providing a rapid and simple approach to identify domains of histocompatibility antigenic structures which have been conserved through phylogenetic evolution and are therefore likely to be essential for the function of these molecules.

Conclusion

Xenoimmunization in conjunction with the hybridoma technique is likely to develop large amounts of HLA antibodies with high affinity which will overcome the limitations imposed by the shortage and low affinity of alloantisera on the immunochemical characterization of HLA antigens. Furthermore and more importantly, xenoimmunization may enlarge the specificities of antibodies and provide reagents directed to determinants not recognized by alloantibodies. Development of a large library of xenoantibodies will facilitate the mapping of immunological domains on histocompatibility antigen molecules and will define the role of these domains in disease susceptibility, allograft rejection and cell-cell interactions.

Acknowledgements. This work was supported by NIH Grants•A1 13154, CA 16071 and CA 24329, a Research Career Development Award *(M.A.P.)* and an American Heart Association Established Investigatorship Award *(S.F.).* This is publication 2301 from Scripps Clinic and Research Foundation. The authors wish to acknowledge the skilled secretarial assistance of *Ellen Schmeding.*

References

1 Allison, J.P.; Ferrone, S.; Walker, L.E.; Pellegrino, M.A.; Silver, J.; Reisfeld, R.A.: Transplantation *26:* 452–454 (1978).

2 Ferrone, S.; Cooper, N.R.; Pellegrino, M.A.; Reisfeld, R.A.: J. Immun. *107:* 939–947 (1971).

3 Moller, G. (ed.): Immunol. Rev. *47:* 1–252 (1979).

4 Quaranta, V.; Pellegrino, M.A.; Ferrone, S.: J. Immun. (in press).

5 Quaranta, V.; Walker, L.E.; Ruberto, G.; Pellegrino, M.A.; Ferrone, S.: Submitted.

6 Quaranta, V.; Tanigaki, N.; Ferrone, S.: Immunogenetics (in press, 1980).

7 Quaranta, V.; Walker, L.E.; Pellegrino, M.A.; Ferrone, S.: J. Immun. *125:* 1421–1425 (1980).

8 Wilson, B.S.; Ng, A.K.; Quaranta, V.; Ferrone, S.: in Reisfeld, Ferrone, Current trends in histocompatibility (Plenum Press, New York, in press).

M. Belvedere, MD, Institute of Medical Genetics, University of Torino,
Via S. Epifanio, 14, I–10100 Torino (Italy)

The Immune System, vol. 1, pp. 234–237 (Karger, Basel 1981)

The Impact of Transplantation Surgery on the Development of Immunology

H.D. Brede[1]

Paul-Ehrlich-Institut, Frankfurt am Main, FRG

Immunology as a biological discipline developed in phases. Its roots lead back into the mystical irrational world of ancient medicine. The first scientific approaches can be traced back into the 18th century in Europe, propagated enthusiastically by Lady *Montagu.* These ideas influenced the *Hunter* brothers and their pupil *Jenner,* leading to the introduction of vaccination.

The second step of scientifically sound approaches followed during the era of *Louis Pasteur.* The breakthrough was achieved by *Paul Ehrlich* and his contemporaries, *Behring, Bordet, Metchnikow, Roux,* and *Landsteiner.* At the end of World War I the scene of activity shifted to the United States, mainly to the Rockefeller Foundation, where there were many developments in blood group serology, specific antibodies, and serotherapy. Only outsiders, like the Japanese research worker *Shinoi,* kept themselves busy with questions related to cellular immunology, and this, in the case of *Shinoi,* in connection with transplantation experiments.

The fascinating development of chemotherapy, mainly of sulfonamides, followed by antibiotics, forced immunology for some decades into the position of a scientific subject of secondary importance. Then, rather unexpectedly, the sudden activation of transplantation surgery surprised immunologists and found them mostly unprepared to fulfill their assignments. I myself was caught in such a position in the area of Cape Town in 1967, and I would like to report briefly about the impact of transplantation surgery on the development of immunology. Since the mid-sixties kidney transplants had gained more and more general acceptance; this was the

[1] *H.D. Brede* is the successor of *Niels Jerne* as president of the Paul-Ehrlich-Institut in Frankfurt.

reason for the foundation of a working group concerned with kidney transplants, and all that goes with such an activity, at the University of Stellenbosch, north of Cape Town, in cooperation with the Johns Hopkins University in Baltimore. I was one of the members of this group in my position as the head of clinical microbiology of the medical faculty of the University of Stellenbosch. In cooperation with other disciplines concerned with transplantation, we started a baboon colony, arranged for traps in different areas of the Cape Province, and pedaled happily along in experimental surgery. We tried different preparations of antisera, studied complement systems, used irradiation and immunosuppressive drugs, and found ourselves involved mainly in theoretical approaches to gain insight into the mechanism of rejection, tolerance, and tissue typing. In December 1967 – quite unexpectedly – one morning we received the news that *Chris Barnard,* at that time an associate professor of surgery at the neighboring University of Cape Town, had performed the first heart transplant in a human being. From then on, our scientific work was confronted with many unexpected effects, which resulted in hectic developments in immunology.

On December 3, 1967, *Barnard* transplanted the heart of a young woman, who was fatally injured in a street accident, into a recipient (Washkanski) with a cardiac ailment of long standing. The kidneys of the donor were offered to our group and hastily transported in a plastic bag with some ice cubes to our hospital. They were transplanted into a child experiencing complete kidney failure. The child was being kept alive by the use of artificial kidneys. The problems of the kidney transplant recipient could be foreseen. He was treated with cortisone and azathioprine. Tissue typing was only performed after both operations had been finished. Although the kidney transplantation did not cause unforeseen difficulties, the first human cardiac transplantation kept everyone alert and was accompanied by much excitement. First of all the recipient was treated with cortisone, azathioprine, irradiation, and actinomycin D. The patient developed secondary infections, mainly *Pseudomonas,* and died 18 days after the operation. Shortly afterwards the second human cardiac transplantation followed. This time the recipient (Dr. Blaiberg) survived for more than 1 year. Activities on the medical laboratory level were tremendous and ranged from various biochemical investigations to the preparation of anti-lymphocytic sera. The whole world was taking part; one conference followed another. Research work concentrated on methods for detecting rejection and on the study of immune complexes, tissue typing, tolerance induction, and organ preservation.

Our own group turned out more than 40 papers within 12 months. They dealt with our baboon research colony, different methods of immunosuppression, antigen determination, immunoglobulins, metabolic alterations, organ perfusion, organ storage with oxygen and helium, significance of blood groups in the survival of untreated baboon renal allotransplants, and surgical techniques, to name only the main fields of activity [1–3, 6]. These were only the activities of one group. The whole scientific world was influenced, and the first International Congress of Immunology in 1971 was a demonstration of the explosive development of transplantation immunology. Few scientific disciplines have undergone the rapid evolution that immunology has experienced since the formulation of the theories of natural and clonal selection, the induction of tolerance, the development of organ and tissue transplantation, and, related to it, the renaissance of tumor immunology. *Jerne* [4] provided convincing support for selective theories of immunity and created new techniques to prove them.

Over the last 20 years, clinical transplantation has been based on the idea that transplantation antigen itself is the major barrier to the grafting of foreign tissues. Therefore, tissue typing and host immunosuppression have been seen as the main rational means of encouraging allograft acceptance by an immunocompetent host. Now, overriding emphasis is placed on the way antigen is presented to the recipient's immune system. Transplantation antigen will only be highly immunogenic when presented to recipient T cells on the surface of donor stimulator cells of lymphoreticular origin. Tissue parenchymal cells, even when they express transplantation antigens, do not behave as strong T cell immunogens. A detailed theoretical explanation of this effect has been presented by *Lafferty and Woolnough* [5]. According to this theory the barrier to tissue allotransplantation is not simply transplantation antigen, but rather antigen carried on the surface of active stimulator cells, i.e. blood cells, passenger leukocytes. Therefore, it should be possible to facilitate allotransplantation by treatments directed at washing out or eliminating such cells before grafting. We have already achieved better results by dipping and rinsing baboon kidneys in heparin solutions. Blood elements within the graft constitute a major barrier to tissue transplantation. It is possible and advisable to eliminate the immunogenicity of a foreign graft by treatment of the tissue rather than the recipient.

From the universally disappointing results of intestinal and lung transplants with their high load of lymphoid cells to the frequently successful transplantation of kidney, heart, and liver, success is inversely correlated with the passenger lymphocyte load. Spectacular successes may be expected from organ culture techniques, especially in the case of thyroid transplantation and of pancreatic islets.

Recapitulating the last 14 years of transplantation history, there has been an extraordinary rapid progress. When I remember my own first struggles to detect transplant rejections apart from obvious functional failures, I tried to achieve results with the measurement of hetero-agglutinins, hetero-hemolysins, and changes in the complement system [2]. Today an overwhelming number of tests are theoretically available. The blastogenic, the cytotoxic, and the macrophage migration inhibition tests have been employed most often in the immunodiagnosis of organ allografted patients. The blastogenic test has been the least reliable. Direct lymphocyte-dependent target cell cytotoxicity assays have been more successful. The most widely applied in vitro technique and the most successful in humans has been the macrophage migration inhibition system.

Not only transplantation immunology but also tumor immunology, immunogenetics, preservation techniques, prospective immunodiagnosis, etc. have benefited from basic research in immunology. *Jerne* was and is one of the most outstanding pioneers in the development of basic immunology.

References

1 Baboons in Organ Transplantation Research: S. Afr. med. J. *17:* suppl., August (1968).
2 Brede, H.D.; Murphy, G.P.: S. Afr. med. J. *42:* special suppl., pp. 22–26 (1968).
3 Groenewald, J.H.; Brede, H.D.; Weber, H.W.; Murphy, G.P.: Medical Primatology 1970. Proc. 2nd Conf. Exp. Med. Surg. Primates, New York 1969, pp. 153–164 (Karger, Basel 1971).
4 Jerne, N.K.: Proc. natn. Acad. Sci. USA *41:* 849 (1955).
5 Lafferty, K.J.; Woolnough, J.: Immunol. Rev. *35:* 231 (1977).
6 Murphy, G.P.; Weber, H.W.; Brede, H.D.; Retief, F.P.; Retief, C.P.; Zyl, J.A. van; Zyl, J.J.W.: Am. Surg. *35:* 292–300 (1969).

Prof. H.D. Brede, Paul-Ehrlich-Institut, Paul-Ehrlich-Strasse 42–44,
D-6000 Frankfurt am Main (FRG)

The Immune System, vol. 1, pp. 238–244 (Karger, Basel 1981)

Membrane Biology and the B Lymphocyte

Charles L. Sidman [1]

Basel Institute for Immunology, Basel, Switzerland

This paper, written for *Niels Jerne,* discusses three areas of membrane biology as they apply to the B lymphocyte.

Membrane Receptors and Architecture

One advantage in studying the B lymphocyte is that several of the membrane receptors which control the cell's function are well-known and molecularly defined. These are surface immunoglobulin (sIg) molecules and class II major histocompatibility complex (MHC) antigens (Ia) [21]. (In comparison, no T lymphocyte membrane receptors are generally agreed upon.) Both Ig and Ia are multi-chain glycoproteins.

Membrane molecules can be classified as either 'peripheral' (not inserted directly into the lipid bilayer) or 'integral' (inserted into the lipid bilayer). Integral membrane proteins may be further subdivided into those which are exposed at only one face of the membrane (either external or cytoplasmic) and those which extend on both sides. Signal transmission from the exterior to the interior of a cell is thought to require the latter, or 'transmembrane', class of integral membrane structures.

Surface Ig heavy chains (both μ and δ) have been shown to be integral membrane proteins while the light chains are peripheral [29]. Nucleic acid and protein studies of μ [15, 25] indicate an amino acid sequence compatible with transmembrane orientation, but a cytoplasmic extension has not been experimentally demonstrated [34]. Even if the orientation proposed from the sequence data is accepted, only two amino acids would be exposed to the cytoplasm [15,25]. This would appear insufficient to define specific

[1] *Charles Sidman* has been a member of the Basel Institute for Immunology since 1978.

interactions with cytoplasmic components (see below), and suggests that associations with other transmembrane molecules, perhaps within the membrane, may be important in the signalling function of sIg molecules. We might speculate further that lipophilic mitogens such as LPS may influence normal lymphocyte triggering mechanisms at just such an intramembrane level of molecular interactions.

The other well-defined class of B cell surface receptors, Ia antigens, is composed of three different polypeptide chains [14]. Two, the α and β chains, show variation among haplotypes, presumably carry the actual receptor combining site(s), and are integral and transmembrane polypeptides [14, 29, 34]. The third Ia chain is the i (for invariant) polypeptide. It does not show variation among haplotypes, is very abundant within the cell, and is weakly and non-covalently associated with the other two chains [14, 22]. Radiolabelling studies have failed to show the i polypeptide as either integral or external-peripheral [29]. However, preliminary proteolysis results [24] may indicate an external disposition of the i chain, which would be consistent with its carbohydrate content [22]. Suggestions as to its function have included roles in signal transmission through the membrane (if internal), or in biosynthesis and membrane deposition (if external).

A final, and already alluded to, aspect of membrane architecture in B cells is the existence of specific associations among membrane components within the plane of the lipid bilayer. The first such demonstration was that complexed sIg molecules bind F_c receptors, but that complexed F_c receptors do not bind free sIg molecules [1]. More recently, an intricate hierarchy of affinities among IgM, IgD, F_c receptors, and LPS receptors has been reported, which changes with the B cell's state of activation [11]. Numerous papers also report an association between class I MHC antigens and viral membrane components [8, 18]. Associations among membrane components will be an active field of research in the future in terms of both cell triggering and immune regulation.

Signal Transmission and Cell Triggering

On an overall cellular level, B cell activation appears to involve two stages of interactions. The first is with sIg molecules; the second, normally involving cooperation with T cells, is where Ia (and other) interactions occur [21]. In B cells [and apparently T cells as well; 19], the first phase of interactions generates or activates a membrane receptor to receive the second round of signals [17].

When multivalent ligands interact with sIg molecules under physiological conditions, a complex and striking series of reactions occurs within minutes [27]. Ligand-receptor complexes coalesce into small microaggregates, then into a large 'cap' at one end of the B cell. The cell undergoes a series of contractile, and motile, episodes. Finally, the surface complexes are both shed and endocytosed. Several hours later, sIg molecules reappear on the B cell membrane.

Except for the motility response, this series of reactions is a perfect example of the general membrane phenomenon of 'receptor-mediated endocytosis' (RME) [13, 33]. As in all other examples of RME, there are a finite (and saturable) number of cellular receptors (here, sIg molecules), which serve to concentrate the target ligand onto the cell. After binding, ligand-receptor complexes are gathered into clathrin 'coated pits' and endocytosed in 'coated vesicles' [26].

What functions might this series of membrane reactions play in B cell triggering? A direct linkage between these early surface events and later activation and differentiation has not been established. However, one consequence of the early events, in and of themselves, is to polarize and orient the B cell, thus allowing directed chemotactic movement in a ligand gradient [35]. This might play a role in lymphocyte homing to antigen in vivo. Another suggestion [31] is that the capping/clearing process might be a means for the B cell to evaluate a potential triggering event. Most functional B cell responses require the presence of ligands through several cycles of receptor exposure/removal. Thus, if the B cell can terminate and remove a signal by a single such cycle, that signal is not an effective inducer. Accessory cells may function in part by presenting non-removable, multivalent, membrane-bound ligands for B cell induction. Finally, many hormone or metabolite receptors undergo 'down regulation' or 'biosynthetic feedback' [13, 33]. The first is when a cell becomes unresponsive to ligands due to the clearance of receptors from its membrane; the second is when repeated receptor cycles eventually lead to a diminution in the rate of receptor synthesis. Normal B lymphocytes temporarily undergo the former, and easily tolerized immature B cells are susceptible to both [30].

Further comparison of the ligand-induced B cell cycle and other examples of RME [13, 33] suggest additional questions for those studying the B lymphocyte. One such point is the intracellular fate, and possible function, of ligand which is internalized complexed to sIg. In several other RME systems, internalized ligand is utilized for further signalling or metabolic purposes. We might ask whether, within the B cell, any antigen escapes degradation to affect later cellular function. A second and related question is the fate of the sIg molecules themselves. Other RME receptors recycle from plasma membrane to coated vesicle to lysosome and back again on the order of 10^4 times per molecule. No studies on the fate of the sIg itself, rather than the ligand, have addressed this point.

One of the first intracellular consequences of membrane interactions is the physical association of particular membrane structures and cytoskeletal elements (such as actin and myosin) after ligand binding to the former. This has been demonstrated both biochemically [10] and by double immunofluorescence [4]. Again, the relationship of the membrane-cytoskeletal linkages to functional cell activation is only conjectural at this time. The occurrence of cytoskeletal linkages [and their pharmacologic disruption; 5] serves to distinguish two classes of membrane molecules, however. Surface Ig, F_c receptors, and thymocyte TL antigens all respond actively to ligand binding and form cytoskeletal linkages, while class I MHC and Thy-1 antigens do not. Since functional receptor roles have been shown for B cell sIg and F_c receptors, such a role also seems possible for TL antigens on immature T cells.

Another rapid intracellular consequence of specific ligand-receptor interactions is the release of Ca^{++} ions stored within the B cell, perhaps on the inner face of the plasma membrane, into the cytoplasm [7]. This may be the actual trigger for the cytoskeletal linkages and movements, and could control various enzymes too.

Changes in at least two enzyme systems have been shown after ligand-sIg interactions. One is an esterase [3] which is essential for cell motility but not for capping, and which may play a role in further signal transmission to the cell nucleus (see below). A second enzyme activity which can be inferred after sIg interactions is a methylase, as methylation-inhibiting drugs block capping at a point distal to Ca^{++} mobilization [6].

Finally, how might signals at the B cell membrane be transmitted to the nucleus? As a criterion for such transmission, the phosphorylation of non-histone nuclear proteins has been used [23]. This occurs after anti-Ig treatment of intact cells, and can be duplicated in the isolated nuclei of non-anti-Ig-treated cells by exposure to the cytoplasmic contents of anti-Ig-treated cells. The responsible cytoplasmic factor is a protein, which is activated from a precursor by an induced membrane protease [16]. At present, this concept of a '2nd messenger' protein activated by a plasma membrane protease is the best model of how membrane interactions might be transmitted to the B cell nucleus.

Membrane Protein Biosynthesis

The control of the expression and intracellular traffic of different classes of cellular proteins are important issues in cell biology today. Most splenic B cells bear membrane receptor IgM [32] and are activatable to IgM secretion [2]. From physical and sequence data on various protein and nucleic acid species [for review see 28], it is now clear that membrane (μ_m) and secretory (μ_s) μ chains (and other Ig heavy chain classes as well) are different polypeptide species. In the case of μ chains, the difference is due to the two proteins having different C-terminal portions. Approximately 95% of the molecules are identical, however.

I have studied the biosynthesis of these two polypeptides [28]. Both are translated with cleavable leader peptides, as are most other proteins synthesized on membrane-bound polysomes. Both are ASN-linked core-glycosylated from dolichol-lipid intermediates. After this stage the processing of the two forms diverges. Membrane μ appears to directly gain its final carbohydrate components and be placed in the cell membrane, while μ_s undergoes at least two other intermediate stages of glycosylation before finally being secreted.

The first major conclusions from these studies concern the levels of control on the cellular expression of membrane and secretory μ chains. The two μ mRNAs are thought to be normally generated by differential RNA splicing of larger primary transcripts, which contain coding for both of the alternative C-termini [9]. Thus, generation of the two mRNA species depends on the balance of the RNA splicing mechanism. It also clearly depends on the DNA sequences for the two alternative C-termini being present and transcribed. As the exon for the μ_m C-terminus is downstream from that for the μ_s terminus and the rest of the V and Cμ coding exons [25], it may be possible that the μ_m C-terminal exon could be deleted during differentiation and still leave expression of μ_s intact. Either of these mechanisms, therefore (imbalanced RNA splicing or a μ_m DNA deletion), could explain the situation in IgM-secreting hybridoma cells, which express the entire μ_s pathway, but totally lack that for μ_m at the protein level [28].

In contrast to the just-described pre-translational control of μ_m expression in hybridoma cells, resting immunocompetent normal B lymphocytes show another type of control over IgM expression [28]. These cells possess and translate mRNA for both types of μ chains, but only complete the pathway for μ_m. Secretory μ molecules appear to receive the core carbohydrates, but not to go through the further carbohydrate modifications leading to secretion. Only after activation do the further carbohydrate modifications, and actual secretion, occur. Resting B cells thus show post-translational control of μ_s expression.

Since carbohydrates are at least implicated in the post-translational control of μ_s expression, one might ask if the carbohydrates of μ_s and μ_m are in fact different. At one time [20], it was thought that μ_m entirely lacked the terminal sugars (galactose, fucose and sialic acid) which are present in μ_s. Now, however, we know that μ_m has at least some galactose and sialic acid [12]. Although the question is not yet definitively resolved, the biosynthetic intermediates give strong evidence of different carabohydrate modifications in μ_m and μ_s chains.

This raises the further issue of how the carbohydrate structures of the two μ chains are differentiated, which is particularly puzzling since the same amino acid residues appear to be glycosylated in all but one case [15]. One possibility is that the two μ forms are segregated into separate subcellular compartments, and that the observed glycosylation differences reflect different enzymes in the two compartments. Alternatively, the small difference in C-terminal amino acids might somehow influence the glycosylation of the large portion of polypeptide which is common to the two μ forms. It is clear that the glycosylation pattern itself does not determine the membrane vs secretory fate of the polypeptides, since the two chains show their different physical and membrane-attachment properties even when not glycosylated [28].

While ample precedent exists for the carbohydrate portion of both secretory and membrane glycoproteins exerting profound effects on the proteins' expression [28], we cannot yet say whether the different glycosylation patterns of μ_s in secreting vs non-secreting cells exert any direct control over μ secretion. They may just reflect some other process. Hopefully, work now in progress will illuminate the role of carbohydrate and other mechanisms, in the control of membrane and secretory IgM glycoproteins specifically, and in various classes of cell products in general.

References

1 Abbas, A.; Unanue, E.R.: J. Immun. *115:* 1665–1671 (1975).
2 Andersson, J.; Coutinho, A.; Lernhardt, W.; Melchers, F.: Cell *10:* 27–39 (1977).
3 Becker, E.L.; Unanue, E.R.: J. Immun. *117:* 27–32 (1976).
4 Braun, J.; Fujiwara, K.; Pollard, T.D.; Unanue, E.R.: J. Cell Biol. *79:* 409–418 (1978).
5 Braun, J.; Fujiwara, K.; Pollard, T.D.; Unanue, E.R.: J. Cell Biol. *79:* 419–426 (1978).

6 Braun, J.; Rosen, F.S.; Unanue, E.R.: J. exp. Med. *151:* 174–183 (1980).
7 Braun, J.; Sha'afi, R.I.; Unanue, E.R.: J. Cell Biol. *82:* 755–766 (1979).
8 Callahan, G.N.; Allison, J.P.: Nature, Lond. *271:* 165–167 (1978).
9 Early, P.; Rogers, F.; Davis, M.; Calame, K.; Bond, M.; Wall, R.; Hood, L.: Cell *20:* 313–319 (1980).
10 Flanagan, J.; Koch, G.L.E.: Nature, Lond. *272:* 278–281 (1978).
11 Forni, L.; Coutinho, A.: Nature, Lond. *273:* 304–306 (1978).
12 Goding, J.W.; Herzenberg, L.A.: J. Immun. *124:* 2540–2547 (1980).
13 Goldstein, J.L.; Anderson, R.G.W.; Brown, M.S.: Nature, Lond. *279:* 679–685 (1979).
14 Jones, P.P.; Murphy, D.B.; Hewgill, D.; McDevitt, H.D.: Immunochemistry *16:* 51–60 (1978).
15 Kehry, M.; Ewald, S.; Douglas, R.; Sibley, C.; Raschke, W.; Fambrough, D.; Hood, L.: Cell *21:* 393–406 (1980).
16 Kishimoto, T.; Kikutani, H.; Nishizawa, Y.; Sakaguchi, N.; Yamamura, Y.: J. Immun. *123:* 1504–1510 (1979).
17 Kishimoto, T.; Miyake, T.; Hishizawa, Y.; Watanabe, T.; Yamamura, Y.: J. Immun. *115:* 1179–1184 (1975).
18 Kvist, S.; Ostberg, L.; Persson, H.; Philipson, L.; Peterson, P.A.: Proc. natn. Acad. Sci. USA *75:* 5674–5678 (1978).
19 Larsson, E.-L.; Coutinho, A.: Nature, Lond. *280:* 239–211 (1979).
20 Melchers, F.; Andersson, J.: Transplant. Rev. *14:* 76–130 (1973).
21 Moller, G. (ed.): Immunol. Rev. *52* (1980).
22 Moosic, J.P.; Nilson, A.; Hammerling, G.J.; McKean, D.J.: J. Immun. *125:* 1463–1469 (1980).
23 Nishizawa, Y.; Kishimoto, T.; Kikutani, H.; Yamamura, Y.: J. exp. Med. *146:* 653–664 (1977).
24 Reske, K.; Rude, E.: Immunobiology *157:* 266–267 (1980).
25 Rogers, J.; Early, P.; Carter, C.; Calame, K.; Bond, M.; Hood, L.; Wall, R.: Cell *20:* 303–312 (1980).
26 Salisbury, J.L.; Condeelis, J.S.; Satir, P.: J. supramol. Struct. *4:* suppl., p. 203 (1980).
27 Schreiner, G.F.; Unanue, E.R.: Adv. Immunol. *24:* 37–165 (1976).
28 Sidman, C.L.: Cell *23:* 379–389 (1981).
29 Sidman, C.L.; Bercovici, T.; Gitler, C.: Mol. Immunol. *17:* 1575–1583 (1980).
30 Sidman, C.L.; Unanue, E.R.: Nature, Lond. *257:* 149–151 (1975).
31 Sidman, C.L.; Unanue, E.R.: J. exp. Med. *144:* 882–896 (1976).
32 Sidman, C.L.; Unanue, E.R.: J. Immun. *121:* 2129–2136 (1978).
33 Sly, W.S.: in Svenerholm, Mandel, Dreyfus, Urban, Structure and function of the gangliosides, pp. 433–451 (Plenum Press, New York 1980).
34 Walsh, F.S.; Crumpton, M.J.: Nature, Lond. *269:* 307–311 (1977).
35 Ward, P.A.; Unanue, E.R.; Goralnick, S.; Schreiner, G.F.: J. Immun. *119:* 416–421 (1977).

Dr. C.L. Sidman, Basel Institute for Immunology, Postfach, CH-4005 Basel (Switzerland)

The Immune System, vol. 1, pp. 245–251 (Karger, Basel 1981)

The Biophysics of Membrane Proteins: Vertical Displacement, Aggregation, Patching, and Cell-Cell Recognition

Donald F. Gerson[1]

Basel Institute for Immunology, Basel, Switzerland

Speaking on the past and future state of immunology at the gathering commemorating the 10th anniversary of the Friedrich Miescher Institute, in Basel, *Niels Jerne* described B and T cell interactions as they were then understood. He remarked that the cells must recognize one another by a means which had not yet been considered. His comment remained with me, and I offer this new model for membrane-protein and cell-cell interaction as a handsel to *Niels Jerne.*

Membrane proteins reside in a fluid, lipid layer having a thickness close to the dimensions of the proteins themselves. In order for a protein to be a stable constituent of such a phase, it must be lipophilic, or hydrophobic. For instance, membrane IgM has a long stretch of hydrophobic amino acids which attach to the membrane. Membrane proteins move laterally and coalesce on the membrane surface (patching and capping) following receptor-antigen or receptor-hormone interaction. Membrane proteins can also move in a direction normal to the membrane surface (vertical displacement), and this is most pronounced when the lipid composition of the membrane changes, thereby changing the relative hydrophobicities of the bulk membrane and the membrane proteins. Consideration of the quantitative measure of hydrophobicity, surface energy, leads to a simple model which has many of the properties associated with membrane proteins. Newly developed methods for measuring the surface and interfacial energies of biological materials using biphasic mixtures of poly(ethyleneglycol) and dextran [3–7] provide some data with which to test this model, and there is

[1] *Donald Gerson* has been a member of the Basel Institute for Immunology since 1979.

a qualitative correspondence between expectation and observation. In the following, I will describe first the model, and then several aspects of its behavior.

Surface Energies and the Vertical Location of Membrane Proteins

At every interface between immiscible phases, there is an associated interfacial free energy, γ, the units of which are erg/cm^2. There are 4 relevant interfaces when considering membrane proteins, each with an associated interfacial energy: γ_{mex} between the membrane and the exterior medium; γ_{pex} between the membrane protein and the exterior medium; γ_{pm} between the membrane and the membrane protein; and γ_{min} between the membrane and the interior of the cell. For simplicity, the protein is considered to be homogeneous and spherical, with radius *r*.

How do the relative hydrophobicities (interfacial energies) of the protein and membrane affect the vertical placement of the protein at the outer interface of the membrane? Qualitatively, if the protein is as hydrophobic as the membrane, or more so, it will be totally immersed in it; if the protein is much more hydrophilic than the membrane, it will be totally displaced from it. Between these extremes, the protein will lodge in the membrane, but protrude into the external (or internal) medium a little. Considering only the external medium and the membrane, the protein will reach an equilibrium position when the total surface energy at the protein-membrane interface equals the surface energy at the protein-external medium interface, with correction for the lost area of membrane-external medium interface. The equality of these surface energies gives the following expression (eq. 1), in which h is the height of that part of the

$$H = \frac{h}{r} = 1 + \frac{\gamma_{pex} - \gamma_{pm}}{\gamma_{mex}} \tag{1}$$

spherical protein protruding into the external medium [1]. If H is between 0 and 2 (i.e. if the protrusion is less than the diameter of the protein), the protein remains attached to the membrane surface. The larger the value of H, the more the protein protrudes. Negative values of H imply complete submersion of the protein into the bilayer, while high (>2) values of H imply detachment from the membrane. The calculation of H requires a value for γ_{pm}, which, unlike γ_{pex} and γ_{mex}, cannot be measured experimentally. The value of γ_{pm} is estimated from the correlation of *Neumann and Sell* [8] (eq. 2a):

$$\gamma_{mp} = \frac{\sqrt{\gamma_{pex^2} - \gamma_{mex^2}}}{1 - 0.015\sqrt{\gamma_{pex}\gamma_{mex}}} \tag{2a}$$

or from the more generally applicable relation developed in this laboratory (eq. 2b):

$$\frac{\gamma_{mex} + \gamma_{pex} - \gamma_{mp}}{2\sqrt{\gamma_{mex}\,\gamma_{pex}}} = \exp\left[\gamma_{mp}\,(0.00007\,\gamma_{pex} - 0.01)\right]. \tag{2b}$$

Figure 1 diagrams the relative vertical positions of proteins and membranes of various hydrophobicities (or values of γ_{pex} and γ_{mex}). Assuming here that the external medium is an aqueous solution, as would be the case for single cells, the higher the value of γ the more hydrophobic the structure. For instance, γ between water and oil is about 40 erg/cm^2, while γ between water and wet filter paper is about 10^{-2} erg/cm^2. The proteins A to D in figure 1 are increasingly hydrophobic (γ_{pex} = 0.1, 1.0, 3.0 and 10.0), and the 3 horizontal rows represent membranes of increasing hydrophobicity (γ_{mex} = 1, 2 or 10). In a relatively hydrophilic membrane (γ_{mex} = 1), the proteins with $\gamma_{pex} \geqslant 1$ (B, C, D) are submerged, and only the hydrophilic protein, A, rests at the membrane interface. As membrane hydrophobicity increases to 2 and 10 erg/cm^2, the proteins begin to protrude more and more.

Shinitsky [10] and *Shinitsky and Souroujon* [11] have demonstrated that the vertical protrusion of certain membrane proteins depends on the lipid content of the membrane. The D blood group antigens of Rh+ human red blood cells protruded from the cell membrane much more in membranes of high cholesterol: phospholipid ratios than in ones of low ratios. Measurements of lipid bilayers have demonstrated that membranes of relatively high cholesterol are much more hydrophobic than membranes of relatively high phospholipid composition. Thus, according to the model presented above, the relatively hydrophilic D antigen would be expected to rise out of the membrane to some extent. These results were attributed to alterations in microviscosity [10, 11]. Although microviscosity does change, there is no simple physical model which can causally relate viscosity, a resistance-to-flow parameter, to the equilibrium position of proteins in a

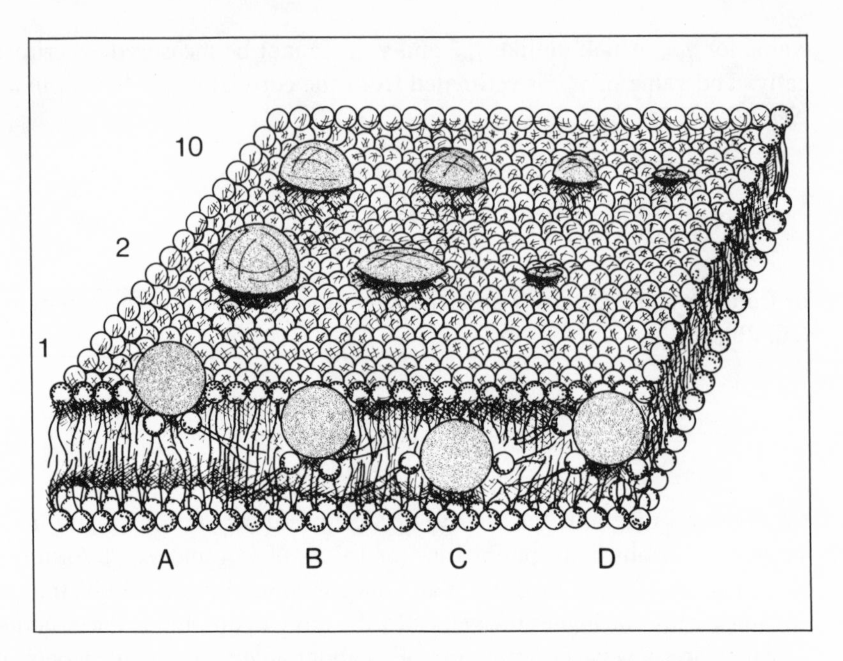

Fig. 1. Relative vertical positions of protein molecules in a cell membrane for various hydrophobicities of proteins (increasing from A to D) and of membranes (increasing from 1 to 10).

membrane. It is, however, undoubtedly true that alterations in the lipid composition of a membrane will lead to changes in both microviscosity and interfacial free energy, and thus the correlation observed by *Shinitsky* [10] and *Shinitsky and Souroujon* [11] is quite reasonable.

Surface Energies and the Aggregation of Membrane Proteins

Membrane proteins aggregate in response to a variety of triggering substances such as mitogens, antigens, and hormones [9]. Consideration of surface energies can also lead to an interesting model of this phenomenon.

The free energy of aggregation ΔGa, per area of contact, of like particles in a uniform medium is given by equation 3, where γ is the surface energy of the particles and Keq is

$$\Delta Ga = 2\gamma = -RT \cdot \ln (Keq), \quad (3)$$

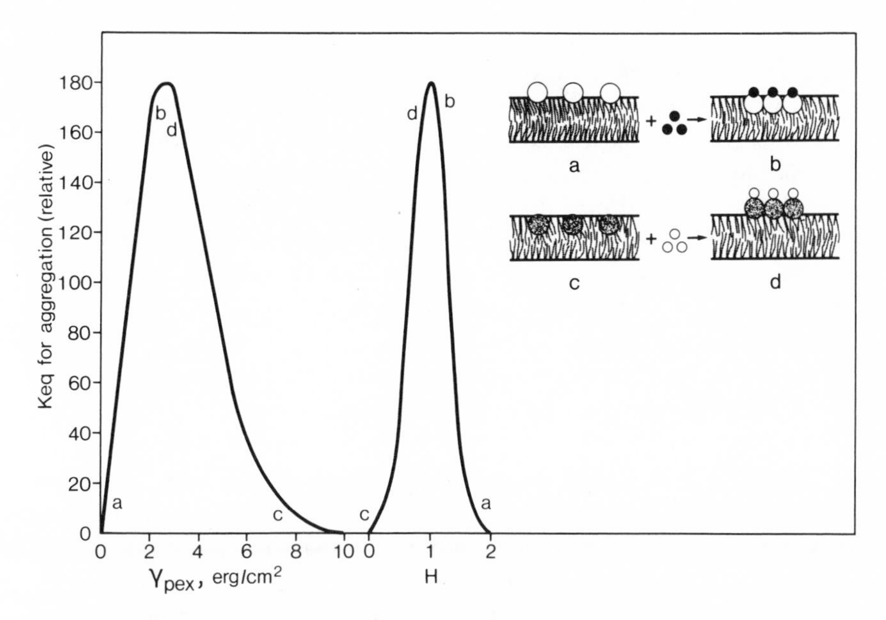

Fig. 2. Equilibrium constants (Keq) for the aggregation of free membrane proteins as a function of the hydrophobicity of the protein (γ_{pex}, left), and the corresponding vertical height (H) above the membrane surface (middle). Possible consequences of receptor-ligand interactions involving hydrophobicity changes are diagrammed on the right.

the equilibrium constant between individual and paired particles. To apply this to the situation described by figure 1, we can make an approximation that the free energy of interaction will be the weighted average (in terms of H) of the free energies of aggregation of the protein in the membrane, γ_{pm}, and external medium, γ_{pex}. If we do this, and plot the resulting values of Keq against γ_{pex}, for constant γ_{mex}, a curve with a relatively sharp peak results (fig. 2, left), indicating that there is an optimal γ_{pex} for the aggregation of membrane proteins in a given membrane. There is also an associated change in H (fig. 2, middle). If ΔGa is small it could be counteracted by Brownian motion, but if it is large, the aggregate will grow. The kinetics of the process undoubtedly depend on membrane viscosity. Some aspects of this problem have been discussed by *Gershon* [2].

The relation between this result and the patching which occurs following receptor-ligand interaction will now be discussed. If a receptor is very hydrophilic, γ_{pex} is low, the attachment of a hydrophobic ligand to it could increase its hydrophobicity (fig. 2, right, a $\rightarrow$ b), and this could increase the degree of aggregation enormously. This result could also be obtained from conformational change following attachment due to recognition at a specific site. This membrane protein will also sink into the membrane. A converse process is also possible (fig. 2, right, c → d) in which a hydrophobic protein becomes hydrophilic, rises out of the membrane and aggregates. The relatively large aggregates may then be phagocytized as a result of surface energies [3, 12]. This seems to be precisely the behavior of hormone receptors (e.g. insulin) [9]. It would be interesting to know if coated pits, the apparent sites of phagocytosis following patching, are more or less hydrophobic than the rest of the membrane.

Surface Energies and Cell-Cell Recognition

When 2 cells touch, the interfacial energy in the area of contact becomes very low compared to γ_{mex} (eq. 2). This can have a large effect on both the protrusion of membrane proteins in that area and on their local aggregation. If the membrane protein is more hydrophilic than either membrane, it may be pushed aside as the 2 cells meet, or it may completely lodge in the more hydrophilic membrane. This could result in protein transfer following cell-cell contact. If the membrane protein is more hydrophobic than either membrane, transfer could also occur by partitioning, although originally the protein would have been completely submerged in its own cell membrane. If the membrane protein has a hydrophobicity between that of either membrane, it will sit at the cell-cell interface. This last case is interesting since it suggests the possibility of receptors which are invisible when the cell is in an aqueous medium, but which come to the cell-cell interface following contact by another cell. The appearance of membrane proteins in tight-junctions in freeze-fracture electron micrographs is in accordance with this model. The mutual recognition of different cell types could result from the transfer or interaction of membrane proteins by these mechanisms.

Conclusions

Consideration of the interfacial free energies associated with cell membranes and the proteins which reside in them has led to model situations with clear experimental and biological analogues. Patching, the phagocytosis associated with capping, changes in the vertical positioning of membrane proteins, and some aspects of cell-cell recognition could result from changes in the interfacial free energies of cell surface components.

References

1 Albertson, P.A.: Partitioning of cell particles and macromolecules (Wiley, New York 1971).
2 Gershon, N.D.: Proc. natn. Acad. Sci. USA *75:* 1357–1360 (1978).
3 Gerson, D.F.; Meadows, M.G.; Finkelman, M.; Walden, D.B.: in Ferenczy, Advances in protoplast research, pp. 447–456 (Akadémiai Kiadó, Budapest 1980).
4 Gerson, D.F.: Biochim. biophys. Acta *602:* 269–280 (1980).
5 Gerson, D.F.; Akit, J.: Biochim. biophys. Acta *602:* 281–284 (1980).
6 Gerson, D.F.; Scheer, D.: Biochim. biophys. Acta *602:* 506–510 (1980).
7 Gerson, D.F.: in Pernis and Lefkovits, Immunological methods, vol. 2, pp. 105–138 (Academic Press, New York 1981).
8 Neumann, A.W.; Sell, P.J.: Z. phys. Chem. *227:* 187–195 (1964).
9 Schlessinger, J.: in DeLisi, Blumenthal, Physical and chemical aspects of cell surface events in cellular regulation, pp. 89–115 (Elsevier, Amsterdam 1979).
10 Shinitsky, M.: in DeLisi, Blumenthal, Physical and chemical aspects of cell surface events in cellular regulation, pp. 173–181 (Elsevier, Amsterdam 1979).
11 Shinitsky, M.; Souroujon, M.: Proc. natn. Acad. Sci. USA *76:* 4438–4440 (1979).
12 Van Oss, C.J.; Gillman, C.F.; Neumann, A.W.: Phagocytic engulfment and cell adhesiveness (Dekker, New York 1975).

Dr. Donald Gerson, Basel Institute for Immunology, Postfach, CH–4002 Basel (Switzerland)

The Immune System, vol. 1, pp. 252–255 (Karger, Basel 1981)

What Do Membrane Potentials Have to Do with Immunology?

Hansruedi Kiefer[1]

Basel Institute for Immunology, Basel, Switzerland

A basic requirement for the proper functioning of even a simple cellular system is communication between its components. In this sense communication means transmitting information to and receiving information from the environment, then transforming it into signals or instructions which the cell can understand and respond to. This exchange of information can be mediated by soluble agents (transmitters) or by direct cell-cell interaction. Whatever the case may be, the cell membrane is intimately involved in the transformation of messages.

How can a *soluble* mediator interact with the membrane of a lymphocyte and what are the first events triggered by such an interaction in a well balanced resting cell? Four different modes of interaction can be envisaged for a soluble mediator:

(1) *Passive permeation* requires that the mediator be of low molecular weight. The mediator must also be lipophilic or at least, as in the case of some proteins, be able to assume a configuration which will give the appearance of a lipophilic molecule. Since passive permeation is rather nonspecific it, per se, will be of little value in mediating a complex system like the immune system.

(2) *Active transport,* the usual way of regulating the uptake of nutrients, requires energy and specific acceptor sites, but is an unlikely candidate for communication between cells.

[1] *Hansruedi Kiefer* has been a member of the Basel Institute for Immunology since 1971.

(3) *Nonspecific binding* of a mediator to the plasma membrane can occur by either electrostatic or hydrophilic/hydrophobic interactions. Binding to or insertion into lipids may lead to fluidity changes of the lipid bilayer and to surface energy or hydrophobicity changes. It might also be the method of choice for passively acquiring and expressing a 'foreign' antigen.

(4) *Binding to specific receptors* is certainly the most attractive possibility when it comes to giving specific messages to specialized cells. The *immediate* consequences of the binding of a soluble agent like an antigen or a transmitter to its receptor are changes in the tertiary structure of the receptor molecule, and/or aggregation and cross – linking of receptors [7, 8].

Conformational changes of the receptor molecule may in turn lead to: (1) enzyme activation; (2) channel formation (gating for cations); (3) exposure of new binding sites inside or outside; (4) flip/flop transport to the inside; (5) release of, or induction of changes in, receptor-associated structures; (6) perturbation of the lipid bilayer, leading to changes in permeability to ions, energy requirements for permeation, fluidity, hydrophobicity, and surface energy.

Aggregation or cross-linking (patching) may lead to: (1) change in hydrophobicity, surface energy; (2) removal of transmitter/receptor complexes (capping, shedding, endocytosis).

The nervous system resembles the immune system in many ways. As *Niels Jerne* [4] wrote: 'These two systems stand out among all other organs of our body by their ability to respond adequately to an enormous variety of signals. The cells of both systems can receive as well as transmit signals.' In the nervous system signals are transmitted by binding of transmitter molecules to specific receptors. These will, by conformational changes, initiate Na^+ gating and depolarization of the membrane. This electrical signal then propagates as an action potential along the axon membrane to, e.g. a motor endplate, at which the electrical signal is converted back into a chemical one, by the release of transmitter (acetylcholine) into the synaptic cleft. Acetylcholine will then bind to receptors on the muscle cell and again the membrane becomes permeable to Na^+ and causes a breakdown of the membrane potential, which in turn leads to the contraction of muscle fibers.

Membrane potentials in both the immune system and the nervous system are based on K^+ diffusion potentials:

$$\Delta\Psi = \frac{2.3RT}{F} \log \frac{[K^+]_{out}}{[K^+]_{in}}.$$

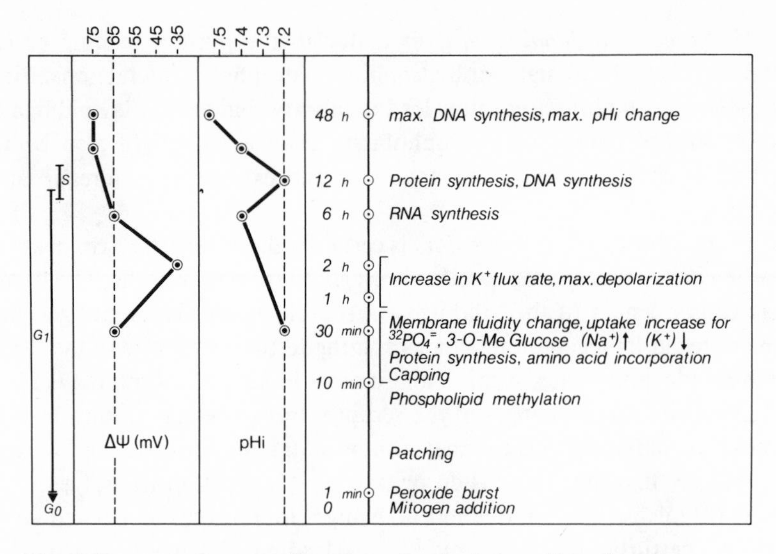

Fig. 1. Sequence of events after mitogenic stimulation of lymphocytes.

Selective permeabilities of the membrane are maintained by a Na^+/K^+ dependent ATPase. Among the earliest events after the addition of mitogens to lymphocytes in vitro, a signal leading to differentiation and proliferation, is a decrease in K^+ and increase in Na^+ content of the cells [5], as well as an increase in fatty acid turnover in the lipid phase of the plasma membrane [9]. Both events reflect changes in the permeability of the membrane.

It has recently been shown that these events occur simultaneously with a drop in membrane potential [6]. This depolarization of the lymphocytes is a necessary but not sufficient first signal for the proliferation of lymphocytes. It is the product of a specific signal only.

Figure 1 shows an attempt to correlate some biochemical and biophysical events after mitogenic activation of mouse spleen lymphocytes in vitro. It should be kept in mind that all quantitative aspects of this table represent an entire population which is asynchronous and only a fraction of which is capable of responding to a mitogenic signal at all. The time course of the depolarization (i.e. 2 h until maximum depolarization is reached) is therefore unrealistic. It is assumed that an individual cell in vivo reaches its maximum depolarization within minutes after being effectively triggered. However, in vitro, the method of cell preparation influences the early time course considerably. Depolarization or increased permeability seems to throw a well-balanced resting cell into a seemingly uncontrolled state of chaos. But with the help of factors produced by accessory cells, it makes a remarkable recovery. It reorganizes its membrane, adjusts its internal pH [*Gerson* et al., manuscript in preparation], activates enzymes [2], accelerates the transport of nutrients and starts the synthesis of various products needed for its new task [9]. Note that this entire sequence of events closely resembles the events which take place after fertilization of sea urchin eggs [1]. But the first specific signal, to a lymphocyte or to a sea urchin egg, changes the membrane potential by changing the permeability of the plasma membrane [3, 6].

References

1 Epel, D.: The molecular basis of cell-cell interaction, pp. 377–388 (Liss, New York 1978).
2 Hirata, F.; Toyoshima, S.; Axelrod, J.; Waxdal, M.J.: Proc. natn. Acad. Sci. USA *77:* 862–865 (1980).
3 Jaffe, L.A.: Nature, Lond. *261:* 68–71 (1976).
4 Jerne, N.K.: Annls Inst. Pasteur, Paris *125C:* 373–389 (1974).
5 Kaplan, J.G.: A. Rev. Physiol. *40:* 17–36 (1978).
6 Kiefer, H.; Blume, A.J.; Kaback, H.R.: Proc. natn. Acad. Sci. USA *77:* 2200–2204 (1980).
7 Koshland, D.E., Jr.: Trends Biochem. Sci. *5:* 297–302 (1980).
8 Schlesinger, J.: Physical chemical aspects of cell surface events in cellular recognition, pp. 89–118 (Elsevier, New York 1979).
9 Wedner, H.J.; Parker, C.W.: Prog. Allergy, vol. 20, pp. 195–300 (Karger, Basel 1976).

Dr. H. Kiefer, Basel Institute for Immunology, Postfach, CH-4005 Basel (Switzerland)

The Immune System, vol. 1, pp. 256–261 (Karger, Basel 1981)

Polyenes as Membrane Probes for Lymphocytes, Macrophages and Tumor Cells

J.R. Little[1], *G. Medoff, S.F. Shirley, T.E. Shine*

Departments of Medicine, Surgery, and Microbiology and Immunology, and the Department of Medicine at The Jewish Hospital, Washington University School of Medicine, St. Louis, Mo., USA

Amphotericin B (AmB) and certain other polyene antibiotics are effective antifungal agents [12], and their mechanism of action is related to binding the ergosterol in cell membranes of fungi [7]. Until recently relatively little was known about the effects of AmB on the host except for the renal, hematopoietic and other acute toxicities encountered during parenteral antifungal therapy. In studies carried out mainly in mice it has been shown that AmB and related polyenes can produce striking augmentation of humoral and cell-mediated immunity [1] and they are also highly effective in combination chemotherapy of tumors [10]. The synergism of antitumor drugs in combination with AmB in chemotherapy is clearly related to two effects: (a) enhancement of tumor cell permeability to antitumor agents [11], and (b) stimulation of the host's immune response. In the AKR system using the drug combination of AmB and BCNU, the immune effects are probably dominant [9].

The success of AmB in the treatment of tumors led to a study of the mechanisms of the immunoadjuvant action of polyenes. Early studies showed that AmB and its water-soluble derivative, amphotericin methyl ester (AME) provided potent stimulation of antibody responses to hapten-protein complexes [1] and several other antigens. Other immunization studies in mice have shown that AME is a potent adjuvant in AKR, A/J, DBA/2 and several other strains when stimulated by T cell-dependent antigens, but antibody responses to three different T cell-independent antigens (type III pneumococcal polysaccharide, phosphorylcholine, or TNP-*Brucella abor-*

[1] *J. Russell Little* was a member of the Basel Institute for Immunology in 1972 and 1973.

tus) were not enhanced [unpublished experiments]. These results agree with others suggesting a critical role for T cells in the mechanism of the polyene adjuvant effects.

A comparison of AME potency with more traditional adjuvants such as complete Freund's adjuvant, *pertussis* and BCG showed a comparable activity for AME [8]. In studies with AME, visible light irradiation of aqueous solutions resulted in disruption of the polyenic structure, loss of antifungal and membrane permeabilizing activity and parallel loss of adjuvant activity [8]. These results indicate that the adjuvant effects of AME depended on sterol binding.

In addition to the preceding studies in immune potentiation, it has been shown that AmB can stimulate certain macrophage functions. *Chapman and Hibbs* [3] showed that AmB augmented macrophage tumoricidal effects and studies from this laboratory have shown that AmB can increase the antifungal effect of murine macrophages on *C. albicans* [*Brajtburg* et al., in preparation].

The studies reported here compare the mouse strain-specific adjuvant potency of AME in vivo with the intensity of polyclonal splenic B cell activation (PBA) induced by AME in culture. The results show genetically determined differences in murine responses to AME which are related to differences in serum lipoproteins.

Materials and Methods

Mice. C57BL/6, C57BL/6-H-2^k (B/6-H-2^k), high cholesterol strain (HC) and BALB/c mice were obtained from the mouse colony at The Jewish Hospital, and AKR mice were from the Division of Radiation Biology, Washington University School of Medicine. Breeding pairs of HC mice were generously provided by Dr. *Andrew Kandutch,* Jackson Laboratories, Bar Harbor, Maine.

Immunization. Mice were immunized by two intraperitoneal (i.p.) injections of 100 μg of trinitrophenylated human serum albumin (TNP-HSA) in 0.2 ml isotonic saline. The secondary immunization was 28 days after the primary. AME was also injected i.p. in 0.2 ml 0.08 *M* Tris-acetate, pH 7.0. The AME dosage was 300 μg per mouse administered only at the time of primary immunization and freshly made stock solutions were shielded from light with aluminum foil prior to injection.

Plaque Assay. The method of *Yamada and Yamada* [18] was used with minor modifications. Trinitrophenylated sheep erythrocytes (TNP-SRBC) were prepared by the method of *Rittenberg and Pratt* [14]. Splenic cell suspensions were prepared from individual mice sacrificed for assay 6 days after the secondary immunization.

Polyclonal B Cell Activation. Spleen cell suspensions prepared from 4 to 6 normal mice were pooled and incubated 48 h in RPMI 1640, 10% fetal calf serum, and varying concentrations of AME. Each culture was then washed and scored for direct plaque forming cells (PFC) using TNP-SRBC as target cells. Cell viability, measured by trypan blue exclusion, was routinely 25–35% after 48 h in the presence or absence of AME.

Results

During testing of several mouse strains with several different immunogens plus AME as an adjuvant, it became apparent that not all strains responded equally. These studies were complicated by the difficulty in excluding genetically determined differences in the response to the immunogens. However, from studies with antigens showing little or no genetic control of the immune response – e.g. sheep erythrocytes (SRBC) and fowl γ-globulin – it became evident that C57BL/6 and other strains on the C57BL background yielded only meager or insignificant responses to AME or AmB as an adjuvant. It was also determined that an inbred mouse strain with a genetically determined high serum cholesterol level [17] responded minimally if at all to AME as an adjuvant. Figure 1 gives results of splenic PFC responses to immunization with TNP-HSA in this high cholesterol (HC) strain (H-2^k) as well as two other H-2^k strains, AKR and B/6-H-2^k. Despite identical antigen dosage and immunization schedule, there was a marked disparity in the immunostimulation provided by AME.

In addition to these in vivo effects, results from cell culture systems provided direct evidence for AmB stimulation of lymphocytes. Polyclonal B cell activation (PBA) by polyenes was first described by *Hammarstrom and Smith* [6]. The present studies confirm and extend their results to show that AME is a more potent activator than AmB. It is particularly interesting in regard to the adjuvant studies that AME showed selective stimulation of B cells from certain mouse strains. Those strains showing enhanced PBA results (fig. 2) were the same strains exhibiting adjuvant effects of AME as shown in figure 1. Direct PFC assays with TNP-SRBC were performed to quantitate changes in PBA stimulated by AME. AKR spleen cells responded vigorously to AME, and AKR was also the strain that responded best to the adjuvant activity of AME (fig. 1). C57BL/6 spleen cells were almost unresponsive to the effects of AME on PBA and this correlates with a nearly uniform absence of AME adjuvant action in this mouse strain. BALB/c mice gave intermediate responses in both systems. The PBA responses of HC mice were very weak, similar to the response of C57BL/6 (fig. 2).

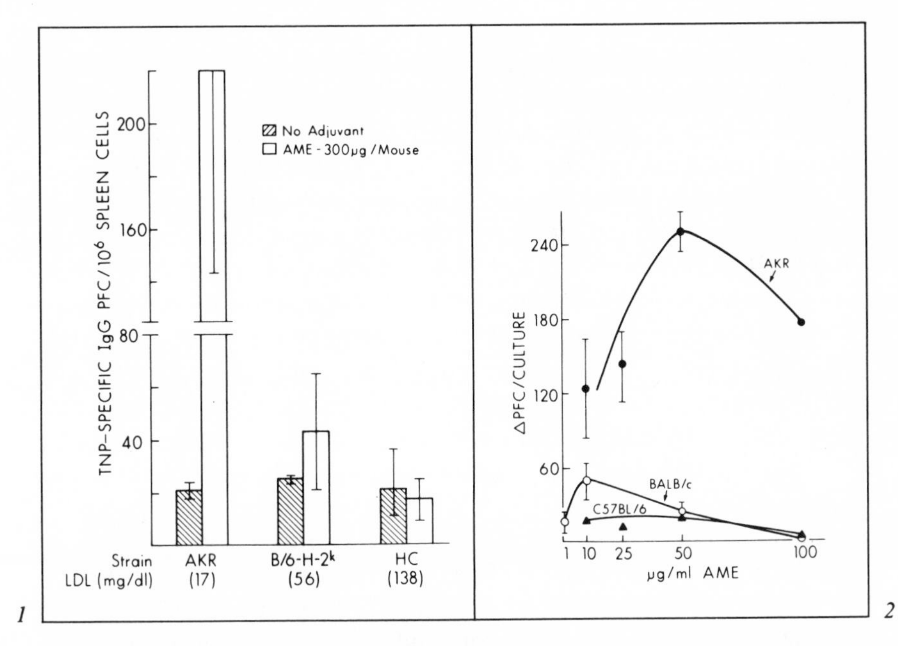

Fig. 1. Mouse strain-specific adjuvant effects of AME. The immunization schedule used in each strain was the same. The HC mice are an inbred strain with genetically determined high serum cholesterol and LDL levels [17]. All 3 strains share the H-2^k haplotype. The fasting LDL concentrations for each mouse strain are indicated below the abscissa.

Fig. 2. Polyclonal B cell activation of murine spleen cells by AME. Spleen cell suspensions from nonimmune mice of the indicated strains were cultured 48 h in the presence of varying concentrations of AME (abscissa). At the end of this interval the contents of each spleen cell culture were scored for IgM PFC using TNP-SRBC. Control cultures were separate aliquots of the same spleen cell suspension containing no AME and the values shown on the ordinate represent the difference between PFC/culture for control and experimental cultures.

Discussion

The mouse strain-specific effects of AME provide an important lead for future studies directed at elucidating the adjuvant mechanisms of the polyenes. Since polyene binding to membrane cholesterol is assumed to be the primary event in all lymphoid cell responses, it is suggested that AKR and C57BL/6 mice express significant differences in membrane lipids. Different membrane cholesterol-phospholipid ratios may result in differences in polyene binding as well as dissimilar lateral diffusion in the plasma membrane even in the absence of polyenes. Since cell membrane diffusion measurements are possible in single living cells [4], AmB and other polyenes may comprise a group of valuable ligands for studies of the regulation of biological processes occurring at the surface of lymphocytes and tumor cells.

In addition to enhancing murine humoral immunity it has been shown that AmB augments cell mediated immune responses to dinitrofluorobenzene (DNFB), oxazolone [15] and SRBC [unpublished results]. When AmB was administered to AKR mice at the time of skin sensitization with dilute DNFB, a significant enhancement of ear swelling occurred following a challenge application of DNFB. Other mice sensitized in the same way with DNFB plus AmB showed augmented lymph node cell proliferation when these cells were restimulated in culture with dinitrobenzenesulfonate (DNBS) [16]. This result supports other data which show that AmB acts by enhancing the immune reactivity of lymphoid cells rather than by amplifying the inflammatory reaction that might have resulted from a generalized increase in cellular permeability. These immunostimulant effects of AmB in AKR mice were in marked contrast to results obtained in C57BL/6 mice sensitized in an identical fashion. AmB failed to augment contact sensitivity in C57BL/6 mice and sometimes suppressed it.

In the analysis of the probable mechanism of the immunoadjuvant effects of polyenes, it is important that AmB partially ablated tolerance to DNFB induced by a single injection of DNBS in AKR mice [15]. Immune tolerance in this system has been shown to be mediated by suppressor T cells [13] and tolerance ablation by AmB leads to the conclusion that AmB may be selectively toxic for suppressor cells. This suppressor cell toxic effect could account for the immune enhancement produced by AmB as well as the interference with tolerance induction.

Therefore, a working hypothesis is suggested for the mechanism of the immunoadjuvant effects of AmB based on the general features of the cell circuit model of *Cantor and Gershon* [2]. The proposed locus for the adjuvant effects of AmB is the T cell subset bearing the Lyt-1,2,3 surface antigens. This relatively immature cell type gives rise to the Lyt-2,3 suppressor/killer cells which in turn inhibit Lyt-1 helper cells and B cells. The implied mechanism of immune augmentation by AmB is through inactivation of a normal suppressor regulatory cell, leaving the positive regulatory cells and effector cells to act unopposed. This model predicts that AmB might also be *immunosuppressive* through interference with the several effector functions of Lyt-2,3 (or Lyt-1,2,3) bearing cells including host protective antiviral immune effects. Recent publications which have documented immunosuppressive effects of AmB [5] are important in this regard. Assuming that a single cell type is the target for immune augmentation by AmB, this model is attractive in its simplicity. However, it may be that multiple cell types are affected and the adjuvant mechanisms are much more complex.

Acknowledgments. The authors wish to express their gratitude to Drs. *V. Hauptfeld* and *D. Shreffler* for their typing studies on the major histocompatibility complex of the HC mice and to Dr. *Virgie Shore* who determined the levels of serum LDL in several inbred mouse strains. This work was supported by an NIH Program Project Grant No. 5 P01 CA 15665 and Training Grants No. 5 T01 GM00371 and No. 5 T32 CA 09118.

References

1 Blanke, T.J.; Little, J.R.; Shirley, S.F.; Lynch, R.G.: Cell. Immunol. *33:* 180–190 (1977).
2 Cantor, H.; Gershon, R.K.: Fed. Proc. *38:* 2058–2064 (1979).
3 Chapman, H.A.; Hibbs, J.B.: Proc. natn. Acad. Sci. USA *75:* 4349–4353 (1978).
4 Elson, E.L.; Schlessinger, J.; Koppal, D.E.; Axelrod, D.; Webb, W.W.: in Membranes and neoplasms, p. 137 (Liss, New York 1976).
5 Ferrante, A.; Rowan-Kelly, B.; Thong, Y.H.: Clin. exp. Immunol. *38:* 70–76 (1979).
6 Hammarstrom, L.; Smith, E.: Scand. J. Immunol. *5:* 37–43 (1976).
7 Kinsky, S.C.: A. Rev. Pharmacol. *10:* 119–142 (1970).
8 Little, J.R.; Plut, E.J.; Brajtburg, J.K.; Medoff, G.; Kobayashi, G.S.: Immunochemistry *15:* 219–224 (1978).
9 Medoff, G.; Kobayashi, G.S.; Kwan, C.N.; Schlessinger, D.; Venkov, P.: Proc. natn. Acad. Sci. USA *69:* 196–199 (1972).
10 Medoff, G.; Valeriote, F.; Lynch, R.G.; Schlessinger, D.; Kobayashi, G.S.: Cancer Res. *34:* 974–978 (1974).
11 Medoff, J.; Medoff, G.; Goldstein, M.; Schlessinger, D.; Kobayashi, G.S.: Cancer Res. *35:* 2548–2552 (1975).
12 Medoff, G.; Kobayashi, G.S.: New Engl. J. Med. *302:* 145–155 (1980).
13 Phanuphak, P.; Moorhead, J.W.; Claman, H.N.: J. Immun. *113:* 1230–1236 (1974).
14 Rittenberg, M.B.; Pratt, K.L.: Proc. Soc. exp. Biol. Med. *132:* 575–581 (1969).
15 Shirley, S.F.; Little, J.R.: J. Immun. *123:* 2878–2882 (1979).
16 Shirley, S.F.; Little, J.R.: J. Immun. *123:* 2883–2889 (1979).
17 Weibust, R.S.: Genetics, N.Y. *73:* 303–312 (1973).
18 Yamada, H.; Yamada, A.: J. Immun. *103:* 357–363 (1969).

Dr. J. Russell Little, Washington University School of Medicine, The Jewish Hospital of St. Louis, 216 South Kingshighway, St. Louis, MO 63110 (USA)

The Immune System, vol. 1, pp. 262–269 (Karger, Basel 1981)

Immunological Tolerance and the Single Cell Tradition

G.J.V. Nossal[1]

The Walter and Eliza Hall Institute of Medical Research, Melbourne, Australia

Two conceptual approaches have shaped cellular immunology as we know it today, and *Niels Jerne's* contribution to each has been quite critical. The first we may term the single cell approach. This is an example of the classical reductionist mode of science through which complex systems are reduced to component parts that are then analyzed. Being able to study single cells in the immune system turned out to be more important than had been expected because a clonal repertoire exists as the basis of the cellular immune system. There is thus a major thread coming from the discovery that one cell always makes only one antibody [13]; to the invention of the Jerne hemolytic plaque technique [7], which made the study of antibody formation by single cells universally accessible; to the scaling up of 'one cell – one antibody' through hybridoma technology [9]. This approach, supplemented by technologies for growing B lymphocytes as functional clones [reviewed in 11] has taught us much of what we know about B cells and antibody production. The similar conceptual thread for the T lymphocyte lineage, begun by *Gowans'* studies on small lymphocyte transformation, has taken a more winding pathway but has led to the cytotoxic T cell clones and immortalized T lymphocyte lines that are so enriching our understanding of

[1] *Gus Nossal* was a member of the Board of Advisors of the Basel Institute for Immunology from 1970 to 1975.

T cell physiology. The second conceptual approach, more recent than the first, sees lymphocytes not as quasi-independent units of a complex system, but as functionally connected through a series of positive and negative feedback loops. *Jerne's* experiments on the paradoxical role of specific IgM antibody in augmenting antibody-production [3] was an early example of this second approach. However, the importance of cellular interactions in immunology was not fully appreciated until the discovery of helper and suppressor T lymphocytes. It took the genius of *Jerne* to realize that these two major approaches to cellular immunology not only could be, but in fact had to be, married if the immune network was to be understood [6].

Scientific tides are given to the same pendular swings as most human movements. At the moment, the tendency is to regard every regulatory event within the immune system as another example of the subtlety of T cells. Indeed, the recent proof that Interleukin 2 is itself a T cell product adds weight to this standpoint. Important though the network approach indubitably is, there is a certain danger in forsaking the single cell tradition as being able to shed light on immunoregulation. It is that subtle variances in receptiveness to signals, properties inherent to the single lymphocyte as such, will be lost sight of. In the field of immunological tolerance, we can see how important it is to keep a foot in both camps: to study suppressor T cells, but also to be aware of the effects of antigen on single lymphocytes. Given the number of laboratories eager to emphasize the former, we have chosen to place our main effort into the latter area over the past 5 years.

It took *Jerne* only a very short period to realize that a key advantage of *Burnet's* [1] approach to selective theories of antibody formation over his own [5] was the ready way in which clonal selection could provide for immunological tolerance. A quarter of a century later, we are still arguing over just how important a silencing of repertoire elements is in the phenomenon of self-nonself discrimination. Like most great biologists, *Jerne* tenaciously hangs on to elegant grand visions until irrefutable evidence against them turns up. Thus, while the debate about T cell recognition of antigen was at its height, he steadfastly refused to believe that nature would have bothered to invent the V gene repertoire twice over, once for B cells and once for T cells, and in this perception it looks as though he will be proved right, although I wish the final resolution of this problem would come more swiftly. Similarly, he has been supportive of the general position that silencing of self-reactive clones early in their ontogeny is the *leitmotif* of self-tolerance. In penning these few lines in *Jerne's* honor, I express the hope that our present view of tolerance does not displease him.

The Concept of Clonal Anergy in B Lymphocyte Tolerance

The time gap between *Lederberg's* [10] clear articulation of the notion for which we coined the phrase clonal abortion [14] and its rigorous testing by experimentation proved to be very long, chiefly because of methodological constraints. Early work on tolerance amongst antibody-forming cells involved injecting antigen into immature animals and challenging them as adults with immunogenic antigen. This approach cannot reveal detailed cellular mechanisms, given the complication of T cell-B cell interactions and antibody-mediated feedback effects. Adoptive transfer methods were an advance but their usefulness is limited by (1) the fact that frequently the assay is nonlinear in terms of donor cell number:host response characteristics; and (2) the observation that the cells read out in adoptive assays represent a small and atypical sub-set of early B lymphocytes termed pre-progenitors [4]. In vitro assays avoid the latter difficulty but, as usually performed, fail to offer a direct readout of B cell precursor numbers, as bulk cultures are subject to regulatory controls in much the same manner as are whole animals. So it turns out that an experimental answer to the question: 'how many anti-A B cells are there present in an animal tolerant to antigen A?' is not easy to obtain.

A key development in the field was that of limiting dilution analysis of precursors of antibody-forming cells [reviewed in 11]. *Marbrook* and *Lefkovits* were the two pioneers here, and *Metcalf and Klinman* [12] and our group [15, 19] were the first to realize how critical these techniques were to prove to the resolution of tolerance problems. Using the enumeration of antibody-forming clones as a final readout, one can at last determine how many functionally competent B cells there are present with reactivity to a test antigen. The study alone cannot differentiate between true clonal deletion, i.e. an actual destruction of the immature B cell; or clonal silencing through receptor modulation, blockade or some other mechanism. The limiting dilution approach solves the read-out problem, but leaves the question of how to recognize the target cell for tolerance induction. Varying the age at which tolerogen is introduced into an animal, or the organs source studied (e.g. bone marrow versus spleen) allows certain broad conclusions to be reached, but the best way to gain final understanding is to fractionate B lymphocytes and their precursors into subsets based on biophysical characteristics and/or cell surface markers. These more homogenous subpopulations can then be placed in tissue culture, already at limiting dilution to minimize influences of other cells, and incubated with various concentrations of putative tolerogens. Then the tolerogen can be removed and a strong antigenic or mitogenic challenge substituted. Later analysis for plaque-forming cells or antibody production constitutes the clonal readout. We have found this two-stage limiting dilution tissue culture approach to be a very powerful tool for both T independent and T-dependent responses. Its only limitation is that the tolerance induction period must be restricted to a maximum of 48 h because of nonspecific cell death.

Using this experimental design [17, 18, 21], we have established the following facts about B lymphocyte tolerance using hapten-protein conjugates as model tolerogens:

(1) There is not an absolute, all-or-none difference between various B cell subpopulations in their capacity to be negatively signalled by nonimmunogenic antigen. All B cells can react provided the stimulus is strong and long enough.

(2) There are, however, profound quantitative differences in the concentrations of tolerogen required to silence the B cell; the differences are 1,000- to 100,000-fold in some experimental situations, suggesting that in 'real life' they would approach all-or-noneness.

(3) The most important variable determining susceptibility is the degree of maturity of the B cell. Cells first encountering antigen as they move from the pre-B cell (surface Ig-negative) stage to that of the immature B cell are the most sensitive. This, of course, is the circumstance that prevails with respect to potentially self-reactive cells, which must 'see' their antigen for the first time as the s-Ig receptors are being placed into the plasma membrane. Cells that have reached s-Ig-positivity but are derived from newborn spleen or adult bone marrow are next in susceptibility, and mature B cells are the least susceptible. The immature B cell is in the ambivalent situation that it can be triggered by antigen plus an appropriate second signal, but that it is very sensitive to signal 1 alone. We do not know what metabolic event betokens final maturity. It is certainly not the acquisition of IgD as such, as s-IgD-negative small lymphocytes from spleens of mice aged 2 weeks or older exhibit adult-type tolerance resistance and thus behave as mature.

(4) For the immature cells, the concentrations of antigen which suffice for an effective negative signal are far too low to saturate the cell's s-Ig receptors, or to cause complete s-Ig loss through modulation. Rather, the cell receives and stores a negative signal which negates the effects of later positive signals, whether received via the still intact s-Ig receptors or via the effects of polyclonal B cell activators or mitogens.

(5) Anti-immunoglobulin μ-chain antibody can mimic essentially all the effects of hapten-protein conjugates and acts as a useful model 'universal tolerogen' [17]. Again, negative signalling of immature cells demands far less ligand than is required for irreversible modulation of receptors.

(6) The hapten density is important in determining tolerogenicity of conjugates; the higher the hapten density, the lower the molarity of the tolerogen required for a specific effect.

(7) The nature of the protein carrier is important in determining the degree of tolerogenicity. Immunoglobulins such as human γ-globulin (HGG) make particularly effective tolerogens when used as carriers for haptens such as DNP, NIP or fluorescein. This is only partially accounted for by the increased cell surface binding properties conferred by the Fc piece [18]. This point will be discussed further below.

(8) The differences in tolerance susceptibility between immature and mature cells which can be demonstrated through in vitro studies differ in degree according to the particular conjugate used, being greatest with oligovalent and least with highly multivalent conjugates [18]. Nevertheless, the greater sensitivity of immature cells can be readily demonstrated with all conjugates tested.

(9) The sensitivity of immature cells of the B lineage is not an in vitro artifact, because (provided the quantitative in vitro readout system is used) the concentration differences for threshold tolerance induction can equally readily be demonstrated in vivo [8, 16]. It turns out that tolerogen introduced at 14.5 days' gestation via the placental circulation is effective at 10^5- to 10^6-fold lower concentration than is required for B cell tolerance in adult animals.

(10) Tolerance induced in vivo, whether this be in utero or in newborn life, does not lead to a deletion of the B cell concerned [16]. When appropriately low antigen concentrations are used to induce tolerance, one can detect perfectly normal numbers of antigen-binding B cells, with a normal avidity spectrum, in profoundly tolerant mice, provided suitable quantitative techniques are employed. This is the case even when antigen is introduced into fetal mice 2 days before the first B cells appear in the foetal liver or spleen. In other words, the developing B cells can register and store negative signals which prevent their later capacity to be triggered while at the same time maturing and displaying their full-density surface Ig coat.

It thus appears that our earlier concept of clonal abortion needs to be modified into one more appropriately described as clonal anergy. Cells with receptors for tolerated antigens appear not to be killed through early contact with antigen but rather rendered incapable of responding to T-dependent or

T-independent antigenic signals, or to mitogens or polyclonal activators of various sorts. The question that then arises is why nature bothers to keep such cells around. Would it not be more sensible to arrange for a way of destroying them? The answer may emerge if we regard tolerance as only part of the complex process of immunoregulation. Self-antigens presumably persist at approximately stable molar concentrations in extracellular fluids, and thus do or do not induce clonal anergy amongst B cells dependent on the concentration and the epitope valency. In situations of chronic infections, cancer, autoantibody formation and immune complex diseases, however, it may well 'pay' to have a system where cells are reversibly silenced rather than killed. This may offer another echelon of regulatory possibilities at the single cell level. In any case, if the life span of the virgin B cell is a week or less, the cost of keeping an anergic B cell alive for its natural life span may not be very great.

Clonal Anergy in Self-Tolerance: the Limits

There are three major caveats that must be entered before we can accept the induction of clonal anergy at the pre-B to B transition as the major physiological mechanism for self-tolerance amongst B lymphocytes. The first relates to valency.

If appears that receptor aggregation is in some way involved in the induction of the negative signal. Univalent antigens appear to be ineffective tolerogens in vitro and unimpressive in vivo. Furthermore, the negative signal is generated through some metabolic event in the cell, e.g. is inhibited by inhibitors of protein synthesis [20]. The question then arises as to how the B cell repertoire remains non-reactive to the many univalent antigens present in the body. Admittedly, the clonal anergy mechanism seems suitable for autologous cell surface macromolecules such as major erythrocyte antigens, histocompatibility antigens, etc.; but does it address monomeric serum proteins, hormones, etc.? Perhaps these are rendered operationally multivalent, e.g. by absorption to cell surfaces, or perhaps they are tolerated because of indirect influences such as their own lack of immunogenicity or their capacity to induce tolerance in helper T cells or activation of suppressor T cells. This remains a fertile, albeit difficult, field for further study.

Secondly, we must consider why some molecules are poor tolerogens even if introduced early in life. The most serious challenge to the position outlined in this paper, and one we must take very seriously, comes from

Diener's group [2, 22]. They find that some molecules are completely non-tolerogenic even when introduced via the placental route during foetel life; and that apparently minor alterations in the chemical nature of carriers can alter their properties as tolerogens in hapten-carrier systems. They tend to regard hapten-immunoglobulin models as representing a special case. Their work raises the issue whether associative recognition may be necessary for B cells as well as T cells, and therefore whether tolerogens, before being 'seen' by B cells, must associate with cell membranes in some way. If so, the carrier could well exert profound effects.

Thirdly, any statement about tolerance induction must take note of affinity considerations. Clonal anergy to self constituents could not be allowed to proceed untrammelled by such constraints, as this could lead to anergy in the total repertoire, given the number of cross-reactions amongst antigens and antibodies. Thus, when endotoxin permits the synthesis of some low affinity autoantibodies we should not be surprised, nor regard this as weighing against clonal anergy arguments.

Tolerance in T Lymphocytes

It would be of great interest to determine whether the clonal anergy mechanism also works for the T lymphocyte system. So far, we have confined our studies on T cells to the cytotoxic T lymphocyte precursors (CTL-P) for allogeneic stimulation. This system lends itself well to enumeration of clonable antigen-reactive cells. Using in vivo tolerance induction with living semi-allogeneic cells, it is clear that the number of CTL-P can be grossly reduced within 2 weeks or so. As there is no method for accurate enumeration of antigen-binding virgin T cells, we cannot determine whether this is due to clonal abortion or the induction of clonal anergy. We believe that the clonal approach will have much to offer in the study of T lymphocyte tolerance.

Conclusions and Summary

No student of *Jerne's* network theory would seek to force all the myriad models of immunological tolerance, nor the great variety of physiological situations which they seek to mimic, into a single conceptual straightjacket. It is not a question of whether tolerance in antibody formation is due to suppressor T lymphocytes *or* to clonal anergy. Each of the two great traditions, the study of the single cell, and the study of lymphocyte-lymphocyte and macrophage-lymphocyte interactions, has its role to play in the elucidation of an obviously complex puzzle. We believe the phenomenon of clonal anergy to be of particular interest, because it represents a previously unsuspected level of regulation. We know nothing of the nature or duration of the negative signal so effectively registered and stored by the immature cell, nor of the influences, metabolic or pharmacological, which might reverse it. Nor can we see clearly the nature of the gradual change in the maturing B cell which renders it progressively more refractory to anergy induction. I suspect the existence of a regulatory network right within the single cell, and I hope this concept, wedding the two traditions, might pique *Niels Jerne's* interest.

References

1 Burnet, F.M.: Aust. J. Sci. *20:* 67–69 (1957).
2 Diner, U.E.; Kunimoto, D.; Diener, E.: J. Immun. *122:* 1886–1891 (1979).
3 Henry, C.; Jerne, N.K.: in Killander, Nobel symposium 3: gamma globulins, pp. 421–428 (Interscience, New York 1967).
4 Howard, M.C.; Fidler, J.M.; Baker, J.; Shortman, K.D.: J. Immun. *122:* 303–319 (1979).
5 Jerne, N.K.: Proc. natn. Acad. Sci. USA *41:* 849–857 (1955).
6 Jerne, N.K.: Annls Immunol. *125C:* 373–389 (1974).
7 Jerne, N.K.; Nordin, A.A.: Science *140:* 405 (1963).
8 Kay, T.W.; Pike, B.L.; Nossal, G.J.V.: J. Immun. *124:* 1579–1584 (1980).
9 Köhler, G.; Milstein, C.: Eur. J. Immunol. *6:* 511–519 (1976).
10 Lederberg, J.: Science *129:* 1649–1653 (1959).
11 Lefkovits, I.; Waldmann, H.: Limiting dilution analysis of cells in the immune system (Cambridge University Press, Cambridge 1979).
12 Metcalf, E.S.; Klinman, N.R.: J. exp. Med. *143:* 1327–1340 (1976).
13 Nossal, G.J.V.: Br. J. exp. Path. *39:* 544–551 (1958).
14 Nossal, G.J.V.; Pike, B.L.: J. exp. Med. *141:* 904–917 (1975).
15 Nossal, G.J.V.; Pike, B.L.: Immunology *30:* 189–202 (1976).
16 Nossal, G.J.V.; Pike, B.L.: Proc. natn. Acad. Sci. USA *77:* 1602–1606 (1980).
17 Nossal, G.J.V.; Pike, B.L.; Battye, F.L.: Immunology *37:* 203–215 (1979).
18 Pike, B.L.; Battye, F.L.; Nossal, G.J.V.: J. Immun. *126:* 89–94 (1981).
19 Stocker, J.W.; Nossal, G.J.V.: Contemp. Top. Immunobiol. *5:* 191–210 (1976).
20 Teale, J.M.; Klinman, N.R.: Nature, Lond. (in press).
21 Teale, J.M.; Layton, J.E.; Nossal, G.J.V.: J. exp. Med. *150:* 205–217 (1979).
22 Waters, C.A.; Diener, E.; Singh, B.: J. exp. Med. (in press).

Prof. G.J.V. Nossal, The Walter and Eliza Hall Institute of Medical Research, Post Office, Royal Melbourne Hospital, Victoria 3050 (Australia)

The Immune System, vol. 1, pp. 270–277 (Karger, Basel 1981)

Aspermatogenic, Agglutinating and Immobilizing Antigens: Extraction, Separation, Purification and Molecular Weight Determinations by Equilibrium Ultracentrifugation

Seymour Katsh[1], *Henry Klostergaard*[1], *Grace F. Katsh*[1], *Jimmy Klostergaard*

Department of Pharmacology, University of Colorado Medical School, Denver, Colo.; Department of Chemistry, California State University, Northridge, Calif.; Department of Molecular Biology and Biochemistry, University of California, Irvine, Calif., USA

One can destroy an entire cell line, the seminiferous epithelium, selectively and specifically with but a single injection of aspermatogenic antigen (ASA). ASA has been extracted from guinea pig testes, epididymides containing sperm, and spermatozoa flushed from epididymides. After using extracts, a fraction was isolated which induces aspermatogenesis in microgram amounts and only the seminiferous epithelium is deleted [11]. The origin of this antigen has been traced to the idiosomic granules which develop in association with the Golgi apparatus [5]. Purification and characterization of ASA has been accomplished [10, 13]. More recently, the molecular weight of ASA has been established to be 10,520 dalton by equilibrium ultracentrifugation [12]. The implications of such studies for the control of undesirable growths should be obvious.

Although it has been known for more than 80 years that injection of sperm can induce antibodies which cause specific sperm immobilization

[1] *Henry Klostergaard* and *Grace* and *Seymour Katsh* were colleagues of *Niels Jerne* at Caltech in Pasadena in 1954–55.

and sperm agglutination [15–17], very little information has come to light in the intervening years, demonstrating that these antibodies react with two separate antigens. In this paper we show that these two antigens are separable and are different entities.

We are pleased to dedicate this paper to our friend *Niels Jerne,* who played in the sand at the beach with the most junior author long ago, when he was very junior indeed.

Materials and Methods

Extraction and Purification of Antigen. Crude guinea pig ASA was prepared as described previously [10] with some modifications and additions. After removal of ASA, the remaining material was lyophilized and dissolved in 1.0 ml absorption buffer (1.0 *M* Na Cl; 0.02 *M* - PO_4, pH 7.0; 1 × 10^{-4} $CaCl_2$, $MgCl_2$, $MnCl_2$; 0.02% merthiolate). All of the material does not dissolve in this small volume of buffer and, therefore, it was centrifuged and washed 3 times with 0.5 ml absorption buffer. The supernatants were pooled and lyophilized. The material was placed on a concanavalin A column (16 × 200 mm, void volume 21 ml, 4 °C) followed by 3 washings with 1.0 ml absorption buffer. The rate of elution was 5.0 ml/h. The 20 ml collected after the void volume was labelled CAF 1 (concanavalin fraction 1). The buffer running through the column was changed to absorption buffer plus 1.0 *M* methyl-*D*-glucoside. After 30 ml of glucoside buffer was run through the column, the absorption buffer was used again. The collection of the next 30 ml was labelled CAF 2. (The column was regenerated by washing with 100 ml absorption buffer.) The eluates were collected in 3.0 ml aliquots and read at 280 nm (Beckman spectrophotometer). The plot of the values showed a defined peak at 20–50 ml; this is CAF 1. A much smaller yet well-defined peak was determined at 51–100 ml; this is CAF 2. Both fractions were lyophilized. Both fractions were placed on separate Bio-gel P-100 columns (2.5 × 90 cm, void volume 100 ml, 4 °C) using the buffer 0.3 *M* NaCl; 0.05 *M*-PO_4; pH 7.0; 0.02% NaN_3. Fractions were collected using an automated fraction collector in 9.7 ml volumes and read at 280 nm. Plots of these values showed 5 well-defined peaks. These materials were lyophilized. After biological testing (in male guinea pigs), two active fractions were ascertained. The lyophilized material from the active fractions was electrofocused: electrofocusing column, LKB model 8108, wide range pH 3.0–10.0 ampholines in sucrose density gradients, 0–30%. The cathode was at the top, the ampholines and the sample were protected against reduction at the cathode by 5.0 ml 0.1 *M* NaOH and against oxidation at the anode by 14.0 ml 0.1 *M* N_2SO_4 in 30% sucrose. The temperature was maintained at 4 °C with a Lauda refrigerated circulator, model K-2/RD. The sample, 10 mg, was loaded onto the column at the 20% sucrose gradient and the run was at 900 V for 72 h. 3 ml aliquots were collected and monitored at 250 and 280 nm. Plots showed peaks at pH 2.4, 4.8, 5.7, 7.3. After analysis by the Fluram technique [14] and bioassay, it was determined that the immobilizing antigen (IMA) had an isoelectric point at pH 4.8 and the AGA (agglutinating antigen) had an isoelectric point at pH 5.7. It was these two fractions that were used for molecular weight determinations by equilibrium ultracentrifugal analyses. The extraction, isolation and purification procedures will be reported in extenso elsewhere.

The antigenic materials were bioassayed as follows. Each antigenic preparation (10 μg) was dissolved separately in distilled water (1.0 ml) and then emulsified in Freund's complete adjuvant. The emulsions were injected separately into the nuchal region of 4–6 male guinea pigs. The animals were sacrificed 2 months later, testes were excised and evaluated for aspermatogenic response. Epididymides were excised and reverse flushed for sperm evaluation (agglutination and immobilization). A battery of other tests was applied at each step of the extraction, isolation and purification procedures. These and the bioassay procedures will be reported in extenso elsewhere.

*Molecular Weight Determinations.*The equilibrium ultracentrifugation method of *Nazarian* [18] was used with some modifications. In this method, the solution of the macromolecule is subjected to rotation until no further change is detectable in the concentration pattern across the cell. At such time, the effects of sedimentation and diffusion are equal but of opposite direction. Thus, a particularly simple relationship holds between the concentration pattern and the molecular weight.

To check the partial specific volume of the antigens, the experiments were performed using D_2O and H_2O as solvents. The conformation of the antigens should be the same in both solvents and the molecular weight by the deutrium exchange is increased by 1.55%. We then have two equations and can therefore leave the partial specific volume as unknown along the molecular weight in H_2O. The IMA was investigated using D_2O first and then H_2O. The temperature was 20 °C (293.1 K) and the speed of rotation was 27,690 rpm. Using H_2O as solvent, these experiments were conducted at the same temperature but at a speed of 23,150 rpm. The AGA was investigated first in the D_2O solvent, 20 °C and 27,690 rpm. After this experiment, the antigenic material was lyophilized and then dissolved in H_2O to repeat the study in this solvent.

Results

In the extraction, isolation, and purification of IMA and AGA, dialysis is not possible because of the loss of material. These antigens, particularly AGA, are somewhat thermolabile. Immuno-double diffusion and the immunoelectrophoretic studies revealed a single band.

Animal Studies. As previously found [11], 2 months following immunization with ASA, all animals were aspermatogenic (fig. 1a, b). The seminiferous epithelium had been deleted: only the Sertoli cells and some type A spermatogonia remained.

After 2 months, those animals injected with IMA were observed to have immobilized sperm upon excision of the epididymides and reverse flushing (through the vas deferens) to obtain the spermatozoa stored therein. After the same interval, those animals injected with AGA were observed to have agglutinated sperm which were obtained from flushing the epididy-

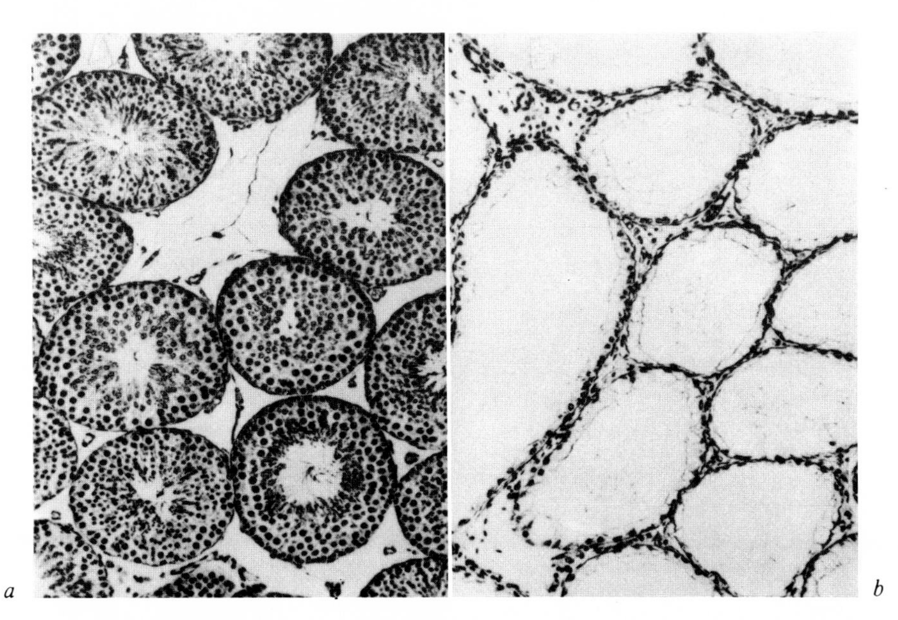

Figure. 1. Cross-section (× 150) of testis of guinea pig injected 2 months previously with (a) bull sperm and Freund's complete adjuvant, or (b) 10 μg purified aspermatogenic antigen (ASA) emulsified in Freund's complete adjuvant. *a* Shows the normal histology and cytology of the guinea pig testis: all stages of spermatogenesis from spermatogonia to spermatozoa are evident. *b* Represents the aspermatogenic response: all spermatogenic cells (with the exception of spermatogonia) are absent.

mides as noted above. Further, antisera obtained from animals immunized with IMA and AGA were observed to cause immobilization and agglutination, respectively, of sperm flushed from the epididymides of non-immunized, control animals injected with saline in Freund's adjuvant. These studies will be described in extenso elsewhere.

Molecular Weight Studies. For IMA in D_2O, we plot in figure 2a the logarithm of the number of fringes counted over the span vs the square of the distance from the center of rotation. A solution of a homogeneous macromolecule yields, in this way, a linear relationship. Since the relationship found is approximated in excellent fashion by the drawn line (least squares, correlation coefficient 0.9986), one observes a macromolecule of nearly perfect homogeneity.

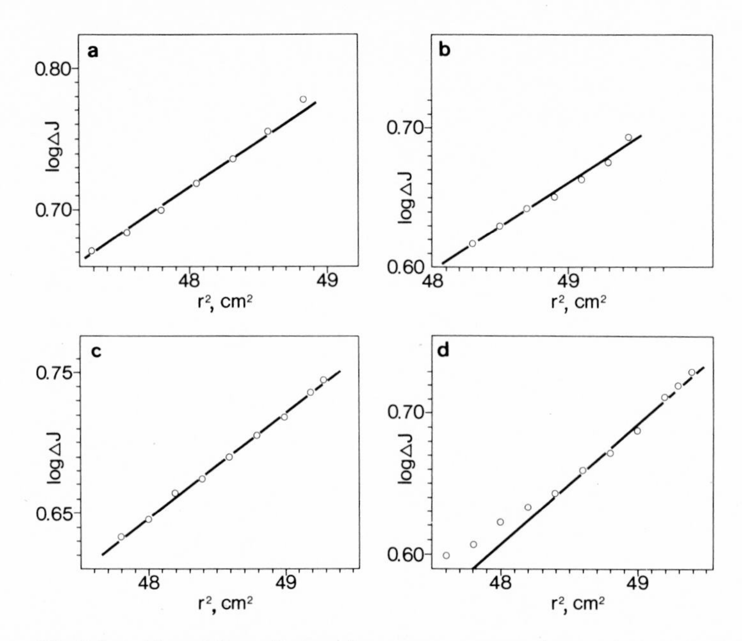

Fig. 2. Log ΔJ vs r^2 from the interference pattern at sedimentation equilibrium. ΔJ is the displacement of a fringe, in fringe numbers, between r^2 and $r^2 + 2.00$ where r is the radial distance from the axis of rotation. *a* IMA in D_2O as solvent; *b* IMA in H_2O as solvent; *c* AGA in D_2O as solvent; *d* AGA in H_2O as solvent.

$MW_{D_2O} \cdot (1-\bar{v} \cdot 1.106) = \text{slope} \cdot 2.303 \cdot 2RT/\omega^2$, where MW_{D_2O} is the molecular weight of the macromolecule in D_2O $(1.0155 \cdot MW_{H_2O}) \cdot \bar{v}$ is the partial specific volume, 1.106 is the density of D_2O at 20 °C, ω is the speed in rad s^{-1}. The slope from figure 2a is 0.06629 cm^{-2}, therefore, $MW_{H_2O} \cdot (1-\bar{v} \cdot 1.106) = 871.4$.

In figure 2b experimental values are plotted in the same fashion with H_2O as solvent. The slope obtained was 0.06079 cm^{-2}, with the correlation coefficient 0.9962; we obtain $MW_{H_2O} \cdot (1-\bar{v} \cdot 0.9982) = 1{,}161$. This result, together with the result of the D_2O experiments, yields: $\bar{v} = 0.699$ ml/g and $MW_{H_2O} = 3{,}843$ dalton.

In figure 2c are plotted the results for AGA as described for IMA in the D_2O experiments. Again, the straight line approximates the findings very well. All points are used, giving a slope of 0.07473 with a correlation coefficient of 0.9989. In the calculation (as above), we obtain MW_{H_2O} = 4,329 dalton.

The H_2O experiments are plotted in figure 2d. Here, the excellent linear approximation found in the D_2O experiments is observed to be weakened, indicating a possible thermal instability of the antigen. It was decided to incorporate the maximun number of points with the highest possible correlation coefficient. This was done by deleting the four results with the lowest r values. Accordingly, this gave the line, as shown in figure 2d, with a slope of 0.8641 cm^{-2} and a correlation coefficient of 0.9952. Proceeding as before, we find $MW_{H_2O} \cdot (1 - \bar{v} \cdot 0.9983) = 1{,}310.3$. If, as before, $\bar{v} = 0.699$ ml/g, we obtain $MW_{H_2O} = 4{,}335$ dalton. This result is in excellent agreement with the D_2O results, though clouded by the change in the sedimentation pattern observed by the low r values. If the two equations involving the molecular weight and partial specific volume are solved together, we find: 0.700 and $MW_{H_2O} = 4{,}332$ dalton.

Discussion

We believe that problems of population and conception as well as restoration of fertility are amenable to resolution using immunologic procedures. These two problems, one of profound significance on a worldwide basis and the other equally profound for the individuals involved, should not be ignored, for the implications are too vast. Much has been written about the immunologic control of fertility/sterility [1, 4, 7–9].

Of all the antigens that may be involved in conception, a sizeable list could be compiled: anterior pituitary hormones, the respective releasing factors, secretions of the reproductive tracts and associated ducts, steroid hormones (when conjugated they become potent antigens) of the gonads and adrenal, and the gametes. We have chosen to examine only the sperm antigens, partly because of their accessibility and the relative ease with which such experimentation could be performed. With this starting point, it is hoped to bring immunoreproduction to its proper place in immunology and biology. There may be a unity in turning the liability of immune reactions inducing sterility toward the advantage of controlling fertility.

We have also noted above that the deletion of a single cell line, the seminiferous epithelium, can be induced by a single injection of aspermatogenic antigen. The implications regarding tumor biology, oncology and the control of non-desirable cellular replication could be worthy of intensive exploration.

Regarding the sperm antigens, it must be emphasized that these antigen-antibody systems are exquisitely specific both as to organ (tissue and cell type) and to species. The organ (tissue, cell type) specificity is readily observable in histological, cytological and electron microscopic sections of the testes of guinea pigs immunized with ASA. Only those cells of the seminiferous cell line containing antigen are deleted. The species-specificity is clearly demonstrable: rooster, mouse, rat, rabbit, opossum, raccoon, bull, ram, horse, goat, etc. sperm antigens do not cross-react with guinea pig sperm antigens in vitro or in vivo. Some of these studies have already been reported [2, 3, 7]. Meanwhile, some progress has been made toward characterization of human testicular and sperm antigens [6].

The merits of this contribution lie not in the procedures and techniques for the extraction, isolation, and purification of the antigens. Such techniques and procedures might even be primitive and most certainly will be superseded. But the prospect of the induction of antibodies to the flagella or cilia of infectious organisms using the model of the immobilizing and agglutinating antigen-antibody systems described herein brings new avenues of preventive and therapeutic medicine. With the low molecular weights for IMA and AMA that we have found, hopes for sequencing and even synthesis of antigenic materials become reasonable and feasible.

Acknowledgements. These studies were supported by grants from the National Science Foundation (USA) (PCM 76–01125) and the Population Council (M73.67).

References

1 Jones, W.R.: Fert. Steril. *33:* 577–586 (1980).
2 Katsh, S.: J. exp. Med. *107:* 95–103 (1958).
3 Katsh, S.: Int. Archs Allergy appl. Immun. *15:* 172–188 (1959).
4 Katsh, S.: Am. J. Obstet. Gynec. *77:* 946–956 (1959).
5 Katsh, S.: Int. Archs Allergy appl. Immun. *16:* 241–275 (1960).
6 Katsh, S.: J. Urol. *87:* 896–902 (1962).
7 Katsh, S.: in Advances in obstetrics and gynecology, vol. I, pp. 467–485 (Williams & Wilkins, Baltimore 1967).
8 Katsh, S.: Immunological aspects of reproduction. Ovum implantation, pp. 307–373 (Gordon & Breach, New York 1969).
9 Katsh, S.: Life Sci. *20:* 761–774 (1977).
10 Katsh, S.; Aguirre, A.R.; Leaver, F.W.; Katsh, G.F.: Fert. Steril. *23:* 644–656 (1972).

11 Katsh, S.; Katsh, G.F.: Fert. Steril. *12:* 522–537 (1961).
12 Katsh, S.; Klostergaard, H.; Katsh, G.F.: Fert. Steril. *27:* 971–974 (1976).
13 Katsh, S.; Leaver, F.W.; Katsh, G.F.: Mt. Sinai J. Med. *XLII:* 405–409 (1975).
14 Katsh, S.; Leaver, F.W.; Reynolds, J.S.; Katsh, G.F.: J. immunol. Methods *5:* 179–187 (1974).
15 Landsteiner, K.: Centralbl. Bakt. *25:* 546–549 (1899).
16 Metalnikoff, S.: Annls Inst. Pasteur, Paris *14:* 577–589 (1900).
17 Metchnikoff, E.: Annls Inst. Pasteur, Paris *14:* 1–12 (1900).
18 Nazarian, G.M.: Analyt. Chem. *40:* 1766–1769 (1968).

Dr. Seymour Katsh, Department of Pharmacology, University of Colorado Medical School, Denver, CO 80262 (USA)

The Immune System, vol. 1, pp. 278–283 (Karger, Basel 1981)

Control of the Cellular Immune Response of Mice to Sheep Erythrocytes by Specific Antibodies and Low Molecular Weight Surface Antigens

P. Albers[1], *H. Gilsenbach*[1]

Chemotherapeutisches Forschungsinstitut 'Georg-Speyer-Haus', Frankfurt am Main, FRG

This work is dedicated to *N.K. Jerne* on his 70th birthday. Almost 15 years ago, he aroused in us an enthusiasm for his field of study that is still strong today, 10 years after his departure from Frankfurt.

In an earlier study we showed that cell-mediated immune responses against sheep red blood cells (SRBC) can be completely abrogated by low molecular weight surface antigens [2]. In this paper we examine the effect of specific antisera, both alone and in combination with low molecular weight surface antigens.

Materials and Methods

Animals. Female colony-bred NMRI mice (purchased from Ivanovas, Kisslegg, Allgäu), aged 6–8 weeks and weighing 25–28 g, were used. In general each group of mice consisted of 10 animals and the experiments were repeated as mentioned in the results.

Antigens. SRBC were obtained from blood collected in Alsever solution. Before use the SRBC were washed twice in phosphate buffered saline pH 7.0 (PBS). To obtain serum, the sheep blood, with no addition, was centrifuged at 3,000 rpm (1,700 *g*), the serum was decanted and stored at –20 °C. This serum contains a small quantity of low molecular weight surface antigens from sheep erythrocytes [2]. Erythrocytes from rabbits were isolated in the same way.

[1] *Paul Albers* and *Heidrun Gilsenbach* were colleagues of *Niels Jerne* at the 'Georg-Speyer-Haus' in Frankfurt from 1966 to 1969.

Antisera. Mice were immunized with 10^8 SRBC and bled after 3 days. The serum of these mice contained specific antibodies mainly of the 19S class. Hyperimmune sera containing 7S antibodies with high avidity were prepared as described elsewhere [7]. The 7S globulin fraction from such anti-SRBC antisera was separated by fractionation on a DEAE-cellulose column [5].

Sensitization. Equal volumes of erythrocytes in saline and incomplete Freund's adjuvant were emulsified just prior to intradermal injection. In general the dose for immunization was 5×10^8 SRBC. The skin was shaved and a total of 0.1 ml (or 2×0.1 ml when immunizing with 5×10^8 SRBC) injected intradermally. If required, 0.1 ml antiserum or 0.5 ml sheep serum were injected i.p. 0.5 h before sensitization.

Challenge and Assay of the Delayed Hypersensitivity Response. Usually 14 days (but sometimes only 7 days) after the sensitization, 10^8 SRBC in 0.025 ml PBS were injected intradermally on the dorsal aspect of one hind paw. If required, 0.1 ml antiserum or 0.5 ml sheep serum were injected i.p. 0.5 h before challenge.

The footpad swelling (FS) measured 22 h later compared with the non-challenged foot was the index of the degree of delayed type hypersensitivity (DTH). The footpad swelling was measured by the method of *Axelrad* [3]. The size increase of the challenged footpad was expressed as a percentage, and for each experimental group the arithmetical mean was calculated. The average size increase of a control group of unsensitized animals, which also receive the challenge antigen dose 22 before the measurement, was subtracted from the value for the experimental group.

Statistics. The results were compared with Wilcoxon's test.

Results

We studied the dependence of FS on the antigen dose used for sensitization. In these experiments the time between induction and expression of the immune response was 7 days. Each group of mice consisted of 20 animals. The results of the experiment: Sensitization dose 5×10^5 SRBC, FS = 6.4%; sensitization dose 5×10^6 SRBC, FS = 15.8%; sensitization dose 5×10^7 SRBC, FS = 29.2%; sensitization dose 5×10^8 SRBC, FS = 31.5%.

Effect of Specific Antisera on the Cell-Mediated Immune Response to SRBC. The sensitization to SRBC may be suppressed significantly by antisera containing 19S antibodies. FS of the antibody-treated animals = 26.8%; FS of the controls = 31.2% ($p < 0.025$). From the results of our previous study, we calculate that their effect corresponds to a 10-fold reduction in antigen dose in the experimental mice. The experiment was repeated 7 times. The sensitization of the immune response is also suppressed by antisera containing 7S antibodies. FS of the antibody-treated animals = 27.8%; FS of the controls = 29.8% ($p > 0.025$). The experiment was repeated 8 times. When the time between induction and expression of the immune response is 7 days, FS of the antibody-treated animals = 27.2%; FS of the controls = 34.0% ($p < 0.001$). The experiment was repeated 11 times.

As shown by the following experiments, the suppression is specific and due to material in the 7S globulin fraction. The sensitization was carried out with 2.5×10^7 SRBC. FS of the animals treated with hyperimmune serum to SRBC = 18.6%; FS of mice treated with hyperimmune serum against rabbit red cells = 24.6%; FS of the controls = 24.9%. The experiment was done with 20 mice per group. The time between induction and expression was 7 days. The difference between the first and the second group was significant ($p < 0.05$). The 7S globulin fraction of a hyperimmune serum against SRBC was separated by fractionation on a DEAE-cellulose column [5]. Application of either hyperimmune serum or an equivalent amount of its globulin fraction resulted in the same suppressive effect. FS of the animals treated with hyperimmune serum = 23.8%; FS of the mice treated with the globulin fraction = 21.7%; FS of the controls = 38.8%.

If injected before the challenge with erythrocytes, antisera containing 19S antibodies were able to enhance the cellular immune response. FS of mice treated in this way = 34.5%; FS of the controls = 30.7%. The experiment was repeated 7 times ($p < 0.025$). Hyperimmune serum against SRBC does not enhance. FS of the antiserum-treated animals = 29.3%; FS of the controls = 30.7%. The experiment was repeated 7 times. With a time of 7 days between induction and expression of the immune response, FS of the antiserum-treated animals = 31.5%; FS of the controls = 30.9%. In neither case was there a significant difference.

Effect of the Treatment with a Combination of Antisera and Surface Antigens on the Cell-Mediated Immune Response to SRBC. There is no influence on the sensitization of the cellular immunity by low molecular weight surface antigens from SRBC contained in sheep serum [2]. As shown above, specific antisera partially suppress sensitization. If we add non-effective sheep serum to the suppressive antisera, we expect that the suppressive effect will be neutralized with sheep serum. FS of the controls = 32.2%; FS of the animals treated only with sheep serum = 32.9%; FS of the animals treated with sheep serum and antiserum = 33.0%. The experiment was repeated 7 times. As expected, we found no significant differences between the groups of mice. With 7S-antiserum, FS of the controls = 30.2%; FS of the animals treated only with sheep serum = 30.8%; FS of the animals treated with non-effective sheep serum and suppressive antiserum = 30.3%. The experiment was repeated 13 times.

If given before challenge with SRBC [2], the surface antigens of SRBC in sheep serum have a strong suppressive effect on a cellular immunity. 7S antibodies have no effect while 19S antibodies enhance the response. To obtain a high level of 19S antibodies at the time of the challenge, we injected sheep serum twice [1]; one injection 2 days before the challenge, the other 0.5 h before. The dose for sensitization was 2.5×10^8 SRBC. FS of the controls = 30.6%; FS of the animals treated with sheep serum once before the challenge = 10.7%; FS of the animals treated twice with sheep serum (as a consequence of this, these animals had a very high level of 19S antibodies before the challenge) = 41.7%. The difference between induction and expression of the immune response was 7 days. To obtain a level of 7S antibodies, 0.1 ml antiserum was injected as usual i.p. into the animals 0.5 h before the challenge with 2.5×10^8 SRBC. FS of the controls = 32.1%; FS of the animals treated with sheep serum = 4.8%; FS of the animals treated with sheep serum and antiserum = 6.5%. The experiment was repeated 7 times. The difference between induction and expression was 14 days ($p \leq 0.025$).

Discussion

In the experiments in this study, DTH was measured and its regulation through specific antibodies and surface antigens from erythrocytes was determined. Cytotoxic T lymphocytes have the same surface markers as the T lymphocytes responsible for DTH [6]. It could be shown that cell-mediated cytolysis can be blocked by surface antigens [4]. Therefore, it would seem reasonable to extend the regulation mechanisms found by us to include cell-mediated immune response in general. In this study it could be proved that specific antisera against sheep erythrocytes suppress the sensitization against this antigen. The expression of the immune response is increased by 19S antibodies, but not by 7S antibodies. *Axelrad* [3] was also able to show that rats treated with hyperimmune serum were partially suppressed for sensitization with SRBC. Antiserum containing 19S antibodies had no effect.

In a previous paper we stated that sheep serum did not suppress the sensitization but that it did suppress the expression of a cell-mediated immune response [2]. Data in the present paper support the assumption that the effective substances in sheep serum are surface antigens from the erythrocytes. Specific antibodies can reduce and even compensate for the suppressive effect of sheep serum on the expression of cell-mediated immunity. On the other hand, sheep serum can cancel out the suppressive effect of specific antibodies on the sensitization of the cell-mediated immune response, due to the surface antigens it contains. We can state this quite simply: specific antibodies suppress the sensitization of a cell-mediated immune response and complementary surface antigens suppress its expression. As antagonists they can mutually impede their activities to such an extent that their effect disappears completely. Because of this strong effect we believe that other control mechanisms are of secondary importance.

Specific antibodies most likely exert their effect by blocking antigenic determinants on the erythrocytes. From 0.01 to 0.001 ml of specific antiserum, placed in the antigen suspension, has approximately the same effect for suppressing sensitization as 0.1 ml of antiserum injected i.p. before sensitization (data not shown). It can be assumed that the surface antigens of the sheep serum block the sensitized lymphocytes. When transmitted to normal syngenic mice, lymphocytes covered with surface antigen trigger a cell-mediated immune response in inbred mice that is much weaker than that triggered by non-covered sensitized lymphocytes (data not shown).

We have also been successful in achieving similar results with other types of cells. For example, a cell-mediated immune response triggered by lymphocytes can be greatly suppressed by means of surface antigens from lymphocytes (which can be easily obtained from lymphocytes with buffer containing Triton) (data not shown). We believe that results obtained for sheep erythrocytes can be applied in the treatment of disease in three situations: (1) tumors, (2) some autoimmune diseases, and (3) allograft rejection. It should be noted here that all three of these cases are influenced by the cell-mediated immune responses of the DTH type [8].

Tumors. Once the tumor has been removed, the cell-mediated immune response must be strengthened. First of all, specific antibodies should be neutralized by surface antigens from the tumor cells, because antibodies suppress sensitization. Finally, tumor cells free of antibodies have to be implanted into the skin for the sensitization. Such cells can be grown in nude mice or in cell culture. It should be stressed that the tumor must not be taken from the original tumor tissue, the cells of which are certain to be covered with specific antibodies against the tumor; therefore, as shown by our results, the original tumor tissue is completely unsuitable for sensitization. Some 1–2 weeks later the humoral immune response against the tumor antigens must be strengthened to overcome the surface antigens of the tumor cells. Immunization is best done with high molecular tumor antigens, which can be prepared from low molecular weight antigens by chemical bonding.

Autoimmune Diseases. Whenever these are caused by a cell-mediated immune response it is necessary to give high doses of surface antigens in order to neutralize the sensitized lymphocytes directed against the diseased tissue. These antigens can be extracted with buffers containing Triton from the tissue or from the diseased cells. At the same time, the production of antibodies must be suppressed, because humoral antibodies – as shown – can remove the effect of these protective antigens.

Allograft Rejection. The same principle as for autoimmune diseases applies to transplantation.

Acknowledgement. We thank Miss *C. Beege* for her excellent technical help.

References

1 Albers, P.; Gilsenbach, H.: in Heymann, Arbeiten aus dem Paul-Ehrlich-Institut, dem Georg-Speyer-Haus und dem Ferdinand-Blum-Institut – Regulation der humoralen Immunantwort durch Antigen und Antikörper, Heft 69, pp. 1–9 (Fischer, Stuttgart 1974).
2 Albers, P.; Gilsenbach, H.: in Brede, Arbeiten aus dem Paul-Ehrlich-Institut, dem Georg-Speyer-Haus und dem Ferdinand-Blum-Institut – Tendenzen der prophylaktischen Medizin, Heft 74, pp. 42–55 (Fischer, Stuttgart 1980).
3 Axelrad, M.A.: Immunology *15:* 159–171 (1968).
4 Bonavida, B.: J. Immun. *112:* 926–934 (1974).
5 Fahey, J.L.; Terry, E.W.: in Weir, Handbook of experimental immunology; 3rd ed., chap. 8, p. 10 (Blackwell, Oxford 1978).
6 Golub, ES.: The cellular basis of the immune response; 2nd ed., p. 269 (Sinauer Associates, Sunderland 1978).
7 Henry, C.; Jerne, N.K.: J. exp. Med. *128:* 133–152 (1968).
8 Sell, S.: Immunologie, Immunpathologie und Immunität; 2. Aufl., pp. 210, 242, 315 (Verlag Chemie, Weinheim 1977).

Dr. Paul Albers, 'Georg-Speyer-Haus', Paul-Ehrlich-Strasse 42/44,
D-6000 Frankfurt/M 70 (FRG)

Allotypes and Idiotypes

The Immune System, vol. 1, pp. 284–290 (Karger, Basel 1981)

The Ideal Rabbit

Andrew S. Kelus[1]

Basel Institute for Immunology, Basel, Switzerland

It is an open secret that *Niels Jerne* has never seen a rabbit in his life, and he was quite certain that this entity was never mentioned by *Kierkegaard* [24] nor by *Wittgenstein* [42, 43]. Nevertheless, *Jerne* had a hunch and postulated: the ideal rabbit (den ideale kanin) exists. After all, it appears on the screen in Hollywood cartoons to please children and grown-ups, and it is more or less real, as far as we can be sure about the reality of existence. The ideal rabbit features all the imaginable immunological goodies. It displays all the polymorphism and diversity of the antibody molecule to such an extent that it makes the immunologist very happy. It has allotypes; it also has idiotypes which *Jerne* [17] predicted in 1960. It obeys the Lord's law of somatic mutations [18], and it has an interesting network [19].

Immunoglobulin Polymorphism

The rabbit has been serving immunology as an experimental animal for over 100 years. But its heyday began about two decades ago when it entered immunogenetics, or to be more precise, when the first genetic markers of rabbit antibodies were described [31]. Until then, rabbit genetics revolved mainly around morphological markers such as those of coat and skeleton, with blood groups being an exception [2, 4–6, 11, 29].

[1] *Andrew S. Kelus* was a founding member of the Basel Institute for Immunology.

Fig. 1. A wooden carving of a rabbit (6 inches high) brought by *P.G.H. Gell* from his trip to Nigeria and given to the author. The carving was made by an unknown folk artist around 1850.

The discovery of immunoglobulin polymorphism in 1956 by *Oudin* [31], and named allotypy by him, has been a strong stimulus, although it caught the imagination of the immunologists with some delay. It was partly a fault of a peculiar habit observed by the French Academy of Sciences to hide findings from the eyes of others. *Oudin* [31] wrote: 'The essentials of the observations reported in this paper and of their interpretation were placed in a sealed envelope with the title "Reaction of specific precipitation between sera of animals of the same species", and deposited at the Académie des Sciences, the 21st December, 1953.'

Actually, the story of allotypy goes back to the very beginning of this century. *Schütze* [38] reported in 1902 that he injected rabbits with blood of other rabbits and obtained so-called 'isoprecipitins', i.e. antibodies reacting with *some* rabbit sera. This important publication appeared too early for the immunological world and went unnoticed. One would agree with *Browning* [3] that 'ignorance is not innocence but sin'.

It is common knowledge that an antibody is an immunoglobulin molecule which consists of heavy and light polypeptide chains. The amino acid sequences of some of these chains have been established. A chain is divided into a variable part, with the antibody combining site at the N-terminal, and a constant part. These parts may be subdivided into several domains and regions. The constant part of the heavy chain determines the class of immunoglobulin: α, δ, γ, ε, or μ. The light chain exists in at least two forms: κ and λ.

Intensive research has revealed many interesting and useful allotypic markers of the immunoglobulin chains of the rabbit. The heavy chains possess allotypes which are determined by a cluster of linked loci: *a*, *x*, and *y* on the variable part of the heavy chains of all classes; *d* and *e* on the constant part of the γ chain; *f* and *g* on the constant parts of α_1 and α_2 chains; and *ms* (or *n*) on the constant part of the μ chain. About 25 alleles belonging to these loci have been reported so far [8, 9, 21, 25–28]. If these loci were independent, this would allow for nearly 10,000 genotypes. Less than 1% have been observed up to now, and only a few recombinants, which were found casually, have been reported. No proper genetic studies have yet been performed to establish the gene order and map distances.

Todd [41], utilizing allotypic markers, found that the IgG and IgM (γ and μ chains, respectively) share the variable part of their chains, in addition to having the same κ or λ chain. This was later extended to other classes of immunoglobulins [10]. Most (95%) rabbit light chains are of the κ type, the rest being λ type. The *b* locus of the κ chain, which determines its allotypes, exists in several allelic forms: *b*4 (*b*4*Ms*3 and *b*4*Ms*7), *b*5, *b*6 and *b*9 [21, 25, 27, 28]. A newly established rabbit strain, called BASILEA, lacks the κ chain and, hence, its allotypes. This chain is replaced by the λ type. BASILEA rabbits are homozygous for *bas*, an allele of the *b* locus [23]. *Ms*3 and *Ms*7 are linked to the *b* locus. They form allotypic determinants on IgM when combined with the μ chains [30].

There are two allotypic specificities of the λ chain: *c*7 and *c*21. The genetics of the *c* locus is obscure; these markers behave like alleles in some rabbit colonies but not in others [25, 28].

The allotypes of the variable part of the heavy chains, determined by the *a* locus, differ from each other by multiple amino acid residue substitutions. The same is true for the *b* locus, where several amino acid substitutions could be associated with the constant part of the κ chains. On the other hand, some investigators ascribed the differences between allotypes of the *b* locus to amino acid interchanges within the variable part of the κ chains [25,

27, 28]. The *d*11-*d*12 allotype differences may be caused by the interchange of methionine-threonine in the hinge region of the IgG molecule, and the differences between allotypes *e*14-*e*15 could be associated with the interchange of threonine-alanine at the position 309. The chemistry of other markers is unknown [9, 25, 27, 28].

The ratio of allelic products in isolated antibodies of a heterozygous rabbit differs remarkably from that measured in its 'normal serum'. In some cases, the product of one allele was practically absent [12, 36]. This led to the conclusion that only cells expressing a single allotypic allele contribute to the antibody pool. Imbalance of the antibody populations preceded directly the discovery of allelic exclusion by *Pernis* et al. [35], who proved that an individual plasma cell of an allotypically heterozygous rabbit expresses only one allotypic allele.

Another important fact was reported by *Dray* [7] who showed that an allotypically heterozygous doe, immunized against an allotype of the buck, delivered offspring which failed to manifest this paternal allotype for many months. This phenomenon was termed 'allotype suppression'. Soon after, a number of researchers confirmed this finding by using passive immunization of newborn rabbits. The most promising method has been to transfer rabbit embryos into pseudopregnant mothers of an entirely different genetic make-up, followed by immunization. In this way 'allotypically blank' rabbits, i.e. individuals expressing no allotypes of the *a* and *b* loci, could be obtained [*Hatton, Kelus and Leibo,* unpublished].

Perhaps the most fruitful discovery made in the rabbit was that of idiotypy [13, 22, 33, 34]. It seemed, at the beginning at least, that each rabbit antibody population of a given specificity possessed unique, non-cross-reacting, antigenic determinants in the variable parts of the polypeptide chains. Cross-reacting idiotypes have since been reported [32]. Recently, so-called common idiotypes shared by anti-allotype antibodies of many individuals have been described [14, 37]. Research on idiotypes extended rapidly to other species and influenced immunological thinking tremendously [cf. 19, 20].

Non-Immunoglobulin Polymorphisms

I am indebted to Dr. *Karen L. Hagen,* University of Illinois Medical Center, for help in preparing this section. Most references to this section may be found in the Biological Handbooks [2].

Blood Groups. Most of the work performed by early investigators concerned the *Hg* locus with its alleles Hg^A, Hg^D, Hg^F, and Hg^N. Subsequently, six other blood group loci have been reported: *Hb, Hc, He, Hh, Hq,* and *Hu.* Two alleles are known at each of these loci. Blood group differences in the rabbit play a role in a haemolytic disease of the newborn (hydrops) and in graft survival time [1].

Histocompatibility. The major histocompatibility complex is similar to that of other mammals; it consists of two polymorphic loci, *RLA-A* and *RLA-D,* and possibly of a third locus. The alleles of the *RLA-A* locus can be grouped into clusters by cytotoxicity tests. This depends on the stimulation of cells with different alleles by each other in the mixed lymphocyte reaction, which is under the control of the *RLA-D* locus. So far, about a dozen *RLA-A* alleles, and only five *RLA-D* alleles, have been found. The histocompatibility complex is important in determining the success of producing allotype chimeras as well as having a strong influence on graft survival [39, 40].

Non-Immunoglobulin Allotypes of Serum Proteins. The products of many genes can be detected by serological methods. For the low-density lipoprotein system one locus with three complex alleles $lpq^{1,3}$, $lpq^{2,3}$, and $lpq^{2,4}$ is known. The locus controlling α_2-macroglobulins has three alleles – $mtz^{1,3}$, $mtz^{1,4}$, and $mtz^{2,3}$. Four other serum protein loci, each with two codominant alleles, have been reported: α_1-acryl esterase *(Ess),* haptoglobin *(Hph),* high-density lipoprotein *(Lhj),* and E serum protein. In addition, there are several other loci with only one known specificity: the A1, H1 and R67 high-density lipoprotein systems, the Ai system of α-globulin, and the uncharacterized J and M protein systems [5, 15; *Hagen,* unpublished].

Other genetic systems are determined by electrophoresis. There is some evidence that the two serum esterases, the cocaine esterase, and the atropine esterase are under the control of one locus with five alleles. Some investigators claim, however, that two very closely linked loci, *Est*-1 *Est*-2, are responsible for these esterases. A third serum esterase locus, so-called 'dependent' prealbumin carboxyl esterase, has also been described. Other serum protein loci, distinguishable electrophoretically, are: the hemopexin locus *(Hx)* with four codominant alleles, and the *Prt* serum protein locus with two alleles. The serum protein locus *Prt* and low-density lipoprotein locus *lpq* are linked to the *b* locus determining allotypes of the κ light chain.

Some red cell enzymes are under the control of a single locus: methemoglobin reductase (or NADH-diaphorase) with two codominant alleles;

adenosine deaminase (ADA) and mannose-6-phosphate isomerase (MPI), each determined by three codominant alleles. Esterases of the red cells and platelets are determined by three loci: the RBC carboxyl esterase-1 (*Es*-1), the platelet carboxyl esterase-2 (*Es*-2), and the RBC carboxyl esterase-3 (*Es*-3). The *Es*-1 locus with its two alleles is closely linked to the *Es*-2 locus with its two alleles. The *Es*-3 locus has three alleles and is not linked to the two other loci.

A polymorphism of the sixth component of rabbit complement has also been reported; individuals without this component are presumably homozygous for a silent recessive allele.

Postscript

The discoveries summed up in this article have been fully appreciated and exploited by *Jerne* [16, 18–20] in his persistent and evolutionary thinking. We may feel happy that we were able to contribute a little to his *Tractatus Immunologiae.* During the last decade, the mouse has taken over the leading part in the immunological drama. The rabbit has been waiting patiently to become a prima donna again. I have no doubt that great days of *Oryctolagus cuniculus* will return.

Let me finish with an anecdote: An international competition was announced one day, the theme of which was the rabbit. An Englishman wrote a brochure, 'The rabbit as a domestic animal'. A Frenchman presented an essay, 'La vie sexuelle du lapin'. A German compiled a 'Kurzes Kaninchenhandbuch' in three volumes. A Pole wrote a political pamphlet, 'The rabbit and the Polish cause'. An Italian prepared a cookbook, 'One hundred ways to cook a rabbit'. But the Danish contribution was: 'Eksisterer kaninen i det hele taget?' ('Does the rabbit exist at all?').

References

1 Adler, L.T.; Adler, F.; Cohen, C.; Tissot, R.G.; Lancki, D.: Transplantation *24:* 338 (1977).

2 Altman, P.L.; Dittmer Katz, D. (eds.): Biological handbooks, III, part 2 (Federation of Am. Soc. for Experimental Biology, Bethesda 1979).

3 Browning, R.: Poems (Chapman & Hall, London 1849).

4 Castle, W.E.: Biblgr. genet. *1:* 419 (1925).

5 Cohen, C.: Genetics in laboratory animal medicine, publ. 1679, p. 34 (Natn. Academy of Science, Washington 1969).

6 Dorn, F.K.: Rassenkaninchenzucht; 2nd ed. (Neumann, Radebeul 1968).
7 Dray, S.: Nature, Lond. *195:* 677 (1962).
8 Dray, S.; Dubiski, S.; Kelus, A.; Lennox, E.S.; Oudin, J.: Nature, Lond. *195:* 785 (1962).
9 Dubiski, S.: Med. Clins N. Am. *56:* 557 (1972).
10 Feinstein, A.: Nature, Lond. *199:* 1197 (1963).
11 Fox, R.R.: in Weisbroth, Flatt, Kraus, The biology of the laboratory rabbit, p. 1 (Academic Press, New York 1974).
12 Gell, P.G.H.; Kelus, A.S.: Nature, Lond. *195:* 44 (1962).
13 Gell, P.G.H.; Kelus, A.S.: Nature, Lond. *201:* 687 (1964).
14 Gilman-Sachs, A.; Dray, S.; Horng, W.J.: J. Immun. *125:* 96 (1980).
15 Hagen, K.L.; Suzuki, Y.; Cohen, C.: Anim. Blood Groups biochem. Genet. *9:* 151 (1978).
16 Jerne, N.K.: Proc. natn. Acad. Sci. USA *41:* 849 (1955).
17 Jerne, N.K.: A. Rev. Microbiol. *14:* 341 (1960).
18 Jerne, N.K.: Eur. J. Immunol. *1:* 1 (1971).
19 Jerne, N.K.: Annls Immunol. *125C:* 373 (1974).
20 Jerne, N.K.: Harvey Lect. *70:* 93 (1976).
21 Kelus, A.S.; Gell, P.G.H.: Prog. Allergy, vol. 11, p. 141 (Karger, Basel 1967).
22 Kelus, A.S.; Gell, P.G.H.: J. exp. Med. *128:* 215 (1968).
23 Kelus, A.S.; Weiss, S.: Nature, Lond. *265:* 156 (1977).
24 Kierkegaard, S.: Enten-eller (Reitzel, Kjøbenhavn 1843).
25 Kindt, T.J.: Adv. Immunol. *21:* 35 (1975).
26 Knight, K.; Hanly, W.C.: Contemp. Top. mol. Immunol. *4:* 55 (1975).
27 Mage, R.: Contemp. Top. mol. Immunol. *8* (in press).
28 Mage, R.; Lieberman, R.; Potter, M.; Terry, W.D.: in Sela, The antigens, vol. 1, p. 299 (Academic Press, New York 1973).
29 Nachtsheim, H.: Z. Tierzücht. Züchtungsbiol. *14:* 53 (1929).
30 Naessens, J.; Hamers-Casterman, C.; Kelus, A.S.: Immunogenetics *8:* 571 (1979).
31 Oudin, J.: C.r. hebd. Séanc. Acad. Sci., Paris *242:* 2489, 2606 (1956).
32 Oudin, J.; Cazenave, P.-A.: Proc. natn. Acad. Sci. USA *68:* 2616 (1971).
33 Oudin, J.; Michel, M.: C.r. hebd. Séanc. Acad. Sci., Paris *257:* 805 (1963).
34 Oudin, J.; Michel, M.: J. exp. Med. *130:* 595, 619 (1969).
35 Pernis, B.; Chiappino, G.; Kelus, A.S.; Gell, P.G.H.: J. exp. Med. *122:* 853 (1965).
36 Rieder, R.F.; Oudin, J.: J. exp. Med. *118:* 627 (1963).
37 Roland, J.; Cazenave, P.-A.: C.r. hebd. Séanc. Acad. Sci., Paris *288:* 571 (1979).
38 Schütze, A.: Dt. med. Wschr. *28:* 804 (1902).
39 Tissot, R.G.; Cohen, C.: Fed. Proc. *37:* 1275 (1978).
40 Tissot, R.G.; Lancki, D.W.; Blaesing, M.: Immunogenetics *8:* 509 (1979).
41 Todd, C.W.: Biochem. biophys. Res. Commun. *11:* 170 (1963).
42 Wittgenstein, L.: Annln Naturphilosophie *14:* 185 (1921).
43 Wittgenstein, L.: Philosophische Untersuchungen (Blackwell, Oxford 1953).

Dr. A.S. Kelus, Basel Institute for Immunology, Postfach, CH–4005 Basel (Switzerland)

The Immune System, vol. 1, pp. 291–298 (Karger, Basel 1981)

Control of Expression of Immunoglobulin Genes in Rabbits

C.L. Martens, A. Gilman-Sachs, K.L. Knight[1]

Department of Microbiology and Immunology, University of Illinois, Chicago, Ill., USA

Rabbit immunoglobulins (Ig) have been extensively studied by immunogeneticists since 1956 when *Oudin* [11] identified genetic variants, or allotypes, of these molecules. Allotypic specificities have been found on both light and heavy chains, and within the heavy chains, genetic variants are found on both the variable and constant regions. Three separate loci, V_Ha, V_Hx, and V_Hy, have been shown to control the synthesis of variable regions of heavy chains; from 70 to 90% of serum Ig molecules carry the V_Ha determinants [1, 2], and approximately 6 and 4% carry the V_Hx and V_Hy determinants, respectively [4]. The allotypic specificities controlled by these loci are on heavy chains of each Ig class, IgM, IgG, IgA, and IgE, independent of antibody specificity. The loci controlling the constant regions of heavy chains are the *n* locus for the Cμ of IgM, the *de* locus for Cγ of IgG, and the *f* and *g* loci for Cα regions of each of two subclasses of IgA (for review, see [5, 6]). All loci coding for heavy chains are closely linked to each other, and the allelic forms, which are inherited as a Mendelian unit, appear together in specific gene combinations or haplotypes (table I) [13]. Of 4,800 possible combinations of all the different heavy chain allotypes, only twelve heavy chain haplotypes have been identified in laboratory rabbits. This suggests that little recombination occurs in the heavy chain chromosomal region.

[1] *Katherine Knight* was a member of the Basel Institute for Immunology in 1975.

Table I. Rabbit heavy chain haplotypes[a]

Designation[b]	Genotype[c]						
	variable region			constant region			
	a	x	y	μ	α	α	γ
A (Agp[1])	a^1	x^-	y^-	n^{81}	f^{73}	g^{74}	$de^{12,15}$
B (Agp[7])	a^1	x^-	$y^{33,30}$	$n^{83,80}$	f^{71}	g^{75}	$de^{12,15}$
C (Agp[3])	a^1	x^-	$y^{33,30}$	$n^{83,80}$	f^{72}	g^{74}	$de^{11,15}$
I (Agp[10])	a^1	x^-	$y^{33,30}$	$n^{83,80}$	f^{69}	g^{77}	$de^{12,14}$
J (Agp[8])	a^1	x^-	y^-	n^{81}	f^{70}	g^{76}	$de^{12,15}$
E (Agp[2])	a^2	x^{32}	$y^{33,-}$	$n^{82,80}$	f^{71}	g^{75}	$de^{12,15}$
F (Agp[9])	a^2	x^{32}	$y^{33,-}$	$n^{82,80}$	f^{69}	g^{77}	$de^{12,15}$
K(Agp[12])	a^2	x^{32}	$y^{33,-}$	$n^{82,80}$	f^{69}	g^{77}	$de^{12,14}$
M(Agp[11])	a^2	x^{32}	$y^{33,-}$	n^{81}	f^{73}	g^{74}	$de^{12,15}$
P (Agp[14])	a^2	x^{32}	$y^{33,-}$	n^{81}	f^{69}	g^{77}	$de^{12,14}$
G (Agp[6])	a^3	x^{32}	y^-	$n^{84,80}$	f^{71}	g^{75}	$de^{12,15}$
H (Agp[5])	a^3	x^{32}	y^-	$n^{84,80}$	f^{72}	g^{74}	$de^{11,15}$

[a] From *Roux* et al. [13].
[b] The letter designation (A thru P) is used for reference to the various haplotypes; the Agp designation refers to those numbers previously published.
[c] The order of the *a, x, y, n, f, g* and *de* genes is arbitrarily chosen and does not imply a known gene order.

Control of Expression of Allelic $V_H a$ Genes

The $V_H a$ allotypes in a heterozygous rabbit are expressed unequally, so that the serum sometimes contains a larger proportion of Ig molecules of one $V_H a$ allotype than the other. This skewed distribution occurs in a predictable way [5, 14] and has been quantitated by several authors [3, 6–8, 16]. In general, there is a hierarchy of $V_H a$ expression of a1>a3>a2, but allotype ratios among individual rabbits vary widely.

Serum samples obtained from 8- to 14-month-old a^1a^2 heterozygous rabbits of known pedigrees were assayed for a1 and a2 Ig by radial immunodiffusion in agar gel [9]. In these adult rabbits we found a1:a2 Ig ratios ranging from 2.5 to 22. Rabbits in which one chromosome carried the A heavy chain haplotype ($a^1x^-y^-n^{81}f^{73}g^{74}de^{12,15}$) consistently had a higher proportion of a1 serum Ig (mean a1:a2 ratio of 12) than was found in rabbits in which other haplotypes encoded a1 (mean a1:a2 ratio of 5). The haplotype of the chromosome encoding a2 did not appear to affect the a1:a2 Ig ratio. Thus when a1 is encoded by the A haplotype, a larger proportion of Ig molecules bear a1 than when a1 is encoded by other haplotypes (B, C, D, J, or I). The levels of total serum Ig are essentially the same in all haplotype combinations, so the difference in ratios of expression of allelic genes is not due simply to increased expression of the A haplotype.

To determine whether the ratio of serum a1 to a2 Ig was under genetic control, we tested sera of 2-month-old rabbits from several families (table II). These a^1a^2 rabbits were progeny either of a^2a^2 homozygous rabbits (heavy chain haplotype E/E) mated with a^1a^1 rabbits heterozygous for heavy chain haplotypes (A/B), or of a^1a^2 (A/E haplotypes) rabbits mated with a^1a^2 (B/E haplotypes) rabbits. The ratios of a1 to a2 Ig were lower in these young rabbits than in adults, but the results were otherwise similar. Serum from a^1a^2 (A/E) rabbits contained a higher proportion of a1 Ig than did serum from their a^1a^2 (B/E) siblings (table II). In one cross between an a^1a^1 (A/B) rabbit and an a^2a^2 (K/K) rabbit, higher a1:a2 Ig ratios were found in A/K progeny than in their B/K siblings (table II). Thus, the higher a1:a2 ratio segregates with the A haplotype.

The correlation of higher a1:a2 ratios in heterozygous rabbits having the A heavy chain haplotype suggests that production of the allelic forms may be under the control of a gene(s) linked to the heavy chain chromosomal region. Alternatively, a larger number of a1 V_H genes may be associated with the A haplotype than with other haplotypes.

Other heavy chain allotypes inherited on the same chromosomes as the predominant V_Ha allotype are also expressed to a greater extent than their alleles. For example, in an $a^1f^{73}g^{74}/a^2f^{71}g^{75}$ rabbit there is more IgA of the f73 and g74 allotypes than of the f71 and g75 allotypes [7]. Thus, allelic imbalance is maintained in all Ig isotypes that have been studied (IgM, IgG, and IgA). Even if the suggested regulator gene were very closely linked to the V_Ha locus, one would expect to find recombinant rabbits in which a high level of, e.g. a2 is present with a low concentration of a1. Such recombinants have not been reported, but the crossover event would probably be very rare.

Coexpression of V_H and C_H Genes

In most cases the V_Ha allotype inherited from one parent is expressed in conjunction with the C_H genes inherited as a chromosomal unit from the same parent (*cis* expression). However, *Pernis* et al. [12] reported that in

Table II. Serum a1:a2 Ig ratios of seven rabbit families

Mating	Progeny No.	Genotype[a]	a1:a2 Ig ratio[b]
A/B × E/E	1	A/E	6.2
	2	B/E	2.2
	3	B/E	2.1
	4	A/E	4.5
A/B × E/E	1	B/E	2.5
	2	A/E	3.7
	3	B/E	2.3
	4	B/E	1.8
A/B × E/E	1	B/E	2.8
	2	A/E	3.8
	3	A/E	4.4
	4	B/E	2.7
	5	A/E	4.7
	6	B/E	2.4
	7	B/E	1.7
A/B × E/E	1	B/E	1.4
	2	B/E	1.8
	3	A/E	3.2
	4	A/E	4.5
	5	A/E	3.8
	6	B/E	2.6
	7	A/E	3.6
	8	B/E	1.9
E/E × A/B	1	B/E	2.0
	2	B/E	1.3
	3	A/E	3.4
	4	B/E	2.6
A/B × K/K	1	A/K	2.7
	2	A/K	2.9
	3	B/K	1.0
	4	B/K	1.3
	5	A/K	2.9

[a]For allogroup designations, refer to table I.
[b]Concentrations of a1 and a2 Ig in serum of 2-month-old rabbits were determined by radial immunodiffusion in agar gel [9].

1–2% of cells synthesizing IgM or IgG, one chromosome contributes the information for V_H and its homolog contributes the information for C_H of a single heavy chain. We have detected small percentages of IgG and secretory IgA molecules bearing V_H and C_H regions encoded by genes *trans* to each other. From 0.5 to 2% of the IgG molecules and from 3 to 8% of sIgA molecules had *trans*-encoded V_H and C_H determinants. Similar percentages of recombinant-like molecules have been detected by cytoplasmic immunofluorescence (table III). Cells containing recombinant-like IgM molecules were very rare; when corrected for allelic ratios, 0.2% or fewer of the cells expressed V_Ha and $C\mu n$ allotypes encoded by different chromosomes (table III). The few double-stained cells detected with anti-*a* and anti-*n* antibodies could have been artifacts of the staining process, but this must be ascertained by examination of a large number of cells. Thus, the percentage of recombinant-like Ig molecules made by V_H and C_H genes *trans* to each other is lowest or nonexistent for IgM, higher for IgG, and highest for IgA. The simplest explanation is that the recombinants we have detected arise from somatic gene recombination. If gene order in the rabbit is like that in mouse, $C\alpha$ is farther than $C\gamma$ from $C\mu$ on the DNA. Thus, more chromosome breakage and reunion can occur between $C\mu$ and $C\alpha$ than between $C\mu$ and $C\gamma$. There is another possible explanation. Some investigators hypothesize that all rabbits carry genes for all V_Ha allotypes. This is based on reports of animals producing small amounts of Ig of an allotype not found in the animal's pedigree [10, 17] and on reports of rabbits which produce high levels of a1, a2 and a3 Ig simultaneously [15, 16]. If these so-called latent allotypes occur because all three V_Ha genes exist in each rabbit, then, for example, the a1g75 sIgA molecules found in an $a^1f^{73}g^{74}/a^2f^{71}g^{75}$ rabbit could be encoded by a rarely expressed a^1 gene *cis* to the a^2, f^{71} and g^{75} genes. To test this possibility, g75 sIgA molecules were isolated from $a^2f^{71}g^{75}/a^2f^{71}g^{75}$ homozygous rabbits whose parents were $a^1f^{73}g^{74}/a^2f^{71}g^{75}$ heterozygotes, and these g75 IgA molecules were assayed for the a1 allotype specificity by radioprecipitation analysis (table IV). Anti-a1 antiserum precipitated no more ^{125}I-labeled g75 sIgA than did normal rabbit serum, but anti-a2 bound to 83% of the IgA [*McNicholas and Knight,* unpublished data]. Thus, it is unlikely that a latent a^1 gene encoded the 3–8% of recombinant-like IgA molecules detected in heterozygous rabbits.

The production of recombinant-like molecules also appears to be regulated by the V_Ha gene expressed. Thus, in an a^1g^{74}/a^2g^{75} rabbit (A/E haplotypes), most of the recombinant IgA molecules are a1g75 rather than a2g74 IgA [7], just as most of the non-recombinant IgA molecules are a1g74 rather

Table III. Double cytoplasmic immunofluorescence of Ig-containing cells of gut sections, spleen, and MLN from four heterozygous rabbits

Tissue	Combination of reagents	C_H cells number	Double-stained cells number	Double-stained cells, %	Recombinant cells, %[a]
Rabbit 2P84-4 ($a^1e^{14}f^{69}g^{77}/a^2e^{15}f^{71}g^{75}$)					
Spleen	anti-a1 and anti-e15	606	60	9.0	1.4
Gut	anti-a1 and anti-f71g75	240	133	35.7	5.4
Spleen	anti-a2 and anti-e14	2,254	27	1.1	0.9
Gut	anti-a2 and anti-f69g77	2,021	20	1.0	0.9
Rabbit R301-3 ($a^1e^{14}f^{69}g^{77}/a^2e^{15}f^{71}g^{75}$)					
Spleen	anti-a1 and anti-e15	563	39	6.5	1.2
Gut	anti-a1 and anti-f71g75	297	199	40.1	7.2
Spleen	anti-a2 and anti-e14	2,487	12	0.5	0.4
Gut	anti-a2 and anti-f69g77	1,748	9	0.5	0.4
Rabbit S317-4 ($a^2n^{81}f^{73}g^{74}/a^3n^{80}f^{71}g^{75}$)					
Spleen	anti-a3 + anti-n81	4,201	15	0.4	0.2
MLN	anti-a3 + anti-n81	459	2	0.4	0.2
MLN	anti-a3 + anti-f73g74	1,107	124	10.0	4.5
Gut	anti-a3 + anti-f73	532	55	9.0	4.1
Gut	anti-a3 + anti-g74	279	20	6.7	3.0
MLN	anti-a2 + anti-f71g75	640	6	0.9	0.5
Gut	anti-a2 + anti-f71g75	533	11	2.0	1.1
Rabbit S196-1 ($a^2n^{81}f^{73}g^{74}/a^3n^{80}f^{71}g^{75}$)					
Spleen	anti-a3 + anti-n81	748	2	0.2	0.1
MLN	anti-a3 + anti-f73g74	599	44	0.6	3.8
Gut	anti-a3 + anti-f73	843	62	6.8	3.7
Gut	anti-a3 + anti-g74	610	33	5.1	2.8
Gut	anti-a3 + anti-f73g74	518	47	8.3	4.6
MLN	anti-a2 + anti-f71g75	637	16	2.4	1.1
Gut	anti-a2 + anti-f71g75	1,834	33	1.8	0.8

[a] Corrected for allelic ratio [6].

Table IV. Radioprecipitation analysis of ^{125}I-sIgA from an $a^2f^{71}g^{75}/a^2f^{71}g^{75}$ homozygous rabbit obtained from a mating of two $a^1f^{73}g^{74}/a^2f^{71}g^{75}$ heterozygous rabbits

Antiserum	% precipitated[a]
Anti-a1	3
Anti-a2	73
Anti-f71g75	84
Anti-b4	84
Anti-α	88
NRS[b]	3

[a] Percent of total ^{125}I counts in precipitate.
[b] Normal rabbit serum.

than a2g75. This finding supports the hypothesis that a regulatory gene closely linked to V_Ha controls levels of production of Ig molecules.

Discussion

The data we have presented demonstrates the complex controls of expression of Ig genes. We have shown that the heavy chain haplotype of a rabbit governs the extent of allelic imbalance of V_Ha gene products. Because of this finding, we have postulated the existence of a regulatory gene(s) associated with the V_Ha gene, which controls relative amounts of the two V_Ha alleles in a heterozygous rabbit. We expect that this study will be extended by an investigation of the number, arrangement and fine structure of V_H genes in rabbits of several heavy chain haplotypes. We have also demonstrated the expression of the V_H and C_H genes encoded on chromosomes *trans* to each other, and suggested that the mechanism of C_H switching may allow the formation of these recombinant-like molecules. Using flow microcytofluorometry, one may obtain sufficient lymphocytes expressing *trans* genes to analyze the arrangement of heavy chain genes in these cells.

Although much is known about the expression and control of C_H genes in mouse myeloma cells, less information is available on the basic questions of generation of antibody diversity, allelic exclusion, and control of selection of V_H and J_H genes. The V_H allotype markers of rabbit Igs have provided many insights into expression of V_H and C_H genes at the protein level. These markers will also be invaluable in studies of the molecular genetics of Ig genes.

References

1 Bornstein, P.; Oudin, J.: J. exp. Med. *120:* 655 (1964).
2 Dray, S.; Young, G.O.; Nisonoff, A.: Nature, Lond. *199:* 52 (1963).
3 Gilman-Sachs, A.; Eskinazi, D.; Dray, S.: J. Immun. *119:* 1396 (1977).
4 Kim, B.S.; Dray, S.: Eur. J. Immunol. *2:* 509 (1972).
5 Kindt, T.J.: Adv. Immunol. *21:* 35 (1975).
6 Knight, K.L.; Hanly, W.C.: Contemp. Top. mol. Immunol. *4:* 55 (1975).
7 Knight, K.L.; Malek, T.; Hanly, W.C.: Proc. natn. Acad. Sci. USA *71:* 1171 (1974).
8 Knight, K.; Schweizer, M.; Pernis, B.: Eur. J. Immunol. *9:* 36 (1979).
9 Mancini, G.; Vaerman, J.-P.; Carbonara, A.O.; Heremans, J.F.: Protides biol. Fluids *11:* 370 (1963).
10 Mandy, W.; Strosberg, A.D.: J. Immun. *120:* 1160 (1978).
11 Oudin, J.: C.r. hebd. Séanc. Acad. Sci., Paris *242:* 2606 (1956).
12 Pernis, B.; Forni, L.; Dubiski, S.; Kelus, A.S.; Mandy, W.J.; Todd, C.W.: Immunochemistry *10:* 281–285 (1973).
13 Roux, K.; Gilman-Sachs, A.; Dray, S.: Molec. Immunol. (in press).
14 Strosberg, A.D.: Immunogenetics *4:* 499 (1977).
15 Strosberg, A.D.; Hamers-Casterman, C.; van der Loo, W.; Hamers, R.: J. Immun. *113:* 1313 (1974).
16 Wolf, B.; Urbain, R.; Miller, A.B.; Kimball, E.S.; Mudgett, M.; Catty, D.; Daneman, J.: J. Immun. *123:* 1858 (1979).
17 Yarmush, M.L.; Kindt, T.J.: J. exp. Med. *148:* 522 (1978).

Dr. K.L. Knight, Department of Microbiology and Immunology, University of Illinois at the Medical Center, Chicago, IL 60612 (USA)

The Immune System, vol. 1, pp. 299–303 (Karger, Basel 1981)

The Complicated Lives of Dominant Idiotypes

J. Quintáns[1]

La Rabida-University of Chicago Research Institute, Department of Pediatrics, University of Chicago, Chicago, Ill., USA

This story begins when I walked into *Niels Jerne's* office, and he inquired about my vital eagerness (*Guillén* said it clearly: 'Porque la vida corre con la sangre ...') and whether or not I understood the most important distribution in biology (Poisson's). We then settled down for a lengthy discussion of immunological matters during which he convinced me not to believe in three things: macrophages, complement, and Freund's adjuvant. Interestingly enough he was not at all concerned about my complete lack of laboratory experience. So I joined the BII staff and eventually established residence in *Ivan Lefkovits'* laboratory.

I will spare the reader of all that befell *Lefkovits* when I joined him except to say that he endured it with Czech malleability, and I suffered no permanent injury in the process. Together, we worked out the limiting dilution assay for B cells. It was *Charley Steinberg* who explained Poisson's distribution to me, as well as single and multi-hit kinetics and other important ingredients that eventually materialized in the paper on the estimation of precursor B cell frequencies specific for SRC [8].

The First Adventures with Idiotypes

As the work in the microculture system progressed it became necessary to utilize an antigenic system to investigate idiotypically defined responses of restricted heterogeneity. In our early experiments we determined precursor B cell frequencies for thymus-independent (TI) phosphorylcholine (PC) antigens and studied the effects of T cell help on clonal expansion, avidity and idiotype of secreted antibody [7]. We were able to demonstrate that

[1] *José Quintáns* was a member of the Basel Institute for Immunology from 1972 to 1976.

idiotypically identical B cells responding to either TD or TI PC antigens assorted independently under limiting dilution conditions. It was, therefore, concluded that separate B cells are engaged in responses to TD and TI antigens although both populations of B lymphocytes expressed the same V-region determinants on their Ig receptors.

The demonstration of functional subsets of PC-specific B cells responsive to TD or TI antigens stirred our interest in the developmental heterogeneity of B cell subpopulations and its relevance to the immune defect of CBA/N mice. Since CBA/N mice were believed to be selectively unresponsive to TI antigens, we suspected they would be useful to study the B cell subset responsive to TD PC antigens.

On How the Patchy Immunodeficiency Led to a Novel Transplantation Model

The analysis of anti-PC responses in CBA/N mice was our first undertaking in Chicago, carried out in collaboration with my faithful student, *Ruth Benca.* By the end of the summer in 1976 we knew that CBA/N mice could not mount anti-PC PFC responses to a variety of immunizations with TD or TI PC antigens [5, 6].

We felt that the inability of mice with the CBA/N immune defect to mount anti-PC PFC responses could be advantageous for cell transfer studies of B cell development if the mice were found to be adequate hosts for transplanted B cells from normal mice. Therefore, we devised a transplantation model utilizing male (CBA/N $\times$ BALB/c) F_1 (NBF_1) hybrid mice which express the CBA/N defect. Our results showed that responsiveness to PC and control TNP antigens is immediately restored in NBF_1 male mice after transplantation of normal spleen or bone marrow cells to unirradiated recipients. Cytotoxicity studies demonstrated that in semiallogeneic transplants the PFC originate exclusively from the donor cells, even up to 8 months after cell transfer. Furthermore, the NBF_1 recipient offered equal permissiveness to dominant and non-dominant anti-PC clones. Thus, we can monitor idiotype shifts occurring over long periods of time in an environment where host contributions to PC responses can be easily controlled, even in long-term experiments.

Although it is also possible to reconstitute PC responses in NBF_1 mice with fetal or neonatal liver cells, grafting liver cells requires preparative irradiation of the host because of the existence of the transplantation barrier in immunodefective mice described by *Quan* et al. [submitted for publication]. The radiosensitive transplantation barrier retards the development of PC-specific B cell clones. This environmental effect provided the opportunity to observe that the emergence of PC-specific antigen-reactive B cells involves quantal processes which are detected as periodic, graded increases in the fraction of transplanted recipients responsive to PC antigens. These processes are amenable to quantitative analysis and we expect that future work will yield information on the events surrounding stem-cell commitment and on the generation and diversification of B lymphocyte lineages.

The Innumerable Complexities of Idiotypic Clonal Dominance

Although one of our recent manuscripts was arrogantly rejected on the basis of our failure to use 'state of the art techniques (such as RIA, Elisa, etc.)', I proudly proclaim here that all the valuable information I report below was obtained using the plaque-forming cell technique. We have utilized the NBF_1 transplantation model to clarify some of the experimental difficulties encountered in previous attempts to study the generation of anti-PC responses in lethally irradiated BALB/c mice reconstituted with immature immunocompetent cells [1, 3]. In our experiments we failed to produce early (3 weeks post-transfer) anti-PC responses after transplantation of 800 rad BALB/c mice with neonatal liver cells. It was possible, however, to achieve modest early restoration of responsiveness if the hosts had been neonatally suppressed with anti-idiotype antibodies; in contrast, a later emergence of responsiveness (10 weeks) was observed in normal recipients. $T15^+$ PFC predominated in the early responses, whereas in later responses T15 idiotypic dominance was lost. In these studies, it could not be established whether the donor or the host were responsible for the anti-PC responses. The work on the NBF_1 transplantation model has provided information on the following aspects of this problem: (1) The neonatal and fetal liver contain progenitors of PC-specific B cells. (2) Generation of PFC responses from transplanted neonatal liver cells follows a well-defined and highly reproducible sequence; responsivity to TNP precedes responsiveness to PC antigens, and in this group TD responses appear before their TI counterparts. It should be noted that in the early experiments TI responses to the R36a vaccine were investigated. Since there are at least two classes of TI antigens, the responses to which display the developmental heterogeneity of *McKearn and Quintáns* [4], further studies in the NBF_1 system are being undertaken to determine the ontogeny of responsiveness to representative TI PC antigens. So far we have utilized only putative TI-2 antigens and we have not excluded that responsiveness to so-called TI-1 antigens is acquired earlier. (3) The number of cells used to reconstitute the adoptive host plays a critical role in anti-PC responses, higher numbers being required to generate TI than TD responses. (4) The injection of normal mouse serum into NBF_1 recipients is immunosuppressive for developing B cells. Since BALB/c mice are known to contain high levels of circulating anti-PC antibodies, one can propose that the facilitating effects of neonatal suppression with anti-idiotype serum described earlier by *Kaplan* et al. [3] are due to removal of the immunosuppression caused by natural antibodies present in the recipient's serum. (5) The ontogeny of BALB/c anti-PC responses in adoptive transfer is characterized by well-defined idiotype shifts. In experiments to be reported elsewhere we have found that the earliest anti-PC responses detected in reconstituted NBF_1 male mice are predominantly $T15^+$. Non-T15 B cells grow more slowly than $T15^+$ clones, but eventually they become the predominant idiotype in both TD and TI anti-PC responses. Therefore, in general, anti-PC PFC responses in adoptive transfer systems do not display the T15 dominance that is characteristic of intact BALB/c mice. In analyzing early responses, T15 predominance will be found, whereas in 'late' responses non-T15 predominance is operative. Even if we specifically suppress the emergence of the T15 clones with a highly specific hybridoma anti-T15 reagent, we cannot accelerate the development of non-T15 clones. The exact causes of this sequence of idiotype changes are not entirely clear although we know that at least two separate components are involved: intrinsic properties of the $T15^+$ and $T15^-$ B cell clones and the operation of extrinsic regulatory mechanisms which we have not yet been able to elucidate completely.

The inability to reproduce T15 dominance in NBF_1 recipients is not restricted to BALB/c neonatal cells. We found that splenic B cells from adult BALB/c mice lose T15 dominance 3 months after cell transfer to NBF_1 recipients. The shift in idiotype expression originates from Ig^- cells because Ig^+ splenic cells purified in a FACS maintain T15 dominance. The interpretation of these findings is complex and we suspect that idiotype-specific regulatory interactions play an important role. However, since we find that early $T15^+$ dominant B cell responses are eventually superseded by non-T15 clones, an idiotype-specific helper T cell-B cell loop cannot solely account for the clonal competition for idiotypic dominance in the NBF_1 transplantation model.

We are confident that these findings are generally applicable since idiotype shifts in NBF_1 recipients are reproduced in BALB/c mice. For instance, exposure to 100–200 rad initiates events in BALB/c mice which eventually lead to the loss of T15 dominance [2]. Also, *Kaplan* [unpublished observation] discovered that old BALB/c mice spontaneously lose T15 dominance. More recently we found that injection of BALB/c neonates with 0.5 μg of PC coupled to mouse Ig induces unresponsiveness to PC. Partial recovery from tolerance occurs with $T15^+$ clones which are progressively replaced by non-T15 clones. Thus, aging, exposures to irradiation, and neonatal tolerance reproduce the idiotype shifts detected in the adoptive transfer system. Repeated attempts to reverse the shifts have not yet been consistently successful.

Conclusion

Because of the relative simplicity of immune responses to PC we thought it would be feasible to design and interpret experiments involving idiotype clonal patterns. However, the feasibility of the designs has somehow outdone the feasibility of the interpretations. *Lwoff* is quoted as saying: 'Au fond, la recherche scientifique c'est une activité ludique.' For me, the study of idiotype clonal dominance has been 'vraiment ludique'. That is why I like it.

Acknowledgements. This work was supported by NIH Grant AI-14530 and by the National Foundation March of Dimes. I am the recipient of Research Career Development Award AI-00268. I acknowledge the valuable contributions of *R. Benca, D. Kaplan, J.P. McKearn,* and *Z.S. Quan* to the research efforts reported here. I am grateful to *A. Söderberg, S. Davies, L. Gemlo,* and *R.F. Dick* for their technical assistance.

References

1 Augustin, A.; Julius, M.; Cosenza, H.: in Sercarz, Herzenberg, Fox, Regulation of the immune system: Genes and the cells in which they function, pp. 195–199 (Academic Press, New York 1977).
2 Kaplan, D.; Quintáns, J.: J. exp. Med. *148:* 987–995 (1978).
3 Kaplan, D.R.; Quintáns, J.; Kohler, H.: Proc. natn. Acad. Sci. USA *75:* 1967–1970 (1978).
4 McKearn, J.P.; Quintáns, J.: Cell. Immunol. *44:* 367–380 (1979).
5 Quintáns, J.: Eur. J. Immunol. *7:* 749–751 (1977).
6 Quintáns, J.; Benca-Kaplan, R.: Cell. Immunol. *38:* 294–301 (1978).
7 Quintáns, J.; Cosenza, H.: Eur. J. Immunol. *5:* 399–405 (1976).
8 Quintáns, J.; Lefkovits, I.: Eur. J. Immunol. *3:* 392–397 (1973).

J. Quintáns, MD, La Rabida-University of Chicago Research Institute,
Department of Pediatrics, University of Chicago, Chicago, IL 60649 (USA)

The Immune System, vol. 1, pp. 304–310 (Karger, Basel 1981)

The Killer T Cell: Idiotypes, Regulation, and Immune Interferon

Peter H. Krammer[1]

Institut für Immunologie und Genetik, Deutsches Krebsforschungszentrum, Heidelberg, FRG

The immune system consists of a complex mixture of interacting cells generally not sessile in a particular organ but moving freely between lymphatic tissue, blood, and lymph. This implies that complicated regulatory mechanisms must exist which control the system by direct cell to cell contact or by release of soluble factors. Molecules responsible for antigen-specific regulation mark a late development in evolution and might have been preceded by a primitive defense mechanism on a broader specificity level. Multicellular organisms might have found it useful to maintain both principles to assure their survival.

As a possible model for this postulated sequence I take the cytotoxic T lymphocytes (CTL) and discuss their receptors and their antigen-specific regulation. In addition, I shall point out that these cells capable of lysing virus-infected target cells release a soluble lymphokine, immune interferon (IFN-y), which exerts anti-viral activity on a different level. I am pleased to dedicate this paper to *Niels Jerne,* who greatly influenced some of the ideas outlined here.

Idiotypes on Receptors of Major Histocompatibility Complex Restricted Cytotoxic T Lymphocytes

Whereas immunoglobulin receptors on B cells recognize antigens as such, T cells seem to recognize antigens presented on cell surfaces. An additional requirement for efficient triggering of a specific T cell-immune

[1] *Peter Krammer* was a member of the Basel Institute for Immunology from 1973 to 1975.

response is that the antigen-presenting cells are syngeneic to the responder T cells with respect to the major histocompatibility complex (MHC). In particular, murine CTL activated against virus, the hapten TNP, or minor histocompatibility determinants lysed target cells most efficiently when the antigens were presented on cells which shared the H-2K or H-2D region of the MHC with the CTL [3, 11, 16]. Two basic models have been proposed to explain this phenomenon; the one-receptor model postulates one single type of T cell receptor which binds to neoantigenic determinants formed by association of the antigen, e.g. virus, TNP, etc. with H-2K or H-2D antigens, whereas the two-receptor model postulates two different types of T cell receptors, one for the antigen and the other for syngeneic major histocompatibility antigens [17]. Presently, the available experimental evidence does not allow us to decide between these two possibilities. One could infer from the difference of antigen recognition between B and T cells that if the one-receptor model is correct the construction of T cell receptors would reflect this disparity. Similarly, if the two-receptor model is correct, one could postulate anti-self T cell receptors with a repertoire not found among V genes products of B cells which seem to lack apparent anti-self-MHC reactivity. Nevertheless, allotype locus linked idiotypic determinants serologically defined on antibody molecules were found on T helper cells, T suppressor cells, T cells responsible for delayed type hypersensitivity, and alloreactive T cells [for review see 5, 9]. Even though these serological data lack the confirmation that would come from a biochemical analysis, these results, together with the fact that framework markers of the heavy chain of immunoglobulin, were found on T cell receptors and antigen-specific T cell factors, would suggest the interpretation that T cell recognition molecules share at least part or most probably the total variable region of the heavy chain with antibody molecules [for review see 9].

The information on MHC-restricted CTL, however, does not quite fit into this simplified picture. B cell immunoglobulin defined major cross-reactive idiotypes *ARS* and *NP^b^* could not be demonstrated on A/J and C57B1/6 CTL specific for the haptens *p*-azophenylarsonate and (4-hydroxy-3-nitro-phenyl) acetyl [NP], respectively [7, 12]. Similarly, NP-primed C57B1/6 CTL were not heteroclitic, i.e. they did not show better lysis of syngeneic target cells coupled with the cross-reactive hapten (4-hydroxy-5-iodo-3-nitrophenyl) acetyl (NIP) than of target cells coupled with the immunizing hapten NP [1]. Heteroclicity, however, a fine specificity marker, was easily demonstrable for NP-induced primary C57B1/6 IgM antibodies. In addition, antisera reactive against framework determinants of the variable region of antibody heavy chains [2] did not conclusively react with CTL clones from long-term cultures [*v. Boehmer,* personal communication]. Furthermore, in contrast to B cells which rearrange immunoglobulin heavy chain variable region (V_H) genes and genes coding for the heavy chain joining (J) piece, no V_H-J rearrangement was found in CTL clones from long- term cultures using a B cell J-piece DNA probe [*v. Boehmer,* personal communication]. Even though these negative data must be interpreted with caution they seem to distinguish CTL from other T cells.

This stimulated us to develop a new system to define serological determinants of receptors on MHC-restricted 2,4,6-trinitrophenyl (TNP)-specific CTL using these cells directly as immunogens in syngeneic AKR mice. Cell populations enriched for TNP-specific, H-$2K^k$ restricted AKR CTL (obtained from a 5-day culture of cells from the draining lymph nodes of AKR mice 5 days previously sensitized on the abdominal skin by 2,4,6-trinitrophenylchloride) were repeatedly injected into syngeneic AKR mice. After several injections these mice developed AKR anti-(AKR anti-AKR-TNP) antisera, abbreviated as AKRa(AKRaAKR-TNP). These antisera were strongly reactive with a major fraction of AKR anti-AKR-TNP CTL. This activity was specific, since the antisera did not react with alloreactive AKR CTL, H-2 K^k restricted AKR CTL activated by fluorescein-isothiocyanate (FiTC), and H-2 restricted, TNP-specific CTL of strain C57B1/6. Since we were unable to demonstrate any reactivity with the antigen TNP as such, we concluded that the antisera reacted with specificity-associated determinants (idiotypes) on AKR anti-AKR-TNP CTL [8]. Particularly, the fact that AKRa(AKRaAKR-TNP) did not react with AKR anti-AKR-FiTC CTL indicated that the antisera recognized either a receptor for TNP (if the two-receptor model were correct) or a receptor for TNP-self (if the one-receptor model were correct). Our experiments excluded reactivity of the antisera with anti-self receptors as these would be identical on both types of CTL. This situation is suitable for designing experiments which help to delineate the genes coding for the receptors on H-2 restricted CTL and help to decide between the validity of the one or two receptor model.

If the antisera reacted with Balb/c restricted anti-TNP CTL from chimaeric mice AKR→Balb/c where putative anti-self receptors would recognize Balb/c as self, one would have to conclude that the antisera recognize a receptor for TNP. The experiment, however, is only conclusive when AKR→Balb/c anti-TNP CTL are idiotype-positive. If they are idiotype negative one could argue that AKR T cells from chimaeras with receptors for TNP could somatically be selected in a different way from the ones in a syngeneic environment. As it is controversial whether fully allogeneic chimaeras of the type AKR→Balb/c are capable of generating MHC-restricted CTL effector cells at all, AKR→(AKR × Balb/c) F1 (parental into F1) chimaeras, which ought to be able to generate a TNP-specific CTL response, might have to be used for such an experiment.

It should be pointed out that an absolutely conclusive biological experiment which could unambiguously prove the validity of the one-receptor model does not seem to exist. It is essential, therefore, to obtain additional information from biochemical experiments.

The discussion of a possible approach to study the genetic control of idiotype expression of H-2 restricted CTL serologically by using AKRa(AKRaAKR-TNP) still has to take into account both models. If we assume that AKRa(AKRaAKR-TNP) react with CTL receptors for TNP (i.e. the two-receptor model is correct), it is straightforward to test for the genetic linkage of expression of these receptors as the relevant stimulating antigen is TNP alone regardless of the MHC haplotype of the stimulator cells. It is sufficient, therefore, to test MHC or allotype different strains for idiotype expression. If the antisera, however, detect idiotypes on anti-TNP-self receptors (i.e. the one-receptor model is correct) the choice of the haplotype of the stimulator cells for CTL is limited to H-2^k-TNP, AKR-TNP in particular, as one has to account for H-2 restriction of the CTL response. To study the genetic control of T cell receptor expression anti-AKR-TNP CTL from H-2 and allotype different chimaeric mice of the type X→AKR or X→(AKR × X) F1 would have to be tested. It is obvious that this is a complex experimental situation which takes into account only the most obvious genetic loci, the MHC and the allotype locus, which could be assayed for linkage to genetic control of T cell receptor expression.

Regulation of the TNP-Specific Response

Presently, our knowledge of the regulation of the cytotoxic response is derived from the study of the interaction and maturation of different T cell subpopulations upon antigen stimulation [13]. In view of the possibility of manipulating the system it would add a necessary refinement to this information if one would get further insight into the regulation of a cytotoxic response on the idiotype level and follow idiotypically defined clones during the course of this response. This could best be done in an experimental situation where a substantial proportion of a cytotoxic response is due to

CTL of a dominant, cross-reactive idiotype, or a restricted number of idiotypes. I do not know whether this situation applies to the above-described TNP system but the fact that AKRa(AKRaAKR-TNP) always react with a sizeable proportion of AKR anti-AKR-TNP CTL would indicate that this could be a possible interpretation. We are far from combining this idiotypic analysis of AKR anti-AKR-TNP CTL T cell receptors with the regulatory events observed in that system. Nevertheless, I would like to discuss a few findings which lay the groundwork for such a future study.

Cells from the draining axillary and inguinal lymph nodes of AKR mice sensitized 5 days previously on the abdominal skin by 2,4,6-trinitrophenylchloride did not give an anti-TNP cytotoxic response. This could only be obtained when the mice were injected with cyclophosphamide two days prior to sensitization [4], or, alternatively, when the cells were kept for an additional 5 days in culture [8]. These data suggested that suppressor T cells play a critical role in the development of the anti-TNP CTL response. In limiting dilution experiments with *Hamann and Eichmann* [manuscript in preparation] we were able to describe more subtle features of these phenomena. At different times of sensitization of the abdominal skin of AKR mice with 2,4,6-trinitrophenylchloride and subsequent in vitro culture of the draining lymph node cells, samples of cell populations containing antigen-primed cells were further expanded by stimulation with concanavalin A for 2 days and subsequent limiting dilution in the presence of filler cells and T cell growth factors. At the end of the limiting dilution period individual wells were tested for a TNP-specific cytotoxic response. It became evident that at the level of the determination of CTL frequencies similar to the situation in bulk cultures and dependent on the time elapsed after antigen priming, a shift from suppression to an active cytotoxic response could be observed. This could be explained either by a loss of suppressor cells or by resistance to or counteracting of the initial suppressor cells. We think that this system is amenable to experiments which determine whether idiotype–anti-idiotype interactions play a decisive role in regulating the TNP-specific CTL response.

Immune Interferon (IFN-γ)

The connection between IFN-γ and the subject of the two previous sections, namely an antigen-specific idiotype-positive CTL response, is not immediately obvious. I shall, therefore, briefly summarize some new data that we have obtained in collaborative experiments with *Marcucci, Waller,* and *Kirchner* [manuscripts in preparation] which suggest that a connection exists. It was not unambiguously known which cell is responsible for IFN-y production. We approached this question by the following experimental design. Spleen cells were stimulated by concanavalin A for 2 days, then subjected to limiting dilution, and finally cultured in the presence of filler cells and T cell growth factors. After 7 days in the limiting dilution cultures (which had been seeded with 1, 2, 4, 8, etc. cells/well) the cells from individual wells were harvested and aliquots were tested for total non-specific cytolytic activity on EL 4 tumour target cells in the presence of concanavalin A and for their ability to produce IFN-y by induction with T cell mitogens. These experiments yielded the following results: (1) IFN-y production could only be obtained by induction; (2) the producer cells of IFN-y were T cells, a finding which we independently verified in clones of T cells in long-term culture in the presence of T cell growth factors but in the absence of adherent cells; (3) for limiting dilution cultures with sufficiently few input cells where, according to Poisson statistics, the cells in most culture wells were the clonal progeny of a single precursor cell, there was an almost absolute coincidence of IFN-y production and cytolytic activity; (4) occasionally, cells in individual culture wells produced IFN-y and did not kill, and vice versa; (5) cells from cultures with the most potent cytolytic activity gave the highest IFN-y titers, and (6) individual selected cultures containing CTL yielded titres of IFN-y substantially higher than those normally obtained from cell mixture in bulk cultures stimulated by a mixed lymphocyte reaction or T cell mitogens [6, 15].

Since *Sonnenfeld* et al. [14] have shown that helper T cells of the Lyt-1 type did not produce IFN-y, we think that T cells of the cytotoxic lineage release IFN-y at maturation stages which do not entirely coincide with those when they are actively cytotoxic. If we assume that CTL are primarily directed against cellular targets, especially those infected with virus, CTL might most effectively prevent spreading of a virus infection by lysis of infected target cells as well as by release of IFN-y. As speculated in the introduction to this paper, these two mechanisms could have evolved independently but might have been combined to assure a safer way of survival. Whether IFN-y might play an additional role in augmenting cytolytic activity by increasing the density of H-2 target cell antigens [10] or by recruiting precursor CTL into the reactive pool remains to be determined.

Our experiments indicated that CTL clones needed to be triggered by T cell mitogens to release IFN-y. As IFN-y-release also occurs in vivo under physiological conditions and upon antigen stimulation, one could envisage a different physiological signal for IFN-γ induction. This inductive signal could be identical to the one triggering specific cytolytic activity and might act on the antigen-specific CTL receptor directly or on a cell surface molecule involved in a chain event of triggering. Alternatively, antigen-specific stimulation of a second cell type could induce a signal which would lead to IFN-γ release from cells of the CTL lineage. This hypothesis, if verified, would establish the link between a specific signal and the release of a 'non-specific' lymphokine.

References

1 Ando, J.; Kisielow, P.: Eur. J. Immunol. *9:* 211–213 (1979).
2 Ben-Neriah, Y.; Wuilmart, C.; Lonai, P.; Givol, D.: Eur. J. Immunol. *8:* 797–801 (1978).
3 Bevan, M.J.: J. exp. Med. *142:* 1349–1364 (1975).
4 Droege, W.; Suessmuth, W.; Franze, R.; Mueller, P.; Franze, N.; Balcarova, J.: in Quastel, Cell biology and immunology of leukocyte function, pp. 467–483 (Academic Press, New York 1979).
5 Eichmann, K.: Adv. Immunol. *26:* 195–254 (1978).
6 Gifford, G.E.; Tibor, A.; Peavy, D.L.: Infect Immunity *3:* 164–166 (1971).
7 Hurme, M.; Karjalainen, K.; Mäkelä, O.: Scand. J. Immunol. *11:* 241–246 (1980).
8 Krammer, P.H.; Rehberger, R.; Eichmann, K.: J. exp. Med. *151:* 1166–1182 (1980).
9 Krammer, P.H.: Curr. Top. Microbiol. Immunol. (in press).
10 Lindahl, P.; Leary, P.; Gresser, I.: Eur. J. Immunol. *4:* 779–784 (1974).
11 Shearer, G.M.; Rehn, T.G.; Garbarino, C.A.: J. exp. Med. *141:* 1348–1364 (1975).
12 Sherman, J.I.; Burakoff, S.J.; Benacerraf, B.: J. Immun. *121:* 1432–1436 (1978).
13 Simon, M.M.; Eichmann, K.: Springer Semin. Immunopathol. *3/1:* 39–62 (1980).
14 Sonnenfeld, G.; Mandel, A.D.; Merigan, T.C.: Immunology *36:* 883–890 (1979).
15 Wheelock, E.F.: Science *149:* 310–311 (1965).
16 Zinkernagel, R.M.; Doherty, P.C.: J. exp. Med. *141:* 1427–1436 (1975).
17 Zinkernagel, R.M.; Callahan, G.N.; Althage, A.; Cooper, J.; Klein, P.A.; Klein, J.: J. exp. Med. *147:* 882–896 (1978).

Dr. Peter Krammer, Institut für Immunologie und Genetik,
Im Neuenheimerfeld 280, D–6900 Heidelberg 1 (FRG)

Factors, Hormones, Interferon

The Immune System, vol. 1, pp. 311–315 (Karger, Basel 1981)

T Cell Factors – T Cell Receptors

M. Cramer[1]

Institute for Genetics, University of Cologne, Cologne, FRG

Discussing the relationship between antigen-specific T cell factors and antigen-specific T cell receptors may seem a delicate task for two reasons: (a) The issue seems to be solved in the minds of a large number of immunologists but not that of *Niels Jerne* although (b) at present the data are so sparse that no one (except perhaps *Jerne*) can make a good guess of the final answer. In fact, both of these points encouraged me to pick up this topic.

Obviously, this article does not at all intend to review the complex field of antigen-specific T cell molecules and functions. The reader is here referred to recent summaries of this subject [2, 4, 9, 10, 14, 17]. It is also obvious that some of the arguments brought up in this discussion are implicit in the thoughts other people have expressed on various occasions.

The Problem

There is no doubt that T lymphocytes react to antigenic stimuli in a highly antigen-specific way [2, 9, 10, 14]. Therefore, these T lymphocytes must possess an antigen-specific recognition structure (receptor) through which information is transferred from the cell's surroundings into its interior. Such receptor molecules have been studied in isolated form or on the

[1] *Matthias Cramer* was a student at the Basel Institute for Immunology from 1971 to 1974.

cell surface in immunochemical and serological terms [recent reviews: see 2, 4, 9, 10, 14]. In principle, a T cell surface receptor is the only antigen-specific molecule required in the system since, at least a priori, the effector functions the cell may have to carry out do not necessarily involve specific antigen recognition (see below).

Nevertheless, there is accumulating evidence for the existence of soluble antigen-specific T cell-derived factors which are released into culture media or body fluids and are capable of carrying out T cell functions in an antigen-specific fashion. This field was recently and very extensively reviewed by *Tada and Okumura* [17]. These T cell factors are characterized by their T cell-substituting functional properties, while serological and immunochemical data are scarce.

Thus, the question arises how these two types of antigen-specific T cell molecules – the soluble factors and the surface receptors – relate to each other. This problem is not discussed in the literature in any detail [but compare 11, 12, 16, 17, 19], probably because on the one hand some of the most extensively studied so-called T cell factors are not molecules readily secreted by T cells but are recovered from mechanical extracts of T cells [17] and on the other hand some T cell receptor molecules are reported to be found in serum [1, 2]. In other words, the distinction between 'factor' (i.e. biologically active effector molecule) and 'receptor' (i.e. surface recognition structure) is not a sharp one in some experimental systems.

Proposition A: T Cell Factors and T Cell Receptors Are Identical or Derived from Each Other

This proposition seems to be among the most plausible solutions to the problem [6, 7, 16, 17]. In this view the recognition structure(s) on the T cell surface serve(s) for the stimulation of the cell and the same structure or a proteolytic split product of it is then – in a soluble form – also responsible for the effector functions of the T cell. It is immaterial for this model whether the specific factors are secreted or shed by the cell, whether they are short-lived, or to what extent they are broken down in order to be released. In any case, with such models the antigen-specific recognition/effector functions of T cells would 'degenerate' to a copy of the B cell receptor/serum antibody system, which is unique in biology inasmuch as here the receptor and the effector molecules are (almost) identical.

Apart from the fact that immunologists would not have to give up such traditional (B cell) views and that under these circumstances T cells would have to produce only *one* type of antigen-specific molecule, some data can be cited in support of proposition A. The variable regions of immunoglobulin heavy chains (V_H) are found on T cell surfaces [2, 5, 14, 16], on T cell receptors [2, 4, 10, 14], and in some T cell factors [7, 16, 17]; I-J region coded structures are found on T cell surfaces [16, 17] and in some T cell factors [16, 17]. Suggestive as these coincidences are, however, they do not rigorously prove that the various structures are localized in the same molecules on the cell surface or in the isolated materials.

It follows from proposition A that functionally different sets of T cells should also have different sets of T cell receptors.

Proposition B: T Cell Receptors Are Distinct from T Cell Factors

Cytotoxic T lymphocytes recognize target antigens and kill targets via cell-cell-contact [3]. No antigen-specific factor is known to be required in this system. Helper T cells described by *Schreier* et al. [15] are stimulated specifically to produce an antigen *non-specific* factor. No antigen-specific factor was found in this system. A/J mice immunized with keyhole limpet hemocyanin (KLH) or the synthetic polypeptide GT are not able to produce an antigen-specific suppressor factor against these antigens [17, 18, 20]. This incomplete set of findings is listed to point out that the production of antigen-specific factors is by no means a general prerequisite for T cell function while the availability of T cell receptors is mandatory. This by itself is already a very general but admittedly not all too convincing argument in favor of proposition B.

Much stronger support may come from the comparison of biochemical and immunochemical properties of antigen-specific T cell factors and T cell receptors [4, 12]. Isolated T cell receptors have a molecular weight (MW) of 150,000 dalton and carry V_H but not major histocompatibility locus (MHC)-coded structures [1, 2, 4] while most of the T cell factors have a MW of 50,000–70,000 dalton and carry V_H *and* MHC coded structures [16, 17].

As pointed out in the introduction, these data are also ambiguous and do not allow us to draw a firm conclusion since the experimental systems involved differ so markedly that everybody could still be looking at different fragments of the same molecule.

An attractive consequence of proposition B – i.e. of the separation of T cell recognition (receptor) and T cell function (specific factor, unspecific factor, or no factor) – is that all functional T cell subsets could theoretically make use of the same T cell receptor repertoire and structure. The type of effector molecules produced or not produced would then determine the subclass of a given T cell. Such a sharing of a common receptor pool among functionally distinct T cells would make it much easier to envisage models of antigen-binding site-related (idiotypic) regulations of T cell responses in general. The fact that under proposition B a given helper or suppressor T cell would have to produce two different molecules sharing the V_H region does not raise any new problems in the lymphocyte system because the same V_H can also be expressed on different classes of immunoglobulin in B cells [13].

Solutions?

To decide between propositions A and B – or any other of the 'mixed' models, which are certainly numerous – we need more of two types of evidence:

(1) Biochemical studies could tackle the precursor-product relationship of T cell receptors and T cell factors. To this end a set of antigen-specific, functional T helper and/or suppressor cell lines or hybridomas [8, 16, 19] should be constructed and analyzed for expression of surface receptor and factor production. It will certainly be advantageous to raise these lines/hybridomas against antigens which at the B cell and the T cell receptor level induce 'major idiotypes'. In this context it is, however, surprising to learn that up to now very little could be done biochemically with some of the existing monoclonal material [16, 19].

(2) Monoclonal antibodies raised against native or monoclonal receptor or factor molecules should be used not only to isolate enough material for structural analyses but, more importantly, to attempt to interfere with the complex processes involved in receptor-mediated stimulation and subsequent factor-mediated T cell function.

Acknowledgements. I wish to thank my colleagues *G.H. Kelsoe, R. Mierau* and *T. Takemori* for critical discussions of this manuscript. The work was supported by the Deutsche Forschungsgemeinschaft through SFB 74.

References

1 Binz, H.; Wigzell, H.: Scand. J. Immunol. *5:* 559–571 (1976).
2 Binz, H.; Wigzell, H.: Contemp. Top. Immunobiol. *7:* 113–177 (1977).
3 Cerottini, J.-C.; Brunner, K.T.: Adv. Immunol. *18:* 67–132 (1974).
4 Cramer, M.; Krawinkel, U.: in Pernis, Vogel, Regulatory T lymphocytes, pp. 39–55 (Academic Press, New York 1980).
5 Eichmann, K.; Ben-Neriah, Y.; Hetzelberger, D.; Polke, C.; Givol, D.; Lonai, P.: Eur. J. Immunol. *10:* 105–112 (1980).
6 Feldmann, M.; Nossal, G.J.V.: Transplant. Rev. *13:* 3–34 (1972).
7 Germain, R.N.; Ju, S.-T.; Kipps, T.J.; Benacerraf, B.; Dorf, M.E.: J. exp. Med. *149:* 613–622 (1979).
8 Goodman, J.W.; Lewis, G.K.; Primi, D.; Hornbeck, P.; Ruddle, N.A.: Mol. Immunol. *17:* 933–945 (1980).
9 Krammer, P.H.: Curr. Top. Microbiol. Immunol. (in press).
10 Lindahl, K.F.; Rajewsky, K.: Int. Rev. Biochem. *22:* 97–150 (1979).
11 Marchalonis, J.J.: Mol. Immunol. *17:* 795–801 (1980).
12 Munro, A.: in McDevitt, Ir genes and Ia antigens, pp. 569–570 (Academic Press, New York 1978).
13 Pernis, B.; Forni, L.; Luzzati, A.L.: Cold Spring Harb. Symp. quant. Biol. *41:* 175–183 (1976).
14 Rajewsky, K.; Eichmann, K.: Contemp. Top. Immunobiol. *7:* 69–112 (1977).
15 Schreier, M.H.; Anderson, J.; Lernhard, W.; Melchers, F.: J. exp. Med. *151:* 194–203 (1980).
16 Tada, T.; Hayakawa, K.; Okumura, K.; Taniguchi, M.: Mol. Immunol. *17:* 867–875 (1980).
17 Tada, T.; Okumura, K.: Adv. Immunol. *28:* 1–87 (1979).
18 Taniguchi, M.; Tada, T.; Tokuhisa, T.: J. exp. Med. *144:* 20–31 (1976).
19 Taniguchi, M.; Takei, I.; Tada, T.: Nature, Lond. *283:* 227–228 (1980).
20 Waltenbaugh, C.; Thèze, J.; Kapp, J.A.; Benacerraf, B.: J. exp. Med. *146:* 970–985 (1977).

M. Cramer, DSc, Institut für Genetik der Universität zu Köln, Weyertal 121, D–5000 Köln 41 (FRG)

The Immune System, vol. 1, pp. 316–323 (Karger, Basel 1981)

Antigen-Specific T Cell Helper Factor: Perspective and Prospect

Michael J. Taussig[1]

Department of Immunology, Agricultural Research Council, Institute of Animal Physiology, Babraham, Cambridge, England

It is an honour and a pleasure to contribute to this Festschrift, and a welcome opportunity to record my personal thanks for the advice and encouragement, both scientific and personal, which I received from *Niels Jerne* while at his Institute in the period 1975–76. During that time there were plenty of occasions on which to cross swords intellectually with my colleagues, both in Basel and elsewhere, who viewed with a characteristic blend of enthusiasm and scepticism my earlier work in Cambridge and Rehovot on the antigen-specific T cell helper factor and its relationship to the immune response (Ir) genes of the major histocompatibility complex (MHC) [15, 21, 24, 25]. Indeed, in a discussion before I came to the institute, Prof. *Jerne* expressed the challenging hope that my work in Basel would convince him of the existence of antigen-specific factors, as his own prejudice was against the very notion of a non-immunoglobulin antigen-recognition system for T cells. At that time, the technical problems of producing the factors led to widespread controversy over this very central point, which has taken some time to come to a final resolution. Happily, 5 years on, there are at least now no doubts over the reality of the MHC-derived factors. Perhaps the most significant recent development is the introduction of T hybrid cell lines producing the factors, offering a stable supply of material for careful and reproducible characterization. Later in this article I will discuss some of the progress they offer, but I would first

[1] *Michael Taussig* was a member of the Basel Institute for Immunology from 1975 to 1977.

like to review in perspective some of the work in which I was involved in Basel. There are several recent comprehensive reviews of the general field [4, 20, 22].

Helper Factors and the Human Antibody Response

One of the exploitable qualities of the antigen-specific helper T cell factors is their ability to function across strain and species barriers, and an early application of their xenoreactivity was the demonstration in Basel by *Luzzati* et al. [11] that a human anti-sheep red cell (SRBC) response could be triggered in vitro by mouse SRBC-specific helper factor. The experiment was suggested because of the great difficulty in obtaining a human primary response to SRBC in vitro. Addition of the mouse factor (a supernatant of cultured SRBC-primed mouse T cells) to cultures of human peripheral blood lymphocytes (PBL) and SRBC produced a spectacular improvement in their plaque response. The effect was antigen-specific, horse (H) RBC being used as a reciprocal control, and was apparently mediated by the same factor which helped mouse B cells in vivo. However, there was also a strong non-specific component, as an impressive proliferation of human cells, far in excess of that expected of specific clonal proliferation, accompanied the specific response. Thus, an antigen-specific agent, the mouse factor, in collaboration with antigen, appeared to set in motion an essentially non-specific mitogenic response in human lymphocytes. This was confirmed by adding both SRBC and HRBC together to a culture with SRBC-specific factor, whereupon a good response to both red cell types occurred [11].

Recently, the reverse of the Luzzati experiment has been employed to assay human antigen-specific T cell factors on mouse spleen cells [7, 18, 26]. A rhesus monkey factor has also been shown to trigger mouse cells [9]. These experiments raise the exciting prospect of detecting specific immune response defects in humans without recourse to immunization, hence avoiding all sorts of ethical problems. Indeed, a start has been made in this approach to the study of human Ir genes, and variation in the ability to produce specific factors for the polypeptide antigens (T,G)-A—L and GAT have been found among unrelated donors [18, 26, 27]. Family studies are awaited with interest.

How the mouse factor and human cells might interact was a question to which I devoted considerable time in Basel and was most fortunate in being able to draw upon the large pool of HLA-typed donors which had been collected by the Institute tissue typing laboratory. It became clear that human PBL were able to absorb out the factor and that this absorption showed an interesting specificity. While the PBL of all donors could, under standard conditions, absorb the mouse factor specific for SRBC, factors against the synthetic polypeptides (T,G)-A—L and (Phe,G)-A—L were absorbed by some individuals but not others and, moreover, the PBL of some donors could distinguish the factors against these two very similar molecules [23]. Thus, the removal of mouse factor by human cells was not a non-specific absorption, but rather seemed to indicate the presence of factor-specific acceptor sites on human cells. This received additional support from extensive family studies which showed the selective absorption characteristic to be inherited in simple Mendelian fashion and to be linked to the HLA complex. Since it had previously been shown that H-2 gene products contributed to the acceptor sites of mouse B cells for T cell factor [15], it was apparent that the mouse factor was indeed probing the analogous system on human cells, implying an impressive degree of preservation of the active moieties of these cell-interaction molecules across species. Furthermore, since in the mouse the inability to absorb a specific T cell factor correlated in several cases with an Ir gene-controlled defect in response to molecules such as (T,G)-A—L [15], it seemed possible that the HLA-related acceptor sites were a reflection of Ir gene control of human responsiveness. With this in mind, we attempted to map the 'acceptor loci' in the HLA region by studying families containing an intra-HLA recombination. One family was highly informative and indicated the presence of two acceptor loci, one associated with the HLA-D locus, the other with HLA-A; on this basis the presence of two Ir gene loci in the HLA region was suggested [23]. I believe that the ability to produce the factor on the one hand and to respond to it on the other represent useful, if somewhat unorthodox, approaches to the problem of studying human Ir genes and cell interaction molecules – and are potentially applicable to other species as well.

Complementation of Mouse Ir Genes

The suggestion that two H-2-linked Ir genes control the immune response in the mouse was first made through studies with the (T,G)-A–-L-specific T cell factor [14, 15, 24]. In collaboration with my colleagues in Rehovot and Cambridge, it was demonstrated that there are three types of H-2-controlled response defect in low or non-responders to (T,G)-A—L, namely an inability to respond to the factor through lack of the B cell acceptor site (H-$2^{k,a,q,j}$ haplotypes); an inability to produce the factor, an apparent T cell defect (H-$2^{f,s}$); or a combination of both defects (SJL mice). Our interpretation was that Ir genes could be expressed in different cellular loci, in either B cells, T cells or both, and this has been amply reconfirmed in different systems. Thus, the apparent equality of high responder and some low responder helper T cells has been observed independently in factor production and cell cooperation for (T,G)-A—L [5, 6], in cell cooperation using TNP-(T,G)-A—L [8] and in delayed hypersensitivity to (T,G)-A—L [19]. Similarly, the reciprocal, T cell defect in H-2^f (B10.M) mice has been reconfirmed in both antibody production and delayed hypersensitivity [6, 12, 19].

In experiments which have become well known, it was possible to test the hypothesis that the (T,G)-A—L response was under dual genetic control by mating low responder strains of complementary type. Three combinations predicted from our cellular studies showed successful in vivo complementation, in that the F_1 hybrids between low responders were high responders to (T,G)-A—L, namely (B10.M × B10.BR)F_1 and (B10.M × I/St)F_1 [15] and (A.SW × B10.BR)F_1 [*Taussig,* unpublished observation], proving the two-gene hypothesis to be correct. Subsequently, dual gene control, as demonstrated by complementation, has been shown to be the rule for many antigens [2]. However, although the (T,G)-A—L complementation experiments were successful on several separate occasions between 1975 and 1977 in both Cambridge and Basel, problems arose when other groups had difficulty in obtaining the same result with the (B10.M × B10.BR)F_1 [1, 13]. In 1977, we were twice unable to reproduce our own earlier result with this F_1 and duly reported the fact [16]. Since this has attracted widespread attention, a few comments are appropriate here. Firstly, although the response of (B10.M × B10.BR)F_1 was acknowledged to be erratic, complementation was not only observed unambiguously several times, but in a double-blind back-cross experiment with progeny of (B10.M × B10.BR)F_1 × B10.M, heterozygotes (H-2^k/H-2^f) could be distinguished from homozygotes (H-2^f/H-2^f) on the basis of response to (T,G)-A—L with 90% accuracy [unpublished observations]. Secondly, the (B10.M × I/St)F_1 hybrids, retested at the same time as (B10.M × B10.BR) in 1977 and on several other previous occasions, showed reproducible complementation; and with the (A.SW × B10.BR)F_1 complementation was successful on both occasions tested (15/16 mice) [*Taussig,* unpublished observations]. Finally, regardless of the behaviour of (B10.M × B10.BR)F_1s on different occasions,

Table I. Dual gene control of the acceptor for (T,G)-A—L-specific helper factor in mice

Strain	H-2	Factor	Acceptor	Response (PFC/spleen) to (T,G)-A—L	
				IgM	IgG
A.SW	s	–	+	550	250
A/J	a	+	–	1,400	620
SJL	s	–	–	300	830
(A/J × SJL)F_1	a/s	+	+	4,700	34,000

Factor and acceptor for the (T,G)-A—L-specific helper factor were assayed in the three inbred strains (A.SW, A/J, SJL). Apparently, A/J lack an acceptor H-2 gene, while SJL lack an acceptor background gene. Dual gene control over the acceptor was confirmed by the (A/J × SJL)F_1 which showed complementation for the acceptor and the anti-(T,G)-A—L PFC response. The secondary response to (T,G)-A—L was assayed after injection of 10 μg in adjuvant, and a boost of 10 μg in saline 1 month later; PFC were measured 7 days after boost (6 mice/group). Complementation in both IgM and IgG responses is evident.

the conclusions regarding expression of factor and acceptor in different low responders and the demonstration of distinct types of low responder to (T,G)-A—L are unaffected: indeed, the latter point has received independent confirmation as already noted above.

While at the Basel Institute, I was in fact able to extend the complementation studies to show that background (non-MHC) genes also contributed to the factor and its acceptor. Table I summarizes the evidence for control of the acceptor by a background gene as well as an H-2 gene by comparing mice of the same background but different in H-2 (A.SW and A/J) or the same H-2 type on different backgrounds (A.SW and SJL). In the complementation test, the mating of two low responders which both fail to express the acceptor (A/J × SJL) produced responder, acceptor-positive F_1s, confirming dual gene control over the acceptor (table I) [17, 22]. Moreover, back-cross analysis of (B10.M × I/St)F_1 animals showed clearly that non-MHC genes contribute to the expression of both the factor and the acceptor [17]. Hence at least four genes seem to be involved in controlling the (T,G)-A—L response via the specific factor and its acceptor. As this conclusion seems to have quite wide applicability, it may be expressed as a 4-gene factor/acceptor model, as shown in figure 1.

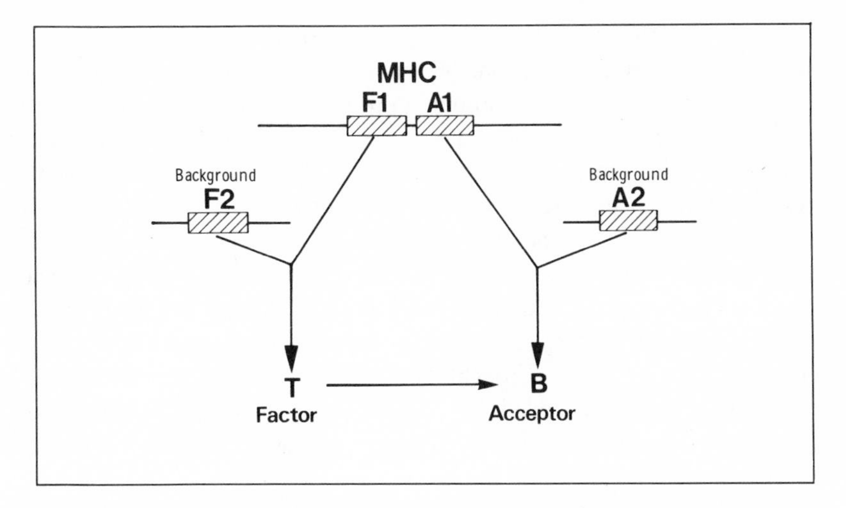

Fig. 1. A four-gene factor/acceptor model. The antigen-specific T cell factor and its appropriate acceptor are coded by closely linked MHC genes (F1, A1 respectively); in addition, the products of background genes (F2, A2) either contribute to their structure or control their production. The background factor gene (F2) may be Ig heavy chain-linked. Other MHC genes (not shown) may also control factor and acceptor production.

The Future: T Cell Hybrids and Factor Structure

Although the conventional methods of preparing antigen-specific T cell factors in culture supernatants or extracts of primed T cells have succeeded in establishing the biological properties and functions of these molecules, they have not so far enabled the structure of the molecules to be studied in detail. This is because of the very low yields of material with high biological activity. However, the factor field has recently been revolutionized by the production of T hybrid lines releasing specific helper and suppressor factors [22]. The T hybrids are produced by fusing antigen-stimulated T cells with a thymoma (BW 5147) to give hybrids which have the permanent growth properties of the thymoma and the functional characteristics of the T cell. The advantages of such permanent culture lines are that they can potentially provide a constant, indeed unlimited, supply of a monoclonal product, the structure and function of which could be thoroughly studied without the complications and frustrations of making tiny amounts of factors from conventional T cells. Recently, lines secreting the specific helper factors for (T,G)-A—L and chicken globulin have been described [3, 10] and it is most exciting to see that their factors have the characteristic properties of the material we originally obtained from in vivo educated T cells, including specificity and H-2 derivation. Moreover, in both cases the helper factors appear to carry the variable domain (V_H) of antibody heavy chain together with I-A determinants, while lacking constant region Ig markers. Several possibilities for the structure of the helper factor are suggested by these results, of which the most likely are the following. (a) The factor is composed of 2 types of polypeptide chain, one of which is Ig-coded and carries the V_H domain, perhaps attached to a 'new' Ig constant region; the second chain is I region coded and might serve a role analogous to a light chain and contribute to the binding site, or rather dictate the function of the molecule (i.e. help versus suppression). (b) Alternatively, the polypeptide portion of the factor is entirely Ig coded (V_H plus a hitherto undefined constant region) and the Ia determinants are carbohydrate groups attached by MHC-coded glycosyltransferases.

The T hybrid lines should provide the ideal material with which to finally put a structure to these rather elusive molecules which may be only a short step from the even more elusive T cell receptor itself. What matters most to me is that these producer lines put the existence of the antigen-specific helper factors, with their unusual and provocative combination of properties, beyond dispute and, though a little late, make a convincing reply to *Niels Jerne's* doubts of 5 years ago.

References

1 Deak, B.D.; Meruelo, D.; McDevitt, H.O.: J. exp. Med. *147:* 599–604 (1973).
2 Dorf, M.E.: Springer Semin. Immunopath. *1:* 171 (1978).
3 Eshhar, Z.; Apte, R.N.; Lowy, U.; Ben-Neriah, Y.; Givol, D.; Mozes, E.: Nature, Lond. *286:* 270–272 (1980).
4 Germain, R.N.; Benacerraf, B.: Springer Semin. Immunopath. *3:* 93–127 (1980).
5 Howie, S.: Immunology *32:* 291–299 (1977).
6 Howie, S.; Feldmann, M.: Eur. J. Immunol. *7:* 417–421 (1977).
7 Kantor, F.; Feldmann, M.: Clin. exp. Immunol. *36:* 71–77 (1979).
8 Kappler, J.W.; Marrack, P.C.: J. exp. Med. *148:* 1510–1522 (1978).
9 Lamb, J.R.; Kontiainen, S.; Lehner, T.: J. Immun. *124:* 2384–2389 (1980).
10 Lonai, P.; Puri, J.; Hammerling, G.: Proc. natn. Acad. Sci. USA (in press, 1981).
11 Luzzati, A.L.; Taussig, M.J.; Meo, T.; Pernis, B.: J. exp. Med. *144:* 573–585 (1976).
12 Marrack, P.; Kappler, J.W.: J. exp. Med. *149:* 780–785 (1980).
13 McDevitt, H.O.: in Katz, Bernacerraf, The role of the histocompatibility gene complex in immune responses, pp. 321–326 (Academic Press, New York 1976).
14 Mozes, E.; Isac, R.; Taussig, M.J.: J. exp. Med. *141:* 703–707 (1975).

15 Munro, A.J.; Taussig, M.J.: Nature, Lond. *256:* 103–106 (1975).
16 Munro, A.J.; Taussig, M.J.: Nature, Lond. *269:* 355–358 (1977).
17 Munro, A.J.; Taussig, M.J.; Archer, J.: in McDevitt, Ir genes and Ia antigens, pp. 487–492 (Academic Press, New York 1978).
18 Rees, A.; Feldmann, M.; Erb, P.; Woody, J.; Kontiainen, S.; Bodmer, J.; Kantor, F.; Zvaifler, N.: Ann. N.Y. Acad. Sci. *332:* 503–515 (1979).
19 Strassman, G.; Eshhar, Z.; Mozes, E.: J. exp. Med. *151:* 265–274 (1980).
20 Tada, T.; Okumura, K.: Adv. Immunol. *28:* 1–87 (1979).
21 Taussig, M.J.: Nature, Lond. *248:* 234–235 (1974).
22 Taussig, M.J.: Immunology *41:* 759–787 (1980).
23 Taussig, M.J.; Finch, A.P.: Nature, Lond. *270:* 151–154 (1977).
24 Taussig, M.J.; Mozes, E.; Isac, R.: J. exp. Med. *140:* 301–312 (1974).
25 Taussig, M.J.; Munro, A.J.: Nature, Lond. *251:* 63–65 (1974).
26 Woody, J.N.; Zvaifler, N.J.; Rees, A.; Ahmed, A.; Hartzmann, R.; Strong, M.; Howie, S.; Kantor, F.; Feldmann, M.: Transplant. Proc. *11:* 382–388 (1979).
27 Zvaifler, N.J.; Feldmann, M.; Howie, S.; Woody, J.; Ahmed, A.; Hartzman, R.: Clin. exp. Immunol. *37:* 328–338 (1979).

Dr. Michael Taussig, Agricultural Research Council, Institute of Animal Physiology, Babraham, Cambridge CB2 4AT (England)

The Immune System, vol. 1, pp. 324–330 (Karger, Basel 1981)

Search for Lymphokines that Regulate Lymphocyte Development

James Watson[1], *Diane Mochizuki, Steven Gillis*

Department of Microbiology, University of California, Irvine, Calif.; Basic Immunology Program, Fred Hutchison Cancer Research Center, Seattle, Wash., USA

Two factors that affect the activation of murine lymphocytes have been distinguished on the basis of biochemical and biological criteria [1]. Both of these factors enhance the mitogenic response of thymocytes to PHA and Con A and stimulate antigen-dependent, cell-mediated and humoral immune responses in culture. They may be distinguished, however, by the presence or absence of the ability to stimulate continuous growth. One of these factors can be obtained from human and mouse macrophages, as well as from culture supernatants of macrophage tumors, and has a molecular weight of about 15,000 dalton. This monokine, originally designated lymphocyte-activating factor (LAF) [4], does not alone stimulate continuous T cell growth. The second factor is a murine spleen cell product described by *Chen and DiSabato* [3] as a 30,000- to 35,000-dalton thymocyte-stimulating factor (TSF). This second factor activity, originally obtained from murine, and then from rat and human sources, differed from LAF in that it was capable of promoting and maintaining proliferation of primary T cell lines in long-term culture.

A new system of nomenclature has been introduced by workers in this field [1]: LAF was designated interleukin 1 (IL-1) and TSF was designated interleukin 2 (IL-2). The biological and biochemical properties of IL-2 isolated from murine, rat, and human lymphoid cells are summarized here.

[1] *James Watson* has, on several occasions, worked as a visiting scientist at the Basel Institute for Immunology.

Many of the experiments described in this paper were initiated as a result of discussions at the Basel Institute for Immunology. They were, in fact, induced by *Niels Jerne's* skepticism about factors that were not biochemically characterized.

Association of Helper T Cell-Replacing Activity with Interleukin 2

We initially began purifying a class of soluble factors that stimulate immune responses to heterologous erythrocyte antigens in T cell-depleted murine spleen cultures [18]. These T cell-replacing factors (TRF) are secreted by mouse spleen cells that have been activated in culture by antigen [18] and the polyclonal T cell mitogen, Con A [9, 21]. We have utilized a microculture assay for quantitation in the purification of TRF from culture supernatants and have focused on the biological activity of a 30,000- to 40,000-dalton protein that exhibits heterogeneity in charge with pIs from 4 to 5 [9].

A class of factors termed 'co-stimulator', also known to be present in these supernatants [15–17], exhibited similar biochemical properties to TRF. Co-stimulator refers to a class of molecules that stimulates thymic lymphocytes of adult mice to proliferate in the presence of the plant lectins PHA and Con A under culture conditions where neither the lectins nor factors alone are mitogenic. Co-stimulator molecules also allow thymocytes to respond to alloantigens and mount cytotoxic lymphocyte (CTL) responses. *Watson* et al. [23] compared the purification procedures and assays, and determined that the molecules responsible for the responses we were measuring may be the same.

As a result of the reports of *Morgan* et al. [13], *Gillis* et al. [5, 6] and *Gillis and Smith* [8] that long-term growth of T cells was stimulated by lymphokines from human [13] and rat [2, 5, 8, 13, 14] conditioned medium, we began to examine murine Con A supernatants for T cell growth stimulatory activity [21]. Our approach was to examine the biological activity of lymphokines in Con A supernatants using different bioassays, and sequentially purifying the stimulatory activities by gel filtration, ion exchange chromatography, and isoelectric focusing (IEF) in an attempt to separate molecules with different lymphokine activities from each other. The biological activities measured included: (a) T cell growth factor (TCGF); (b) the promotion of Con A-induced mitogenesis in thymocyte cultures; (c) helper T cell-replacing factor (TRF) activity in athymic spleen cell cultures, and (d) the induction of CTL in both thymocyte and nude mouse spleen cultures.

The striking molecular homogeneity of the various activities present in such lymphokine preparations led us to conclude that a single class of factors, IL-2, is responsible for the activity in each assay system [7, 9–12, 21, 22].

Molecular Properties of Murine IL-2

After fractionation by gel filtration, ion exchange chromatography and IEF, murine IL-2 is a class of molecules with an apparent molecular size of 30,000 dalton and separable into two components which differ in charge, having pIs of 4.3 and 4.9. Both components support the continuous proliferation of cytotoxic and helper T cells in culture, promote the proliferation of Con A-treated thymocytes, and induce the generation of antibody-forming cells in T lymphocyte-deficient cultures and cytotoxic cells in either thymocyte or nude mouse spleen cell cultures [21].

In addition, within the factor(s) in the 30,000-dalton fraction another subclass of molecules was detected by its ability to induce antibody synthesis specific for heterologous erythrocyte antigens in cultures of nude mouse spleen cells. This material showed considerable heterogeneity in charge, ranging from a pI of 3.0–4.2 (fig. 1). No other lymphokine activity was detected when these molecules were tested in the other assay systems.

Rat and Human IL-2

Con A-stimulated rat spleen cells and mitogen-stimulated human peripheral blood lymphocytes produce culture supernatants which contain a T cell growth factor [2, 5, 6, 8, 14, 20, 21]. Rat and human IL-2 both appear to be single molecular species of 15,000–17,000 dalton (as estimated from gel filtration studies) with pIs of approximately 5.5 and 6.5, respectively [9]. Rat and human IL-2 have identical biological activities when assayed on murine lymphocytes. They enhance mitogenesis by Con A and amplify cytotoxic T cell and humoral antibody responses, under conditions of limiting helper activity. They are inseparable from T cell growth-promoting activity following successive gel filtration, ion exchange chromatography, and IEF. This is strong evidence that all biological activities are mediated by a single class of molecule.

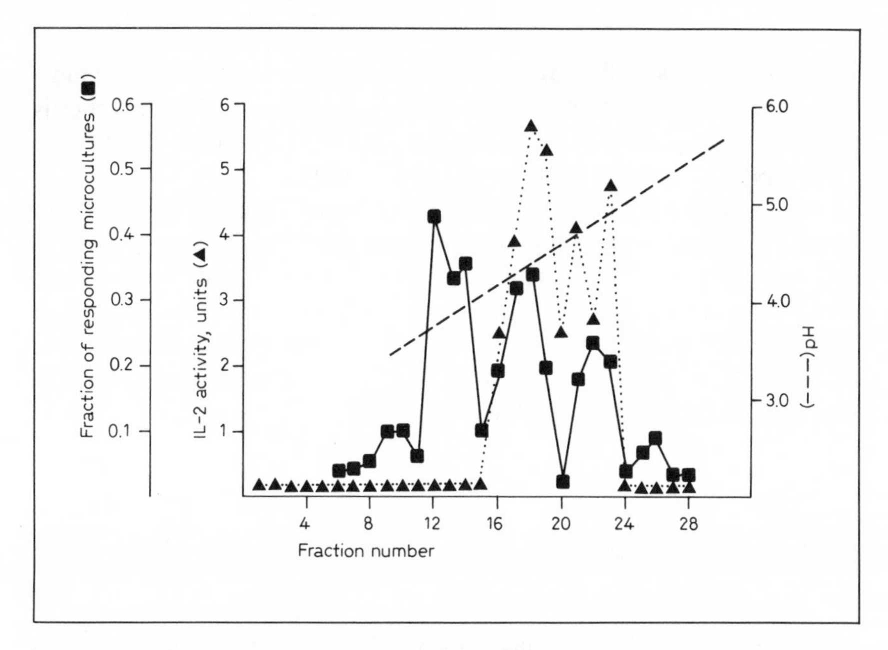

Fig. 1. Lymphokine activities assayed after isoelectric focusing over a narrow pH range (pH 3–6). The lymphokines were prepared from supernatants of Con A-stimulated BALB/c mouse spleen cells. ▲ = IL-2 activity assayed on CTLL cells; ■ = anti-sheep erythrocyte antibody responses in nude spleen cultures.

The Isolation of Cell Lines Producing IL-2

We have screened some 40 murine and human T cell leukemia and lymphoma cell lines for both constitutive and mitogen-induced IL-2 production. Of these, two were found to produce high-titer IL-2 upon mitogen stimulation. Stimulation of murine T cell line LBRM-33 with 1% PHA yielded culture supernatants which contained 1,000–5,000 times more IL-2 activity than the same number (10^6 cells/ml) of optimally stimulated rat or mouse splenocytes [19]. Similarly, PHA stimulation of the human leukemia T cell line Jurkat-FHCRC produced 100–300 times more IL-2 activity than lectin-stimulated human PBL or spleen cells. IL-2 activity from Jurkat-FHCRC supported the in vitro proliferation of both murine and human CTLL, that of LBRM-33 supported growth of only murine CTLL. Culture medium containing 1% PHA was incapable of inducing proliferation of either mouse or human activated T cells [7, 11, 12].

The LBRM cell line was originally derived from a radiation-induced splenic lymphoma in B10.BR mice. The expression of various T cell antigens on LBRM cells establishes this line as a T cell tumor. The Jurkat cells were originally obtained from Dr. *John Hansen,* Fred Hutchison Cancer Center, Seattle, and are a human leukemia T cell line [11].

With both cell lines, supernatants contained maximal IL-2 activity when 10^6 cells/ml were cultured with either 1% PHA or 20 μg/ml Con A and harvested after 24 h. This procedure routinely yields greater than 1,000 U/ml IL-2 activity with LBRM-3 and greater than 300 U/ml IL-2 with Jurkat. The yield was not affected by the presence or absence of serum [7, 11, 12]. Peak production by 24-hour lectin-stimulated cultures was consistently accompanied by poor cell viability.

Comparative Properties of IL-2 Derived from Normal and Malignant Cells

We have recently found that LBRM-33-derived IL-2 is biochemically indistinguishable from conventionally prepared mouse spleen-generated factor. Mouse tumor cell line-derived IL-2 appears to be localized in a 30,000-dalton protein, isolable by net charge into two electrophoretically distinct species with isoelectric points of 4.3 and 4.9, respectively. Both molecular species (1) enhance thymocyte mitogenesis, (2) sustain IL-2-dependent T cell line proliferation, and (3) induce CTL and PFC responses in thymocyte and nude spleen cell populations, respectively. Regardless of the source, murine IL-2 is completely destroyed by exposure to trypsin, subtilisin, and chymotrypsin. For the two preparations tested, LBRM-33-generated IL-2 activity was consistently more resistant to pH, urea and heat treatment than splenic-derived material [12].

While conventional IL-2 stimulated antibody responses in nude spleen cultures, presumably via the clonal expansion of helper T cell precursors, IL-2 from LBRM-33 does not. But we have observed, while purifying murine spleen IL-2, that there is another class of molecules within the IL-2 fraction. This class of molecules may be the true helper T cell-replacing factor; it induces antibody synthesis to heterologous erythrocyte antigens in nude spleen cultures but has no T cell growth factor activity. This material showed considerable heterogeneity in charge ranging from pI values of 3.0–4.2 (fig. 1). LBRM cells do not secrete this class of lymphokine.

Mode of Action of IL-2

Activated T cells, but not resting T cells, readily absorb IL-2 [22]. Furthermore, IL-2 acts on cloned cytotoxic and helper T cell lines. These observations imply that IL-2 interacts directly with T cells. However, there is a strict requirement for antigen in order to observe induction of antibody synthesis or CTL responses with IL-2. We have suggested that T cells or their precursors in thymocyte or nude spleen cultures respond to antigen or mitogen by expressing receptors for IL-2 and that the subsequent clonal expansion of T cells from any effector class (helper, killer, or suppressor) requires only the presence of IL-2. These suggestions are supported by the experimental finding that a brief treatment of thymocytes with Con A renders them responsive to IL-2 [22].

Physiological Significance of Interleukin 2

It is difficult to associate a factor activity in an in vitro response assay with a physiological role in vivo. Studies on the purification of lymphokines with T cell growth factor activity have revealed a class of hormone-like molecules that seems to promote the clonal expansion of antigen or mitogen-activated helper, suppressor, and cytotoxic T cell. However, it should be emphasized that the physiological significance of these molecules is unknown and that IL-2 may represent only one class of regulatory molecule. We have maintained in culture, for several years, murine T cell lines that retain their antigen-specific helper and cytotoxic affector activities and which remain absolutely IL-2-dependent for continued growth [5, 6, 8, 10, 20]. Whatever the physiological role of IL-2 might be, it is of tremendous value as a tool to generate antigen-specific T cell lines.

Acknowledgements. This work was supported by a Research Career Development Award (AI-00182) to *James Watson,* and Grant AI-13383 from the National Institute of Allergy and Infectious Disease, and Grant (1-469) from The National Foundation. *Steven Gillis* is a Special Fellow of The Leukemia Society of America. Supported by NCI Grant 28419 and Grant 1-724 from the National Foundation.

References

1 Aarden, L.A., et al.: J. Immun. *123:* 2928–2929 (1979).
2 Alvarez, J.M.; De Landazuri, M.O.; Bonnard, G.D.; Herberman, R.B.: J. Immun. *121:* 1270–1275 (1978).
3 Chen, D.M.; DiSabato, G.: Cell Immunol. *22:* 211–218 (1976).
4 Gery, I.; Waksman, B.H.: J. exp. Med. *136:* 143–152 (1972).
5 Gillis, S.; Baker, P.E.; Ruscetti, F.W.; Smith, K.A.: J. exp. Med. *148:* 1093–1098 (1978).
6 Gillis, S.; Ferm, M.M.; Ou, W.; Smith, K.A.: J. Immun. *120:* 2027–2032 (1978).
7 Gillis, S.; Scheid, M.; Watson, J.: J. Immun. (in press, 1980).
8 Gillis, S.; Smith, K.A.: Nature, Lond. *168:* 154–156 (1977).
9 Gillis, S.; Smith, K.A.; Watson, J.: J. Immun. *124:* 1954–1962 (1980).
10 Gillis, S.; Watson, J.: Immunol. Rev. *54* (in press, 1980).
11 Gillis, S.; Watson, J.: J. exp. Med. (in press, 1980).
12 Mochizuki, D.; Watson, J.; Gillis, S.: J. Immun. (in press, 1980).
13 Morgan, D.A.; Ruscetti, F.W.; Gallo, R.C.: Science *193:* 1007–1009 (1976).
14 Nabholz, M.; Engers, H.D.; Collavo, D.; North, M.: Curr. Topics Micro. Immunol. *81:* 176–187 (1978).
15 Paetkau, V.; Mills, G.; Gerhart, S.; Monticone, V.: J. Immun. *117:* 1320–1326 (1976).
16 Shaw, J.; Monticone, V.; Mills, G.; Paetkau, V.: J. Immun. *120:* 1974–1980 (1978).
17 Shaw, J.; Monticone, V.; Paetkau, V.: J. Immun. *120:* 1967–1973 (1978).
18 Watson, J.: J. Immun. *111:* 1301–1313 (1973).
19 Watson, J.: Trends biochem. Sci. *4:* 36–40 (1979).
20 Watson, J.: J. exp. Med. *150:* 1510–1519 (1979).
21 Watson, J.; Gillis, S.; Marbrook, J.; Mochizuki, D.; Smith, K.A.: J. exp. Med. *150:* 849–869 (1979).
22 Watson, J.; Mochizuki, D.; Gillis, S.: in Cunninghan, Goldwasser, Watson, ICN-UCLA Symp. on Control of Cell Division and Differentiation (Academic Press, New York 1980).
23 Watson, J.; Shaw, J.; Paetkau, V.; Aarden, L.A.: J. Immun. *122:* 1633–1638 (1979).

Dr. James Watson, Department of Pathology, University of Auckland School of Medicine, Auckland (New Zealand)

The Immune System, vol. 1, pp. 331–339 (Karger, Basel 1981)

The Effect of Interferon on the Immune System: from Skepticism to Confidence?

Iver Heron[1]

Institute of Medical Microbiology, University of Aarhus, Aarhus, Denmark

Interferon (IFN) was discovered more than 20 years ago [25] as a supernatant factor produced by virus-challenged cells which inhibits other viruses from replicating. It was soon demonstrated that IFN was stable on dialysis against buffers of low pH (pH 2), was destroyed by trypsin but not by ribonuclease, and acted on the cells in which virus replicate but did not directly inactivate the virus [30].

The discovery of IFN initially created optimism among virologists, who hoped that a general antiviral reagent would become available for clinical use. Purification and biochemical characterization of IFN of different cellular origins (leukocyte type IFN-α, fibroblast type IFN-β, and immune type IFN-γ), proved to be very difficult, however, and not until very recently was this accomplished. General enthusiasm waned during the long period in which researchers were doing experiments with supernatants containing IFN activity. Furthermore, many other biological effects, apparently not related to the antiviral action, were ascribed to IFN containing preparations over the years. All this made many scientists very skeptical about IFN, and this view is reflected in the following event. A well-known French IFN researcher, *E. DeMayer,* had a paper rejected by a very respectable journal with the comment that: 'he was working with an ill-defined acid stable inhibitor of no great interest and of no heuristic value' [10].

The majority of work on IFN has been done with mouse and human IFN, although the original discovery was made in the chicken. Recently IFN-α and IFN-β of mouse and humans have been completely purified, but

[1] *Iver Heron* was a member of the Basel Institute for Immunology in 1978 and 1979.

the biochemical procedures necessary for the two species, as well as for α and β within a species, proved to be very different, probably due to differences in glycosylation of the proteins.

One of my co-workers *(Kurt Berg)* has spent many years exploring purification protocols with the objective of isolating human leukocyte IFN. This was accomplished [4] and required a combination of several biochemical procedures: precipitation, gel filtration, ligand chromatographies, and affinity chromatography using extensively absorbed antisera. The final preparations were analyzed in SDS-PAGE, and the IFN activity could be eluted from distinct protein bands in at least 5 different molecular weight species in the range of 17,000–21,000 dalton. In retrospect, the quantity of supernatants necessary (several liters), the heterogeneity, as well as the purification factor required (around 200,000-fold), fully explain the difficulties. Modern technology of applied molecular biology and genetic engineering recently took over this part of the field and has in an explosive way yielded insight into IFN gene sequences and the IFN amino acid sequences; a strain of *E. coli* carrying an appropriately constructed plasmid synthesizes high titers of IFN-α [15]. This product proved to be biologically active when assessed for its capacity to protect squirrel monkeys from lethal encephalomyocarditis virus infection [15]. The gene structure studies confirmed the heterogeneity at the protein level, and at least 8 different chromosomal genes for IFN-α have been disclosed in cells originating from an individual embryo [33]. Some 5 different IFN-α genes were identified in an IFN-producing lymphoblastoid cell line [1]. IFN-β produced by human fibroblasts has been found to differ from IFN-α in coding sequences and protein sequences, although some degrees of homology were found [37]. It has been concluded that the genes were derived from a common ancestor. The immune type IFN (IFN-γ) produced by lymphocytes stimulated with mitogens or antigen has not yet been sufficiently characterized.

The functional and evolutionary significance of multiple structural genes for IFN remains unknown, but speculations abound, based on other known multigene family products like β-globin, the immunoglobulins, and the actins. It is not yet known whether the different IFN proteins have different specificity and functional properties apart from their antiviral activity, which defines them.

The so-called non-antiviral effects of IFN preparations include a variety of phenomenologic observations. Among these, the effects on the immune system are prominent; furthermore, among other effects, IFN enhances synthesis of prostaglandins [40] and certain enzymes, e.g. T-RNA methylase. Phagocytosis of carbon by mouse peritoneal cells was reportedly increased [22]. Questions like, 'Are there different receptors for the different IFN proteins on single cells?' and 'Do different cells have the same IFN receptors?' remain to be answered.

Effects of IFN on the Immune System

Inhibition of Replication and Immunosuppression. IFN preparations have been found to inhibit replication of various normal and neoplastic cells in vivo and in vitro. This effect is not due to cytotoxicity but is caused by an increase in generation time. Part of the inhibitory activity on cell growth described in the literature is probably exerted by low molecular weight contaminants [9], but pure IFN material still inhibits to a certain extent multiplication of selected cell lines [3, 11]. It is thus not surprising that IFN has been found to be 'immunosuppressive', when antibody responses have been assessed in vivo as well as in vitro [7, 17, 26]. The B cell has been found in vitro to be the main target for the inhibition via interference with the activation of B cell precursors, whereas B cells once activated and expanding are refractory [6]. This explains the observations that the timing of IFN and antigen administration in vivo is critical. If IFN was given hours after the antigen, no suppression, but rather some enhancement, of the PFC response was observed [39].

IFN exerts suppressive activity on T cell mitogenic responses in vitro and on MLR, but the magnitude of inhibition is not very impressive and relatively high doses of IFN are required. These latter effects have not yet been confirmed using pure IFN proteins. I think it is fair to say that IFN is not a potent agent as far as general immunosuppression in vivo is concerned.

Enhancement of Natural Killer Cell Function. NK function is a new and exciting field, and I will review it briefly [24]. Peripheral lymphoid tissues of mice and of humans contain populations of cells that express cytotoxicity in vitro against neoplastically transformed cells, against certain fetal antigenic structures on cells, as well as against a proportion of cells in normal bone marrow. The cells responsible for this spontaneous effector function have been termed NK cells. They constitute a few percent of the peripheral lymphocyte population and appear to be a heterogeneous population as far as conventional lymphocyte markers are concerned. In humans a T cell subset, as well as null cells, contains the main NK population. In the mouse T cells and non-T cells expressing an Ly 5.1 marker contain NK activity. Both beige mice and patients with Chediak-Higaschi's syndrome are NK-deficient, while athymic nude mice possess increased NK activity. It is controversial whether NK cells are related to the macrophage cell lineages. Most NK cells possess F(c) receptors for IgG, but the receptor can be blocked without loss of the NK function [2].

Successful cloning of cells with NK activity was recently reported [11]. The target cell repertoire of the individual clones was identical to the specificity spectrum of the normal non-cloned NK cells. This means that (at least some) NK cells have receptors that are not clonally distributed, but it does not tell us whether NK cell receptors are heterogeneous in their specificities or whether one receptor recognizes common structures on different target cells.

IFN seem to provide an important regulatory signal for NK cells. Marked augmentation of NK activity has been observed following administration of IFN inducers or IFN material directly, injected in vivo, in mice [14] and in humans [12]. In vitro short term (less than 1 h) treatment of lymphocytes with low IFN doses followed by removal of IFN leads to marked augmentation of NK activity. This effect has proved to be linked to the pure IFN proteins [21, 31]. The process has been analyzed and it was found that IFN increases the number of cytotoxic lymphocytes by recruitment among pre-NK lymphocytes, which were, however, able to bind target cells before IFN treatment [35]. In the mouse the recruitment is also reflected in the acquisition of the Ly 5.1 marker. In the human system it has been found that the IFN-induced NK activation is reversible and that the cells can be reactivated later [35]. In addition to NK cells, other lymphoid cells bind to and form conjugates with NK-sensitive target cells; among these conjugate formers there are cells that secrete IFN upon contact with the target cell [38]. Confrontation of NK cells with virus-infected cells or appropriate tumor cells thus initiates a positive feedback IFN signal resulting in a rapid amplification of NK ativity.

Few malignant cells transferred to athymic mice may produce tumors. Cells of the same origin, but infected with virus (several possible choices) injected in vast excess, failed to grow as tumors [32]. In vitro the uninfected tumor cells were resistant to NK, but the infected cells were sensitive. An infected line induced IFN when injected, and it was capable of preventing growth of certain other tumors. In beige mice, which are NK-deficient, a NK-sensitive melanoma tumor line grew faster and metastasized much faster than in control mice, whereas a NK-resistant line behaved identically in the two strains of mice [36]. Antibodies against IFN made virus-infected tumor cells grow in nude mice and generalized metastasis was observed [36].

From the above there seems little doubt that the NK system plays a role in surveillance against transplantable tumors in the mouse, particularly when other immune functions are deficient. Whether NK and IFN play any role in the prevention of spontaneous tumors in normal mice remains to be determined.

In humans the role of NK and IFN has been investigated less. IFN induction or injection give enhanced NK activity in the peripheral blood [12]. In some patients with multiple sclerosis and lupus erythromatosis, as well as in tumor-bearing patients, lower levels of NK as compared to healthy controls have been observed and relatives of patients with familial melanoma likewise have been found to have diminished NK activity [5].

Effect of IFN on Cytotoxic T Lymphocyte (CTL) Generation. With human lymphocytes [18, 19] when IFN was added to mixed lymphocyte cultures (MLC), the thymidine uptake was slightly inhibited, yet more CTL was generated. IFN treatment of CTL already generated or IFN added late in the culture period had no effect. Most enhancement was found when IFN was present from the initiation of the cultures through the proliferative phase. Both minimal and optimal doses were within 'physiological' levels, and the pure IFN proteins have proved effective. A closer analysis of this IFN effect has excluded trivial, artifactual explanations and shown that the augmented allospecific killer capacity was due to classical T-CTL effectors, not NK-like activity. In vitro generated suppressor cells for CTL were assessed with and without IFN, but IFN did not inactivate such suppressors. Secondary MLCs after primary cycles grown in the presence and absence of IFN showed that primary MLC cells with IFN were more quickly and easily induced by allogeneic stimulators to CTL, indicating that IFN during the primary MLR played an immunoselective role of some sort.

Effects on the Cell Surface. IFN treatment of cells can increase the expression of certain surface antigens. Expression of H-2 antigens on a mouse leukemia cell line were found to be increased [29], and recently [20] we observed that the amount of HLA antigens on human blood lymphocytes increased following a brief exposure to IFN. The cells needed protein synthesis and incubation at 37 °C for at least some hours, after which the anti-HLA antibody-absorbing capacity of IFN-treated cells was enhanced severalfold. Cell size and shape were found to be unchanged. Pure IFN proteins have proved active. Two other antigenic determinants, Ig on B cells and a T cell specific antigen, were unaffected by IFN. The DR antigens (analogous to Ia in mouse) found on a subpopulation of lymphocytes in the blood were not changed by IFN. However, on monocytes the expression of not only HLA but also the DR antigen was changed. The functional implications of this for the antigen presenting role of monocytes has not yet been resolved. Preliminary experiments of our own have shown that when the presenting cell was limiting, IFN pretreated cells were more efficient in presenting PPD to lymphocytes than control monocytes, whereas there was no difference at higher concentrations of monocytes.

IFN has been found to enhance HLA antigens on lymphocytes of different lineages as well as on certain lymphoid cell lines, and normal human fibroblasts in culture react similarly. It thus looks as if this IFN-induced change of the cell surface is expressed not only on lymphoid cells but more widely.

Other effects on the cell surface have been described. *Chang* et al. [8] studied a mouse embryo cell line and showed that IFN treatment increased the density of the plasma membrane and the concentration of some of the major intramembranous glycoproteins. Other reported IFN-induced changes at the cell surface are an increase in net negative charge [27], increased binding of Con A [23], and changes in the binding of TSH [28].

General Discussion

At the present time IFN enjoys some degree of credibility, mainly because the efforts to characterize it and purify it have been successful. The second main area of objection to IFN, the wide breadth of IFN action, remains just as confusing as before. Much work at the cellular level needs to be carried out to clarify whether the multitude of effects represents a reaction pattern of the cell when triggered by one of the IFN proteins. A detailed molecular and cell biological analysis is obviously needed to find some unifying concept.

It is tempting to think of IFN as a hormone when its wide spectrum of action is considered; moreover, many of its non-antiviral effects on cells are antagonistic to those of corticosteroids. But the ability of so many different cell types to produce IFN upon proper induction makes the situation exceptional. That several differentiated cell types have IFN-productive potential could be taken as an indication for the vital importance of the proteins.

There is some evidence that endogenous IFN plays a role in vivo. Inoculation of antibodies against mouse IFN into mice infected with Semiliki Forest virus leads to earlier development of disease and increased mortality [13]. Herpes simplex virus infection was accelerated and Moloney sarcoma virus tumors grew larger in anti-IFN-treated mice [16]. Encephalomyocarditis virus disease progressed more rapidly in antibody-treated mice, and the target organs were different because the mice died from visceral damages instead of from neurological disease [16]. Experiments of this kind clearly show that IFN is of importance for the outcome of certain viral infections, but does not allow distinctions to be made about the role of the antiviral activity and the effects exerted on the immune system. One example from the literature points to an effect of IFN on the defense system. Vaccinia virus infects monkey cells in vitro as well as in vivo, but IFN in vitro did not protect cells from infection, while in vivo IFN effectively prevented the growth of locally inoculated virus [34]. Also herpes viruses are quite refractory to IFN action in vitro but appear to be very sensitive in vivo [16]. These observations support the fact that there is some action other than the antiviral one in vivo.

Finally, I will briefly try to explain how I think the action of IFN on the immune system as mentioned above works in the living animal. I see the NK system regulated by IFN as a surveillance system against transformed cells (by virus or by other means). The specificity of this system is not as exquisite as that of the CTL system, but the NK system is preformed and works very quickly and can be boosted by IFN within less than an hour. At the same time the NK system probably functions as a regulator of hemopoiesis [24].

Some virus-infected cells obviously escape the NK system. IFN induction limits the spread of the infection via the antiviral effect; and the enhanced major histocompatibility antigen (MHC) expression on infected cells and on immunocompetent cells eases the cellular interactions necessary during the initial sensitization phase (in which it is well known that viral antigens and MHC antigens are recognized in association). Virus-specific CTL generation is thought to be augmented by IFN (e.g. alloreactive CTL in vitro), whereas the B cell system is relatively inhibited by IFN. At the stage (some days later) where CTL are generated and react locally with their virus-infected targets, the effect of IFN is not so relevant, since IFN production in vivo occurs only during the first few days of viral infection.

The main effect of IFN on the specific immune response against the virus then is immunomodulation with a preference for the T cell part of the system at the expense of the humoral side. Does this seem plausible in the teleological network? I think it does, since it is well established that the T cell system is particularly important for defense against viral infections (patients with T cell deficiency diseases die from viral infections) and viruses are the most potent IFN inducers.

References

1 Allen, G.; Fantes, K.H.: Nature, Lond. *287:* 408–411 (1980).

2 Barada, F.A.; Kay, H.D.; Emmous, R.; Davis, J.S.; Horwitz, D.A.: J. Immun. *125:* 865–871 (1980).

3 Berg, K.; Heron, I.: J. gen. Virol. (in press).

4 Berg, K.; Heron, I.: Scand. J. Immunol. *11:* 489–501 (1979).

5 Bloom, B.R.: Nature, Lond. *284:* 593–595 (1980).

6 Booth, R.J.; Booth, J.M.; Marbrook, J.: Eur. J. Immunol. *6:* 769–772 (1976).

7 Brodeur, B.R.; Merigan, T.C.: J. Immun. *114:* 1323–1330 (1975).

8 Chang, E.H.; Jay, F.T.; Friedman, R.M.: Proc. natn. Acad. Sci. USA *75:* 1859–1863 (1978).

9 Dahl, H.: Acta path. microbiol. scand. B *5979:* 1–4 (1977).

10 DeMaeyer, E.: Bull. Inst. Pasteur, Paris *76:* 303–323 (1978).

11 Dennert, G.: Nature, Lond. *287:* 47–49 (1980).

12 Einhorn, S.; Blomgren, H.; Strander, H.: Acta med. scand. *20:* 477–483 (1978).

13 Fauconnier, B.: Path. Biol. *19:* 575–578 (1971).

14 Gidlund, M.; Orn, A.; Wigzell, H.; Senik, A.; Gresser, I.: Nature, Lond. *223:* 259–260 (1978).

15 Goeddel, D.V.; Yelverton, E.; Ullrich, A.; Heyneker, H.L.; Miozzari, G.; Holmes, W.; Seeburg, P.H.; Dull, T.; May, L.; Stebbing, N.; Crea, R.; Maeda, S.; McCandliss, R.; Sloma, A.; Tabor, J.M.; Gross, M.; Familletti, P.C.; Pestka, S.: Nature, Lond. *287:* 411–416 (1980).

16 Gresser, I.; Tovey, M.G.; Maury, C.; Bandu, M.T.: J. exp. Med. *144:* 1316–1327 (1976).

17 Gresser, I.; DeMayer-Guignard, J.; Tovey, M.G.; DeMayer, E.: Proc. natn. Acad. Sci. USA *76:* 5308–5312 (1979).

18 Heron, I.; Berg, K.; Cantell, K.: J. Immun. *117:* 1370–1377 (1976).

19 Heron, E.; Berg, K.: Scand. J. Immunol. *9:* 517–526 (1979).

20 Heron, I.; Hokland, M.; Berg, K.: Proc. natn. Acad. Sci. USA *75:* 6215–6219 (1978).

21 Heron, I.; Hokland, M.; Hokland, P.; Berg, K.: Ann. N.Y. Acad. Sci. (in press).

22 Huang, K.Y.; Donahoe, R.M.; Gordon, F.B.; Dressler, H.R.: Infect. Immunity *4:* 581–588 (1971).

23 Huet, C.; Gresser, I.; Bandu, M.T.; Lindahl, P.: Proc. Soc. exp. biol. Med. *147:* 52–57 (1974).

24 Möller, G. (ed.): Immunol. Rev. *44* (1979).
25 Isaacs, A.; Lindenmann, J.: Proc. R. Soc. B *147:* 258–267 (1957).
26 Johnson, H.M.; Smith, B.G.; Baron, S.: J. Immun. *114:* 403–409 (1975).
27 Knight, E.; Korant, B.D.: Biochem. biophys. Res. Commun. *74:* 707–713 (1977).
28 Kohn, L.D.; Friedman, R.M.; Holmes, J.M.; Lee, G.: Proc. natn. Acad. Sci. USA *73:* 3695–3699 (1976).
29 Lindahl, P.; Gresser, I.; Leary, P.; Tovey, M.: Proc. natn. Acad. Sci. USA *73:* 1284–1299 (1976).
30 Lindenmann, J.; Burke, D.; Isaacs, A.: Br. J. exp. Path. *38:* 551–562 (1957).
31 Minato, M.; Reid, L.; Cantor, H.; Lenguel, P.; Bloom, B.R.: J. exp. Med. *152:* 124–137 (1980).
32 Minato, N.; Bloom, B.R.; Jones, C.; Holland, J.; Reid, L.M.: J. exp. Med. *149:* 117–1130 (1979).
33 Nagata, S.; Mantei, N.; Weissmann, C.: Nature, Lond. *287:* 401–408 (1980).
34 Schellekens, H.; Weimar, W.; Cantell, K.; Stitz, L.: Nature, Lond. *278:* 742–743 (1979).
35 Silva, A.; Bonavida, B.; Targan, S.: J. Immun. *125:* 479–484 (1980).
36 Talmadge, J.G.; Meyers, K.M.; Prieur, D.F.; Starkey, J.R.: Nature, Lond. *284:* 622–624 (1980).
37 Taniguchi, T.; Mantei, N.; Schwarzstein, M.; Nagata, S.; Muramatsu, M.; Weissmann, C.: Nature, Lond. *285:* 547–549 (1980).
38 Timonen, T.; Saksela, E.; Virtanen, I.; Cantell, K.: Eur. J. Immunol. *10:* 422–427 (1980).
39 Virelizier, J.L.; Virelizier, A.M.; Allison, A.C.: J. Immun. *117:* 748–764 (1976).
40 Yaron, M.I.; Gurari-Rotman, D.; Revel, M.; Lindner, H.R.; Zor, U.: Nature, Lond. *267:* 457–458 (1977).

Dr. Iver Heron, Institut for Medicinsk Mikrobiologi, Bartholin Bygningen, Aarhus Universitet, DK–8000 Aarhus C (Denmark)

The Immune System, vol. 1, pp. 340–348 (Karger, Basel 1981)

Neuroendocrine Control of the Immune Response

E. Sorkin[1], *A. del Rey, H.O. Besedovsky*
Swiss Research Institute, Department of Medicine, Davos, Switzerland

What does a lymphocyte know about
the waters in which he swims all his life?
A. Einstein (adapted)

The immune system is presently viewed as a network of helper, suppressor and effector lymphocytes and of macrophages, which are integrated in circuits, capable of stabilizing the system after antigen has evoked a distinctive perturbation. Such a view is to a considerable extent based on the crucial ideas on natural selection of antibody formation advanced by *Jerne* [16] in 1955 and his work on feedback control of the immune response by specific antibody molecules [17]. *Jerne's* more recent brilliant concept of the immune system as a dynamic idiotypic-antiidiotypic network of interacting cellular elements [18] has superseded the static view of neatly separated immunological compartments and presents a new point of departure for understanding immunoregulation. Analysis of this complex autoregulated network and its interaction with other integrative neuroendocrine host systems under physiological conditions presents a major challenge.

Despite its seeming autonomy and regulatory characteristics, we propose that the immune system is also subject to other levels of control by the two major integrative host agencies, the central nervous and endocrine system.

The considerable literature on in vitro induced immune responses [22] attests, not surprisingly, to the fact that immune cells perform well in the absence of agents such as hormones and neurotransmitters, which are nor-

[1] *Ernst Sorkin* was a member of the Board of Consultants of the Basel Institute for Immunology from 1970 to 1975.

mal components of their physiological environment. Most host systems behave similarly in that they show to varying extents some degree of autonomy. However, all of these systems, the immune system included, have evident restrictions on their autonomy imposed on them by neuroendocrine mechanisms. In fact, some evidence is already emerging that immunoregulatory circuits integrate information derived from products of the immune system and from hormones and neurotransmitters. As this evidence is still fragmentary, the proposal of neuroendocrine immunoregulation needs close examination and fulfillment of criteria such as are required for any kind of physiological regulatory mechanisms.

Present Information on Neuroendocrine Control of the Immune System

The proposition that immune-neuroendocrine circuits are involved in immunoregulation is supported by experimental evidence fulfilling several basic criteria:

Hormones and Neurotransmitters Interfere with Processes Essential for the Immune Response. Several hormones and neurotransmitters influence a variety of cellular and subcellular processes essential to the immune response, e.g. lymphoid cell proliferation and transformation, metabolism, transport of substances and allosteric changes in membranes, genetic expression, lymphokine synthesis [for summaries see 4, 5, 9].

Lymphocytes Express Receptors for Hormones and Neurotransmitters. Receptors for a number of hormones and neutrotransmitters have been demonstrated on lymphoid cells. This is true for corticosteroids, insulin, growth hormone, oestradiol, testosterone, β-adrenergic agents and acetylcholine [6].

Interference with Neuroendocrine Functions Influences the Immune Response. Numerous experimental findings affirm that hormones and neurotransmitters affect the immune response in a variety of ways [4–6, 10, 12, 25]. Most reports agree that depending on circumstances (i.e. the type, concentration, mode and timing of their application) either immunosuppression or immunopotentiation can be brought about.

Many authors have reported on the consequences of direct manipulation of the brain (e.g. by electrical lesions or stimulation) on immune responsiveness [for review see 23]. While it is evident that these and other related experiments can affect the functioning of the immune system, due to induced gross disturbances, they hardly allow the analysis of the *dynamic flow of information* on which immunoregulation would be based. The same methodological limitations apply to modes of intervention such as organ extirpation, denervation and the administration of pharmacologic doses of hormones and neurotransmitters.

Immune-Neuroendocrine Interactions Exist under Physiological and Pathological Conditions. The evidence for such interactions derives from situations in which endocrine or immunological derangements result in reciprocal functional disturbances. Among examples of such situations the following may be listed: the hypopituitary dwarf mouse (endocrine deficiency – immunological defect); in the germ-free state, in the athymic nude mouse and in the early bursectomized chicken embryos (immunological defects – endocrine alterations) [5].

The Activated Immune System Induces Changes in the Neuroendocrine System and Receives Regulatory Signals. For any control mechanism to be effective, an essential requirement is that it has the capability to perceive changes in the mechanism which should be regulated by it. Accordingly, neuroendocrine alterations should be induced by the activated immune system and this in turn should subsequently receive regulatory signals.

Changes in Blood Hormone Levels Occur during the Immune Response. We ourselves have shown that in the course of the immune response to several otherwise inert, non-replicating antigens, changes occur in corticosterone and thyroxine blood levels [8]. Thus, 5 days after injection of SRBC into rats and mice, corticosterone concentrations in blood rise to levels which are immunosuppressive. The possible immunoregulatory meaning of these hormonal changes will be considered subsequently.

Changes in Local Neurotransmitter Levels Occur in Lymphoid Tissues during the Immune Response. A strong depression of noradrenaline content has been found in spleen but not in non-lymphoid organs of rats 3–4 days after antigenic challenge with SRBC [3]. The decrease in NA content is proportional to the magnitude of the immune response [11]. This decrease in NA splenic level is viewed as the expression of a sympathetic reflex mechanism triggered by the immune response itself. Since denervation of the spleen produces an enhancement of the immune response and α-agonists are immunosuppressive [3], the regulatory implication of this sympathetic reflex most likely is removal of the restraint exerted by NA on immunological cells.

Afferent Signals from the Activated Immune System Elicit a Response in Neurons of the Hypothalamus. Immunoregulation by the central nervous system implies reception of information derived from an ongoing immune response (afferent pathway) and the possible subsequent emission of regulatory signals to immunological cells (efferent pathway). We have approached the problem by determining in rats, challenged with SRBC or TNP-haemocyanin, the firing rates of individual hypothalamic neurons [7]. For example, a highly significant increase in firing rates restricted to the medial hypothalamus was observed on day 5 after SRBC administration. The fact that on day 1 after SRBC injection, a time when no PFC are detectable, no increase in firing rates in the studied neurons occurred, would seem to exclude external artifacts, i.e. handling, stress or some direct action by the antigen itself as being responsible for the neuronal changes.

Lymphokines Act as Afferent Signals for Neuroendocrine Changes during the Immune Response. The endocrine, autonomic and hypothalamic changes described here imply the existence of afferent signals originating in the activated immune system. The nature of these signals remains unknown, but it is a rational speculation that the soluble mediators elaborated by lymphocytes, lymphokines, would be the most likely candidates. Supernatants of Con A-activated lymphocytes were therefore injected into rats. They induced within hours a severalfold increase in concentration of corticosterone in blood attaining immunosuppressive levels analogous to those obtained in antigenically challenged animals by the 5th day following the primary antigenic stimulus [2]. This finding suggests that lymphokines could be afferent signals mediating the kind of neuroendocrine changes observed during the immune response. These results, together with the increase of blood glucocorticoids during the immune response and the data of other laboratories showing inhibitory effects of these hormones on lymphokine production (e.g., LAF, TCGF [14]), lead us to propose the existence of a glucocorticoid-associated immunoregulating circuit.

While the aforementioned examples appear to be indicative for the participation of neuroendocrine mechanisms in immunoregulation, they also lead us to focus on a number of unsolved problems. In particular, the origin and nature of the afferent and efferent signals, their kinetics, and their effect on the different types of immunological cells that receive these signals, are quite unknown. Furthermore, two major conceptual problems need consideration. How could unspecific signals mediated by hormones and neurotransmitters affect immunospecificity? How are autoregulatory immunological circuits to be integrated with neuroendocrine signals?

The Specificity Problem

One main conceptual difficulty is to account for the manner by which non-specific agents such as hormones and neurotransmitters could exert control over a system which involves specific effector cells and products. The early concept of the immune system as being totally compartmentalized into individual clones of cells with different specificities has undergone radical modification. The discovery that distinctive cellular products, presumably lymphokines with polyclonal effects, participate in different stages of the immune response has revealed interclonal influences. In addition, *Jerne's* idiotypic network also involves clones other than those which recognize and respond to a particular external antigen. Specificity seems to emerge from a fine balance between specific clonal expansion evoked by an external immunogen and polyclonal effects of several lymphokines, cross-reactive expanding clones and idiotypic-antiidiotypic reactions.

Taking these several participating components into account, neuroendocrine control could conceivably affect specificity by: (a) selective action on clones which recognize and react to a given external antigen, and (b) stabilization of internal connections of the immune system within a defined range of activity.

Selective Neuroendocrine Control of Clones which Recognize a Given External Antigen. Examples of selective control are those reported by *Hollenberg and Cuatrecasas* [15] and *Bourne* et al. [9]. Thus, only activated lymphoid cells express detectable receptors for insulin [15] and β-adrenergic agents [9]. Such examples provide the clue that those cells which expand clonally after binding external antigens will perceive these agents first, thereby gaining a regulatory advantage as compared with resting and other lymphoid cells which because of their low affinity, their idiotypic-antiidiotypic interaction or their response to non-specific polyclonal signals can only be activated later on.

Another example refers to selective neural effects on lymphocytes responding to external antigens. The immune response develops in well-defined anatomical structures. After immunization of rats with SRBC, PFC first appear in the white pulp of the spleen close to the periarteriolar sheath [13]. This location, which is a T cell-dependent area, may well be optimal for T-B cell interactions. Restricted to the same zone, the neurotransmitter noradrenaline is released by sympathetic nerves [21]. Therefore, SRBC-plaque-forming B cells, as compared with other B cells, would be more influenced by sympathetic signals because of their differential location in the area of the spleen with the highest noradrenaline content. Furthermore, they would be more affected by the phasic changes in splenic noradrenaline content, which we have reported as occurring during the immune response to SRBC.

Stabilization of the Internal Immune Network by Neuroendocrine Mechanisms. It is well known that glucocorticoids are much less immunosuppressive when utilized to intervene in an ongoing immune response than when they are given before or simultaneously with the antigen. We have shown that corticosterone blood levels increase more than 2- to 3-fold several days after immunization with SRBC or 1–2 h after injection of lymphokines. This increase in corticosterone may in turn affect the production of certain lymphokines. Thus, for example, T cell growth factor (TCGF) production was recently shown to be diminished by glucocorticosteroid [14]. In regulatory terms, the possible significance of these phenomena is that while permitting the ongoing immune response to the external antigen (SRBC) to proceed relatively undisturbed, other possibly excessive involvement of lymphoid cells unrelated to the anti-SRBC response is suppressed. Thus, in this instance, specificity would be controlled by reducing 'background noise'. A strong indication for the opertion of such mechanisms is the finding that by adrenalectomy, with consequent ablation of the corticosterone increase, antigenic competition (between HRBC and SRBC, injected sequentially) can be overcome [1].

Common Circuits and Pathways between Internal and Neuroendocrine Signals

We have shown that 'external' neuroendocrine signals restrict the autonomy of the immune system and participate in immunoregulation. However, this concept should not be interpreted as if autoregulation and external regulation are each operating in an absolutely independent manner. Rather, it would be more realistic to assume that common pathways and circuits exist for *both* control mechanisms. During different stages of the immune response the performing lymphoid cell clones are likely to receive both internal and external (neuroendocrine) signals which at times act synergistically, at other times antagonistically. Accordingly, one can assume that autoregulation and neuroendocrine immunoregulation operate in a coordinated fashion. There are a number of experimental findings consonant with this proposition.

(1) Lymphocyte activation by lymphokines, polyclonal or antigenic stimuli seems to be mediated by changes in the intracellular level of cyclic nucleotides [19]. Since many hormones and neurotransmitters also induce changes in the intracellular concentration of these messengers, it is likely that specific hormone binding to lymphoid cell receptors could influence the threshold of lymphoid cell activation by modifying cyclic nucleotide levels.

(2) The previously discussed data on expression of receptors for certain neuroendocrine agents in activated lymphoid cells can be considered to be an instructive example of common circuits between internal and external signals. Internal regulatory signals leading to activation of cells will also make them more sensitive to the action of insulin and sympathetic agonists which thereby might influence their performance.

(3) The regulatory switch from IgM to IgG production seems to be affected by corticosteroid hormones [6].

(4) Lymphokines seem to be fundamental factors controlling lymphoid cell activation, transformation and clonal expansion. There are examples for adrenocortical hormones influencing at physiological concentrations lymphokine production [14, 24]. These findings taken together with the observed increased corticosteroid levels following lymphokine inoculation [2] is suggestive of a hormonally mediated feedback mechanism for regulation of lymphokine production. The physiological significance of these findings remains to be established.

(5) Certain sites within the Fc region of human immunoglobulin molecules have a non-immunological affinity to an unknown constituent on both cell membrane and cytoplasma of ACTH producing cells in the human anterior pituitary [20]. The physiological significance of this finding is unknown, but it is conceivable that the activity of ACTH-producing cells is affected by the binding of the immunoglobulins. If so, this would constitute an example of a product of the immune response being recognized by an

endocrine cell and the possibility that such a product might influence the function of such a highly specialized cell.

Concluding Remarks

The experimental findings alluded to, in particular those on the dynamic changes in the endocrine and sympathetic nervous systems and in the hypothalamus in the course of the immune response, present persuasive indications for physiological external immunoregulation. These changes can be considered as reflecting autoregulatory immune mechanisms subject to and integrated with a complex system of hormonal feedbacks and neural signals. Since the central nervous system coordinates multiple external and internal signals and biorhythms, it can be visualized that several central inputs can affect the basal operation of the neuroendocrine control of the immune response. Numerous questions remain concerning the nature of the afferent and efferent signals, their kinetics and cellular targets and how they are to be synchronized with autoregulatory mechanisms within the immune system. However, it is hoped that as this kind of exploration continues, we will come to understand better how the lymphocytes live in the waters in which they swim all their life.

Acknowledgement. We thank our former institute colleague, Dr. *Maurice Landy,* for editorial treatment of the manuscript and for helpful discussions. This work was supported by the Swiss National Science Foundation, Grant No. 3.213.77.

References

1 Besedovsky, H.O.; del Rey, A.; Sorkin, E.: Clin. exp. Immunol. *37:* 106–113 (1979).
2 Besedovsky, H.O.; del Rey, A.; Sorkin, E.: J. Immun. *126:* 385–387 (1981).
3 Besedovsky, H.O.; del Rey, A.; Sorkin, E.; Da Prada, M.; Keller, H.H.: Cell. Immunol. *48:* 346–355 (1979).
4 Besedovsky, H.O.; Sorkin, E.: in James, Endocrinology, vol. 2, pp. 504–513 (Excerpta Medica, Amsterdam 1977).
5 Besedovsky, H.O.; Sorkin, E.: Clin. exp. Immunol. *27:* 1–12 (1977).
6 Besedovsky, H.O.; Sorkin, E.: in Ader, Psychoneuroimmunology (Academic Press, New York, in press, 1981.
7 Besedovsky, H.O.; Sorkin, E.; Felix, D.; Haas, H.: Eur. J. Immunol. *7:* 325–328 (1977).

8 Besedovsky, H.O.; Sorkin, E.; Keller, M.; Müller, J.: Proc. Soc. exp. Biol. Med. *150:* 466–470 (1975).

9 Bourne, H.R.; Lichtenstein, L.M.; Melmon, K.L.; Henney, C.S.; Weinstein, Y.; Shearer, G.M.: Science *184:* 19–28 (1974).

10 Claman, H.N.: J. Allergy clin. Immunol. *55:* 145–151 (1975).

11 del Rey, A.; Besedovsky, H.O.; Sorkin, E.; Da Prada, M.; Bondiolotti, P.: Submitted for publication.

12 Dougherty, T.F.; Berliner, M.L.; Schneebeli, G.L.; Berliner, D.L.: Ann. N.Y. Acad. Sci. *113:* 825–843 (1964).

13 Fitch, F.W.; Stejskal, R.; Rowley, D.A.: in Fiore-Donati, Hanna, Advances in experimental medicine and biology, vol. 5, pp. 223–321 (Plenum Press, New York 1969).

14 Gillis, S.; Crabtree, G.R.; Smith, K.A.: J. Immun. *123:* 1632–1638 (1979).

15 Hollenberg, M.D.; Cuatrecasas, P.: in Clarkson, Baserga, Control of proliferation of animal cells, pp. 423–434 (Cold Spring Harbor Laboratory, Cold Spring Harbor 1974).

16 Jerne, N.K.: Proc. natn. Acad. Sci. USA *41:* 849–857 (1955).

17 Jerne, N.K.: Annls Immunol. *125C:* 373–389 (1974).

18 Jerne, N.K.: in Cold Spring Harb. Symp. quant. Biol. *XXXII:* 591–603 (1967).

19 Parker, C.W.: in Cohen, Pick, Oppenheim, The biology of the lymphokines, pp. 541–583 (Academic Press, New York 1979).

20 Pouplard, A.; Bottazzo, G.F.; Doniach, D.; Roitt, I.V.: Nature, Lond. *261:* 142–144 (1976).

21 Reilly, F.D.; McCuskey, R.S.; Meineke, H.A.: Anat. Rec. *185:* 109–118 (1976).

22 Schreier, M.H.: in Pick, Lymphokine reports, vol. 2 (Academic Press, New York, in press, 1981).

23 Stein, M.; Schiavi, P.C.; Camerino, M.: Science *191:* 435–440 (1976).

24 Wahl, S.M.; Altman, L.C.; Rosenstreich, D.L.: J. Immun. *115:* 476–481 (1975).

25 Wolstenholme, G.E.W.; Knight, J.: Ciba Found. Study Group No. 36 (Churchill, London 1970).

Dr. Ernst Sorkin, Schweizerisches Forschungsinstitut, Medizinische Abteilung, CH–7270 Davos-Platz (Switzerland)

Lymphoid Tissues

The Immune System, vol. 1, pp. 349–355 (Karger, Basel 1981)

Micro-Environments in Lymphoid Tissues

J.H. Humphrey[1]

Royal Postgraduate Medical School, London, England

Prologue

Niels Jerne has repeatedly been able to discern a wood when most of us looked only at the trees, and as the number of trees has grown the pattern of the wood has become finer and more magical. *Niels* also founded two nurseries, one in Geneva and the other in Basel, where new kinds of tree are planted and old and new are put to practical use. It is a pleasure to contribute to the Festschrift in his honour an essay about some of the tangled thickets which the wood still contains.

Introduction

Much of our present knowledge about the properties, functions and interactions of T and B lymphocytes has been obtained from ingenious in vitro experiments, using mixtures of lymphocytes and adherent cells derived from genetically defined rodents or even humans, and indeed it could not have been obtained in any other way. However, in vivo the cells do not

[1] *John Humphrey* was Chairman of the Board of Advisors of the Basel Institute for Immunology from 1970 to 1973.

occur in mixtures, manipulable by the investigator, but live in or migrate through elaborate and intricate anatomical structures in lymphoid tissues, whose architecture is remarkably similar in all higher and some lower vertebrates.

Although the structure of lymph nodes and spleen used to be regarded as designed primarily to trap microbes and other foreign particles as efficiently as possible, it has also become clear that the delicate architecture also functions to keep cells apart in some sites and to encourage their interaction in others. Thus, although in vitro experiments indicate how T cells, B cells, accessory cells and antigens may interact in principle, the outcome in practice is likely to be determined by the precise micro-anatomical situations in which they actually meet. The purpose of this essay is to show that micro-environments should not be neglected and that we are beginning to know something about them. I shall discuss briefly four different sites where antigen in or on accessory cells and lymphocytes may interact, and the different consequences which ensue. In so doing I confine myself to inert antigens, and shall avoid discussing the complexities associated with the surface antigens of living cells or intracellular micro-organisms capable of replicating intracellularly. I shall also not consider the consequences of direct interaction between free antigen and lymphocytes, even though these have been shown to result sometimes in vitro in tolerization or stimulation of B cells or specific stimulation of suppressor T cells. This may occur also in vivo, but since most antigens (other than proteins for export) are rapidly and extensively taken up by macrophages in vivo, and since it has often been confirmed that immune responses are stimulated more effectively by macrophage-associated than by free antigens, I assume that antigens taken up by or on cells are biologically the more relevant.

The micro-environments to be discussed are identifiable by the nature of the antigen-presenting cells in and of the lymphocyte traffic through them. They are germinal centres, T-dependent areas, marginal zones and sinuses, and spleen red pulp or lymph node medulla, in each of which the outcome of antigen-lymphocyte interaction appears to differ. Persons who work with lymphoid tissues tend to confine their attentions to lymph nodes or to spleens, out of convenience or habit, and most of the experimental data on which my theme is based derive from studies of mouse spleens. In drawing upon other work on similar structures in lymph nodes there is a built-in assumption that generally similar micro-environments in different tissues imply similar influences at work. To argue whether this is strictly true would require more detailed consideration than space allows.

Germinal Centres

Germinal centres in spleen and lymph nodes are characterized by the presence of B and the almost complete absence of T lymphocytes. In the middle of an active germinal centre many of the B cells are blast forms or in mitosis, whereas those in the outer corona are apparently quiescent. Large active macrophages containing nuclear debris are also present, and there is evidence that B cells are being destroyed there. However, the most striking features are follicular dendritic cells (FDC), which have long extended processes in intimate contact with lymphocytes [10] and which bind antigen-antibody complexes to their surface, without ingesting and degrading them. In the spleen binding occurs via activated C3, and only those complexes which activate C3 (by the classical or alternate pathway) become bound [7]; in lymph nodes the evidence that C3 is required is less clear [15]. Radiolabelled antigen in the complexes can remain intact for many months, and is partly displaceable by high concentrations of unlabelled antigen, which suggests that it would be accessible to specific receptors on the surface of lymphocytes in contact with FDC [14]. Active germinal centres appear a few days after administering antigen, and after antibody is first detectable. For this and other reasons it had been suspected by several workers that they were sites of generation of primary B memory cells. Very strong evidence to support this comes from observations that B memory cells are not generated when binding of complexes on FDC is prevented, and that they are generated rapidly and very effectively by injecting minute amounts of preformed complexes when, but only when, these become localized on FDC. Furthermore, not only are B memory cells generated which recognize the antigen in the complexes, but also others which recognize the idiotype of the antibody – an interesting confirmation of the relevance of *Niels Jerne's* network hypothesis, and a reasonable explanation of the cyclical production of specific antibodies which can be shown following administration of certain immunogens. A plausible hypothesis is that an early (T-independent) antibody response is sufficient to cause localization of antigen-antibody complexes on FDC. These stimulate such early B cells as possess specific IgM and C3 receptors, and which migrate so as to come into contact with the FDC, to divide and mature to B memory cells. In the absence of T cell help in the germinal centre they do not differentiate further into antibody producers, but if contact with antigen and T cell help are provided elsewhere they can do so. This evidence is summarized in [8]. Other influences (hormonal, pharmacological) may also act in germinal centres, but they have not been investigated. FDC themselves have recently been isolated by careful enzyme digestion of mouse spleens. They are very easily broken during isolation, and once separated from supporting structures and the lymphocytes which surround them they tend to round up and to aggregate into giant multinucleated cells. As expected they have Fc and C3 receptors, but they do not express detectable surface I-A determinants [4]. This would be unexpected if they were involved in T cell stimulation but not if they interact exclusively with B cells.

T-Dependent Areas (Deep Cortex of Lymph Nodes and Periarteriolar Sheath of the Spleen)

These are characterized not only by the selective migration of T lymphocytes through them but also by the presence of 'interdigitating' cells with which the lymphocytes make intimate contact. The 'interdigitating' cells have been shown to express large amounts of I-A antigen [9]. What is new and exciting is that there is increasing evidence that these cells form part of a system which includes Langerhans cells in the skin, 'veiled' cells in the afferent lymph, and similar cells in the thymus medulla. They share morphological properties, are strongly Ia-positive, stain histochemically for ATP-ase, and (in species which show this feature) may contain Birbeck granules. The latter seems to depend upon whether the cells have recently been in contact with epithelial tissues, and in spleen or mesenteric nodes it is absent. Although these cells may contain small lysosomes and typical lysosomal enzymes, they do not engage in obvious phagocytosis but rather in macropinocytosis. C3 and Fc receptors have been described on some Langerhans and veiled cells, but they are weak and vary with the species, and neither type of cell adheres strongly to glass surfaces [summarized in 1]. The 'dendritic' cells isolated by *Steinman* et al. [12] have so many properties in common with interdigitating cells that, although there is no direct proof, I strongly suspect that they are one and the same. Cells resembling Langerhans cells are prominent in the thymic medulla where they have been assigned a putative role in the education of T cells to recognize self MHC antigens [5], and very recently apparently similar Ia rich cells have been observed in the lamina propria of human intestinal villi, where they are in intimate contact with helper T cells though suppressor T cells lie outside [*Janossy*, personal communication].

One story is as follows: Langerhans cells can take up soluble antigens or skin-sensitizing materials in the skin. Especially when skin is inflamed they migrate into the lymph as 'veiled' cells, and these in turn displace some or all of the interdigitating cells in the T-dependent area of draining nodes and become interdigitating cells in their turn [1]. Here they are ideally placed selectively to stimulate antigen specific helper T cells, and to evoke the delayed type hypersensitivity which results from dermal application of sensitizing antigens. Strong evidence to support this comes from the fact that if mouse skin is depleted of Langerhans cells by UV irradiation, local application of DNCB results not in delayed type hypersensitivity but in specific unresponsiveness [16]. The immunological role of this type of cell under other circumstances, in spleen, central lymph nodes, the gut or peritoneal cavity, may well be similar but clear evidence is not available because most workers who have studied antigen presentation to T or B lymphocytes have not distinguished between interdigitating cells and conventional macrophages. However, if the former are the same as 'dendritic' cells there are two pieces of evidence which favour a similar role. Dendritic cells are extremely effective in vitro at inducing primary mixed lymphocyte reactions [11], and in presenting soluble antigens so as to stimulate syngeneic T cell proliferation. In the latter case antigen presentation was also shown to be under immune response (Ir) gene control [13].

Marginal Zone

The marginal zone of the spleen surrounds the white pulp, and contains large active macrophages which ingest (and if digestible rapidly catabolize) antigen-antibody complexes, particulate materials and many antigens which are not proteins for export and are rapidly removed from the blood stream. This zone is also the main B lymphocyte traffic area in the spleen, through which the cells pass into the red pulp or return to the B cell areas of the white pulp. In studying the capacity of a variety of hapten-conjugated indigestible polysaccharides, which are effective T-independent immunogens in the mouse, to prevent secondary responses in mice primed and boosted with the same hapten on a T-dependent protein carrier, I observed that conjugates with certain uncharged polysaccharides (Ficoll, hydroxethyl starch and Dextran 2000) were much less effective than were conjugates of various anionic polysaccharides [2]. The former also elicited much more prolonged T-independent IgM antibody responses than the latter. Examination of the fate of radiolabelled conjugates showed that the unchanged materials were not taken up by conventional macrophages in lymphoid tissues, liver, and bone marrow but became localized exclusively in macrophages in the marginal zone and marginal sinuses of lymph nodes. By using fluorescent conjugates, which enable live cells containing them to be identified, marginal zone macrophages have been isolated from mouse spleens. Collagenase digestion is needed to release them from what many immunologists consider to be 'debris'. They resemble other macrophages in most respects, but are larger and have big round open nuclei. As isolated they are surrounded by loosely adherent B lymphocytes, probably due to a weak lectin-like interaction, which may be important in returning B lymphocytes into the white pulp, or into the outer cortex of lymph nodes. Marginal zone macrophages adhere well to glass, have Fc and C3 receptors but do not express surface Ia determinants [3]. This is consistent with a function effectively to present T-independent antigens to a responsive subclass of B lymphocytes, without involving T cell help.

Red Pulp and Medullary Cord Macrophages

These make up the bulk of what are normally regarded as typical spleen or lymph node macrophages. They line the spaces through which blood cells leaving via the splenic venous sinuses or lymph cells entering the efferent lymphatic vessels have to pass, and are very effective scavengers of antigens. In view of all the accumulated in vitro evidence that syngeneic Ia-positive adherent auxiliary cells are required to present antigens effectively to helper T cells and in turn to stimulate antibody secretion by B cells, these macrophages would seem to be the likely candidates. They adhere firmly to glass and have Fc and C3 receptors, and most workers who have looked for Ia determinants find them on half or more (but never all) of them. However, I think that their role as effective primary presenters of antigens may require re-examination. When freshly isolated from unstimulated spleens, and identified by uptake of fluorescent acidic polysaccharides, very few express I-A determinants [2]. These only become apparent if the macrophages have been allowed to adhere to glass, or if they are freshly isolated from spleens in which T lymphocytes have recently been activated [3], and the in vitro findings could be misleading. Furthermore, a case can be made, on admittedly circumstantial evidence, that red pulp macrophages actually inhibit T-dependent immune responses by sequestering and degrading antigens [6].

In this contribution I have made no mention of non-specific B and T cell-stimulating factors released by adherent cells, though I am sure that they are important in determining what happens in their neighbourhood. Despite this over-simplification, I hope that two points have emerged. One that some microenvironments can be defined and are important for a full understanding of immune responses, and the other that auxiliary cells should not simply be classed as 'macrophages' and their nature deserves closer examination.

References

1 Drexhage, H.A.; Lens, J.W.; Cvetanov, J.; Kamperdijk, E.W.A.; Mullink, H.; Balfour, B.M.: in van Furth, Mononuclear phagocytes, functional aspects (Nijhoff, Den Haag 1980).
2 Humphrey, J.H.: Eur. J. Immunol. (in press 1981).
3 Humphrey, J.H.; Grennan, D.: Eur. J. Immunol. (in press 1981).
4 Humphrey, J.H.; Grennan, D.: in preparation.
5 Janossy, G.J.; Thomas, J.A.; Bollum, F.J.; Granger, S.; Pizzolo, G.; Brandstock, K.F.; Wong, W.; McMichael, A.; Ganeshaguru, K.; Hoffbrand, A.V.: Immun. *125:* 202–212 (1980).
6 Joshua, D.E.; Humphrey, J.H.; Grennan, D.; Brown, G.: Immunology *40:* 223–228 (1980).
7 Klaus, G.G.B.; Humphrey, J.H.: Immunology *33:* 31–40 (1977).
8 Klaus, G.G.B.; Humphrey, J.H.; Kunkl, A.; Dongworth, D.W.: Immunol. Rev. *53* (in press).
9 Lampert, I.A.; Pizzolo, G.; Thomas, A.; Janossy, G.: J. Path. *131:* 145–156 (1980).
10 Nossal, G.J.V.; Abbot, A.; Mitchell, J.; Lummus, Z.: J. exp. Med. *127:* 277–289 (1968).
11 Nussenzweig, M.C.; Steinman, R.M.: J. exp. Med. *151:* 1196–1212 (1980).
12 Steinman, R.M.; Kaplan, G.; Witmer, M.D.; Cohn, Z.A.: J. exp. Med. *149:* 1–16 (1979).
13 Sunshine, G.H.; Katz, D.R.; Feldmann, M.: J. exp. Med. *152:* 1817–1822 (1980).
14 Tew, J.G.; Mandel, T.: J. Immun. *120:* 1063–1069 (1978).
15 Tew, J.G.; Mandel, T.E.; Miller, G.A.: Aust. J. exp. Biol. med. Sci. *57:* 401–414 (1979).
16 Toews, G.B.; Bergstresser, P.R.; Streilein, J.W.; Sullivan, S.: J. Immun. *124:* 445–453 (1980).

Dr. J.H. Humphrey, Department of Immunology, Royal Postgraduate Medical School, Hammersmith Hospital, London W12 (England)

The Immune System, vol. 1, pp. 356–361 (Karger, Basel 1981)

Some Quantitative Aspects of Lymphocyte Development and Circulation

George I. Bell [1]

Theoretical Division, Los Alamos Scientific Laboratory, University of California, Los Alamos, N. Mex., USA

It is widely believed that, at least in young animals, the rate of lymphocyte generation, chiefly in the bone marrow and thymus, is very large, being sufficient to repopulate the entire lymphocyte population in only a few days. The fate of the vast majority of these rapidly renewed lymphocytes is unknown as are the selection pressures which cause most clones to perish and only a few to prosper and endure. In order to place the consideration of these problems on a more quantitative basis, I was recently persuaded by *Niels Jerne* to devise a mathematical model of lymphocyte kinetics [1], i.e. an accounting scheme for the birth, death, and circulation of lymphocytes. This paper presents some results from application of that model, discusses experiments which would clarify the fate of rapidly renewed lymphocytes and briefly considers some theories concerning the selection pressures. I am indebted to *Jerne* for critical discussions and for the hospitality of the Basel Institute for Immunology where this work was started.

It is important to keep in mind that there are qualitative as well as quantitative changes in lymphocyte kinetics depending on animal age and species. Therefore any conclusions drawn below for young rodents cannot be confidently extrapolated to other animal ages or species. In particular,

[1] *George Bell* was a member of the Basel Institute for Immunology from 1979 to 1980.

recent studies by *Cahill* and colleagues on young lambs, including those in utero, have shown radically different cell kinetics.

Information concerning lymphocyte kinetics has come from labeling of cells in vivo by radioactive DNA precursors, usually tritiated thymidine. Such experiments have shown that in the bone marrow of rodents, the majority of the small lymphocytes are non-dividing cells generated by the rapid proliferation of larger lymphoid cells, sometimes called transitional cells. These small lymphocytes, although non-dividing, are rapidly renewed. For example, in C3H mice, the half-renewal time increases from 14 h in 4-week-old mice to 24 h at 16 weeks [13]. The rate of production of small lymphocytes in the 8-week-old mouse is about 10^8 cells/day [14]. Moreover, in the thymus, non-diving small lymphocytes are being generated at a comparable rate [10]. Since the whole mouse contains only about 10^9 lymphocytes it is apparent that this rate of lymphocyte production is sufficient to repopulate the entire lymphocyte population in only a few days. In fact, this turnover does not occur since many peripheral lymphocytes are relatively long-lived [5, 18].

In particular, our young mouse contains long-lived lymphocytes which circulate from blood to lymphoid tissues to lymph and back again to blood, roughly once per day. This pool of recirculating lymphocytes, largely T cells, is believed by many immunologists [19, 23, 26] to contain 'memory cells', i.e. lymphocytes with circulation patterns and life span which have been qualitatively altered by contact with environmental (or other) antigens. Our mouse contains about 2×10^8 lymphocytes in the recirculating pool [6] and long-term thymidine labeling studies [25] have shown that only about 1% are replaced per day. Thus no more than one percent of the lymphocytes born in the bone marrow and thymus of our 8-week mouse can survive after diverse trials to join the recirculating pool. Indeed the surviving fraction may be much smaller, since lymphocytes in the recirculating pool can also be renewed by mitosis of cells in that pool or in the peripheral lymphoid tissues [24].

What happens to the vast majority of the rapidly renewed small lymphocytes? Let us first consider those produced in the bone marrow. A large proportion are B cells as judged by the presence of immunoglobulin molecules on their surface [15] and/or in the cytoplasm [17]. It is generally believed that these small lymphocytes are mostly exported to the blood which carries them to various lymphoid tissues, chiefly the spleen. If a fraction f, of the rapidly renewed small lymphocytes in the bone marrow is exported to blood, then the rate of export for our young mouse is about 10^8

f/day or 7 × 10^4 f/min. If these cells have a lifetime of τ minutes in the blood, before lodging in tissue, the number of such lymphocytes in blood will be 7 × 10^4 fτ cells. Thus, if the number of these lymphocytes could be measured one could deduce a value for the product fτ.

With present data it seems possible to estimate an upper limit to the number of these lymphocytes as follows. When mice are exposed to long-term thymidine labeling, most of the marrow (and thymus) small lymphocytes become labeled by the third day but only a small fraction of lymphocytes in the blood are labeled. For example, in experiments of *Osmond and Nossal* [16], only 5% of blood lymphocytes are labeled by 3 days. Some fraction of these labeled lymphocytes will be recent emigrants from the marrow and we may obtain an upper limit to the number of emigrants by assuming that they are only labeled blood lymphocytes. If our 25 g mouse has about 3 ml of blood and 2 × 10^6 lymphocytes/ml, then there are about 6 × 10^6 lymphocytes in the blood. Thus about 3 × 10^5 labeled lymphocytes will be in the blood at three days and fτ $\lesssim$ 4 min.

This upper limit to fτ could be refined in a number of ways. Since only a small fraction of the rapidly renewed marrow lymphocytes are T cells [18], while the majority of blood lymphocytes are T cells [20], it is natural to inquire as to the fraction of rapidly labeled blood lymphocytes which are T cells. Such data seem not to have been reported. In addition one could seek to deduce the contribution which other sources of labeled cells including the thymus and thoracic duct are making to the rapidly renewed blood lymphocyte population. Such efforts are underway [1] but handicapped by incomplete data. Finally, one could inject a population of labeled marrow lymphocytes and observe their rate of clearance from blood.

An important fraction of the rapidly renewed lymphocytes of the bone marrow migrate via blood to the spleen [21, 22, 27]. In one experiment [27], nearly all of the injected marrow lymphocytes were recovered from the spleens of recipients. It is thus of interest to see whether the limit on fτ, deduced above, tells us something about how many marrow lymphocytes go to the spleen.

The cardiac output of a 25 g mouse is probably about 12 ml/min. This value is deduced from data on rats and humans [2], assuming log (cardiac output) is proportional to log (animal wt). Moreover, the fraction of cardiac output to spleen has been measured with radioactive microspheres to be about 1% [7], so that blood flow to the spleen is about 0.12 ml/min. Thus, the animal's blood volume (3 ml) traverses the spleen only once every 25 min. It follows that if the lymphocytes were removed from blood only in the spleen [and with the same high efficiency as are the 15-μm microspheres; 7,] then their lifetime in the blood, τ, would be about 25 min.

This value of τ is substantially larger than the upper limit estimated above for $f\tau$. There are two possible reasons for this. Either f is substantially less than unity, implying that most of the rapidly renewed marrow small lymphocytes die in situ, or most of the lymphocytes are removed from blood at locations other than in the spleen. Alternatively, both explanations may be partially true.

Additional information may be gained by considering the labeling of small lymphocytes in the spleen during the first few days of continuous thymidine labeling. During this period, there is a buildup of labeled lymphocytes in the spleen, arising at least in part from the migration of lymphocytes labeled in marrow [3, 5, 18, 21]. By studying the rate of lymphocyte labeling in the spleen, deductions can be made concerning the residence time in the spleen of labeled incoming lymphocytes and of their rate of entry [1]. From these data it can also be concluded that most of the rapidly renewed marrow small lymphocytes do not go to the spleen.

Lymphocyte kinetics in the thymus are similar to the bone marrow in that a large population of rapidly renewed non-dividing small lymphocytes is generated by rapid division of larger precursors. However, in the thymus it is clear that many of the lymphocytes die in situ [9, 12, 18]. In particular, studies on the reutilization of labeled thymidine in the thymus [9, 12] indicate that the majority of rapidly renewed thymocytes die in situ. The rate and fate of the exported lymphocytes can be studied by looking at the rate of appearance of labeled T cells in peripheral lymphoid organs during the first few days after a brief or sustained exposure to tritiated thymidine [18]. In addition, it is possible to preferentially label thymocytes by local injection [11] or application [9] of the label and to observe the subsequent appearance of label in peripheral tissues. Such studies have shown that thymus-derived lymphocytes migrate preferentially to the gut-associated lymphoid tissues within a few days after their generation.

As in the case of marrow derived lymphocytes, additional information on the fate of thymus derived lymphocytes could be gained by measuring the number of labeled cells, in this case T cells, in the blood as a function of time during the first few days after labeling. *Ernström and Larsson* [4] made such measurements in arterial and thymic venous blood to deduce an export rate of rapidly renewed lymphocytes from the thymus of guinea pigs. It would appear that better measurements could be made using local labeling [9].

Discussion

We have presented quantitative estimates of the rate of generation of rapidly renewed small lymphocytes in the bone marrow of young adult mice and of some aspects of their subsequent fate. Few (< 1%) live a long time but where and why the majority die are unclear. It was argued that most of these lymphocytes do not migrate to the spleen. There are two extreme possibilities: they may mostly die in the marrow or mostly migrate to other tissues and die there. Thymus derived lymphocytes appear to mostly die in situ but a significant fraction migrate to gut-associated and other lymphoid tissue. Better insight into the fate of these lymphocytes could be gained from further data on the rate of appearance of labeled T, B, and other lymphocytes in *blood* and in secondary lymphoid tissues during the first few days after the start of continous (or transient) labeling of DNA. Localized labeling has the advantage of concentrating the label in a single population of lymphocytes at the expense of perturbing the system and obtaining, for marrow at any rate, a smaller population of labeled cells.

Why do most of the lymphocytes which are generated so rapidly in marrow and thymus have such a short life span? We may imagine two quite different explanations. (1) Perhaps most of these lymphocytes encounter something which is unhealthy for them and which rapidly terminates their existence. For example, as *Jerne* [8] suggested for thymocytes a decade ago, they might have receptors recognizing self antigens and upon encountering these antigens at some early stage of maturity, die. If this is true for marrow lymphocytes that are largely of the B lineage, it should be detectable by the cultivation of pre B cells and finding that they bind, or can be induced to secrete antibodies to, self antigens. (2) Alternatively, all rapidly renewed marrow and thymus lymphocytes may be destined to die within a few days unless they encounter an appropriate and rare microenvironment. For example, a particular combination of accessary cells, other lymphocytes and antigen might be required in order to permit the survival and further proliferation of a young lymphocyte. If this were true, a fruitful approach might be to study in vitro conditions under which a majority of young lymphocytes die and only a minor and specific fraction survives.

Acknowledgement. This work was supported by the US Department of Energy.

References

1 Bell, G.I.: In preparation.
2 Biol. Data Book III, Fed. Am. Soc. Exp. Biol., Bethesda, p. 1740 (1974).
3 Chan, F.P.H.; Yang, W.C.; Osmond, D.G.: Manuscript submitted.
4 Ernström, U.; Larsson, B.: Nature, Lond. *222:* 279–280 (1969).
5 Everett, N.B.; Tyler, R.W.: Int. Rev. Cytol. *22:* 205–237 (1967).
6 Everett, N.B.; Tyler, R.W.: in Formation and destruction of blood cells, pp. 246–283 (Lippincott, New York 1970).
7 Hay, J.B.; Hobbs, B.B.: J. exp. Med. *145:* 31–44 (1977).
8 Jerne, N.K.: Eur. J. Immun. *1:* 1–9 (1971).
9 Joel, D.D.; Chanana, A.D.; Cottier, H.; Cronkite, E.P.; Laisue, J.A.: Cell Tiss. Kinet. *10:* 57-69 (1977).
10 Joel, D.D.; Chanana, A.D.; Cronkite, E.P.: Ser. haematol. *7:* 464–481 (1974).
11 Joel, D.D.; Hess, M.W.; Cottier, H.: J. exp. Med. *135:* 907–923 (1972).
12 McPhee, D.; Pye, J.; Shortman, K.: Thymus *1:* 151–162 (1979).
13 Miller, S.C.; Osmond, D.G.: Cell Tiss. Kinet. *8:* 97–110 (1975).
14 Osmond, D.G.: J. reticuloendoth. Soc. *17:* 88–114 (1975).
15 Osmond, D.G.: in Immunoglobulin genes and B cell differentiation (Elsevier/North-Holland, New York 1979).
16 Osmond, D.G.; Nossal, G.J.V.: Cell. Immunol. *13:* 132–145 (1974).
17 Owen, J.T.; Wright, D.E.; Habu, S.; Raff, M.C.; Cooper, M.D.: J. Immun. *118:* 2067–2072 (1977).
18 Press, O.W.; Rosse, C.; Clagett, J.: Cell. Immunol. *33:* 114-124 (1977).
19 Raff, M.C.; Cantor, H.: in Progress in immunology, pp. 83-93 (Academic Press, New York 1971).
20 Raff, M.C.; Owen, J.T.: Eur. J. Immunol. *1:* 22–31 (1971).
21 Rosse, C.; Cole, S.B.; Appleton, C.; Press, O.W.; Clagett, J.: Cell. Immunol. *37:* 354–362 (1978).
22 Ryser, J-E.; Vassalli, P.: J. Immun. *113:* 719–728 (1974).
23 Shortman, K.; Howard, M.; Teale, J.; Baker, J.: J. Immun. *122:* 2465–2472 (1979).
24 Sprent, J.: in The lymphocyte : structure and function, vol. 1, pp. 43–112 (Dekker, New York 1977).
25 Sprent, J.; Basten, A.: Cell Immunol. *7:* 40–59 (1973).
26 Strober, S.: Transplant. Rev. *24:* 84–112 (1975).
27 Yang, W.C.; Miller, S.C.; Osmond, D.G.: J. exp. Med. *148:* 1251–1270 (1978).

Dr. George I. Bell, Theoretical Division, Los Alamos Scientific Laboratory, University of California, Los Alamos, NM 87545 (USA)

The Immune System, vol. 1, pp. 362–371 (Karger, Basel 1981)

The Bone Marrow: a Major Site of Antibody Production

R. Benner[1]

Department of Cell Biology and Genetics, Erasmus University, Rotterdam, The Netherlands

The bone marrow (BM) is the central organ in hemopoiesis as well as lymphopoiesis. The newly formed small lymphocytes are mostly of the B cell lineage [44]. Most of the newly formed B cells are disseminated throughout the other lymphoid organs [16], where they become involved in antibody responses or die. In addition to newly formed small lymphocytes, the BM contains a substantial number of long-lived, potentially recirculating, small lymphocytes [41], partly of the B cell lineage, and partly of the T cell lineage [46]. These long-lived lymphocytes are thought to be the B and T memory cells which are produced as a concomitant of immune responses leading to humoral and cellular immunity [44]. In mice, long-lived lymphocytes comprise from 7 to 22% of all marrow small lymphocytes [41]. Most of the long-lived BM lymphocytes are immigrants originating from peripheral lymphoid tissues [49]. Antigenic stimulation can recruit recirculating specific lymphocytes back to the peripheral lymphoid tissues [51]. Thus, the BM is a quantitatively important breeding site of virgin B lymphocytes and a reservoir of long-lived, potentially recirculating B and T lymphocytes.

Apart from these interrelationships between populations of small lymphocytes in the BM and the peripheral lymphoid tissues, the BM B cell compartment produces antibodies. This paper reviews the data on antibody production in the BM. Several crucial pieces of information on this subject have been obtained at the Basel Institute for Immunology, under the stimulating influence of *Niels Kaj Jerne.*

[1] *Robbert Benner* was a member of the Basel Institute for Immunology in 1978 and 1979.

Antibody Formation in Bone Marrow

For the sake of clarity we must distinguish between experiments which demonstrate the production of Ig by BM cells regardless of the antibody specificity of these molecules and experiments which formally demonstrate the production of specific antibodies in the BM upon immunization – we will refer to 'Ig synthesis' and 'antibody synthesis', respectively. Ig and antibody production in the BM has been demonstrated in a variety of animal species, ranging from frogs [21] to higher mammals. Thus, even at the lowest level of phylogenetic development at which the BM occurs as a distinct tissue, the amphibians, antibody-forming cells can be found in this organ.

Experiments Demonstrating Ig Synthesis by the Bone Marrow. All published studies in which it was investigated whether Ig synthesis takes place in the BM have yielded a positive answer. During short-term cultures of guinea pig [1, 3], rabbit [2, 53], monkey [4], and human [37, 57] BM cells, Igs are produced and released into the culture medium. This is apparent from the incorporation of radioactively labeled amino acids. Wherever Ig production by different lymphoid tissues was quantitated, production was found to be largest in the BM, especially when calculated per whole organ [1, 2, 37]. The most extensive study in this field is from *McMillan* et al. [37, 38], who measured the in vitro IgG production by lymphoid cells from five normal human lymphoid tissues: BM, spleen, blood, lymph node, and thymus. Using this data and the total lymphoid cell content of the various tissues, it was calculated that the BM produced more than 95% of all IgG synthesized by the organs which were evaluated.

Ig synthesis also takes place in murine BM. The number of cytoplasmic Ig-positive plasma blasts and plasma cells [23, 24] and the number of Ig-secreting cells [14] in this organ is substantial compared to the numbers in the other lymphoid organs. This distribution pattern, however, is affected by T cell deprivation and antigen deprivation. Thus, Ig-secreting cells are less frequent in the BM of germ-free nude mice [14, 15].

The heavy-chain isotype distribution of the Ig-secreting cells in the BM is different from those in the other lymphoid organs, especially from that in spleen. This has been shown for both mouse [15, 24] and man [26, 27, 55]. IgG-and IgA-secreting cells are relatively frequent in the BM, in contrast to IgM-secreting cells. The V region repertoire, on the other hand, of the IgM secreted by BM cells is the same as that of IgM secreted by other lymphoid tissues [14].

The localization in the BM of the great majority of all IgG- and IgA-secreting cells of an adult mouse and their deficiency in germ-free mice [14, 15] suggest that the IgG and IgA produced are predominantly directed against environmental antigens, which mainly penetrate via mucous membranes and the respiratory and digestive tract. However, it is not clear to what extent antibodies of the IgA class produced by cells within the BM contribute to the secretory IgA in the mouse. In humans the contribution of the BM to the secretory IgA is probably minimal, since the IgA produced outside the secretory sites is almost exclusively of the monomeric form [47]. However, BM-derived IgA might be secreted via the bile. Evidence for this pathway has been presented for rats [43].

In humans there is a striking correlation between the heavy-chain isotype distribution profile of the cytoplasmic Ig-positive plasma blasts and plasma cells (C-Ig cells) in the BM and the levels of the various Ig classes and subclasses in the serum. This has been shown for the following combinations: IgM-IgG-IgA [26, 55], IgGl-IgG2-IgG3-IgG4 [42], IgAl-IgA2 [50], monomeric IgA-polymeric IgA [47], and IgD versus all other isotypes together [56]. Furthermore, the κ/λ ratio of the C-Ig cells in the BM is correlated to a high degree with the ratio of Ig (κ) to Ig (λ) in the serum [26, 55]. The C-Ig cell distribution pattern in the other lymphoid organs did not fit with the serum levels of the various Ig classes and subclasses. In summary, human BM is a major source of all classes and subclasses of Igs in the serum.

Experiments Demonstrating Antibody Production by the Bone Marrow. In 1898 *Pfeiffer and Marx* [45] reported that extracts of BM of rabbits which had been immunized with *Vibrio cholerae* contained specific antibodies. Almost simultaneously *Deutsch* [18] showed that immunization of guinea pigs with typhoid vaccine led to the appearance of anti-typhoid antibodies in the BM. Although these pioneers did not formally prove that these antibodies were actually produced within the BM, their data were suggestive that this was the case. Early in this century *Lüdke* [35] and *Reiter* [48] showed that BM cells from immunized animals release specific antibodies during in vitro cultivation. *Thorbecke and Keuning* [52] were the first to prove that the BM actually produces antibodies upon immunization. Since then several others followed [reviewed in 12]. At present antibody formation has been demonstrated in frog [21], chicken [29], mouse [6, 25], rat [31], mole rat [27], guinea pig [1, 3], rabbit [2, 54], and man [39, 58]. In mice antibodies of the IgM, IgG, IgA [6, 25], and IgE [22, 30] classes can be produced by the BM.

Many studies in which the BM was investigated as a site of antibody formation yielded negative results. This seems to be especially true for mice immunized with T-dependent antigens [summarized in 12]. After primary immunization of mice with T-dependent antigens only small numbers of plaque-forming cells (PFC) appear in the BM [6, 17, 40]. This PFC response in the BM peaks later than in spleen and lymph nodes, usually at least 2 weeks after immunization of the mice. When the antigen is administered together with the adjuvant, PFC in the BM can become quite numerous some months after immunization [13].

Secondary immunization of mice with T-dependent antigens without adjuvant induces a high PFC response not only in spleen and lymph nodes but also in the BM. The first phase of the response, about 1 week in duration, is characterized by much higher numbers of PFC in the spleen and/or lymph nodes than in the BM. But after the first week, the PFC response in the BM is much higher than in all other lymphoid organs combined. This characteristic kinetics is independent of the type of T-dependent antigen and the booster dose and is found for IgM as well as for IgG and IgA antibody production [7, 25]. That is, peripheral lymphoid tissues respond rapidly, but only for a short period, while the BM responds slowly but takes care of a long-lasting massive production of antibodies to antigens which repeatedly challenge the organism.

It has been reported for rat [31], mole rat [28] and rabbit [19] that a single injection of a T-dependent antigen can induce quite substantial antibody formation in the BM. Perhaps the mechanism underlying antibody formation (see below) in BM is different in these species, but the phenomenon can also be explained by cross-reactivity between the antigens used for immunization and environmental antigens. Such environmental antigens might already have primed the animals so that they respond upon antigen injection with antibody formation in the BM. Indeed, certain T-dependent antigens induce antibody formation in the BM of rabbits provided the antigen is injected several times [34, 54].

BM antibody formation against T-independent antigens has not been studied as extensively as after immunization with T-dependent antigens. So far, only a few T-independent antigens have been used: bacterial lipopolysaccharide (LPS), trinitrophenylated LPS (TNP-LPS), and pneumococcal polysaccharide type III (SIII). LPS and TNP-LPS can induce antibody formation in the BM of mice [9], rabbits [33] and frogs [21]. SIII was only tested in mice and did not induce a BM response [5]. LPS and TNP-LPS induce BM antibody formation not only during the secondary, but also during the primary response.

For obvious reasons, the human being is not normally accessible for studies on antibody formation in BM. However, some data are available about auto-antibody production in various lymphoid tissues of patients suffering from rheumatoid arthritis and idiopathic thrombocytopenic purpurea (ITP). In rheumatoid arthritis PFC producing rheumatoid factor are much more numerous in the BM than in the other lymphoid tissues [58]. In chronic ITP, the BM was also found to be the major source of autoantibodies. In this disease the IgG anti-platelet antibody production in the BM was calculated to be tenfold greater than that in the other lymphoid tissues [39].

The Mechanism Underlying Antibody Formation in Bone Marrow

Studies on the mechanism underlying the appearance of antibody-forming cells in the BM have been done almost exclusively with mice. The key observations in this field are the weak PFC activity in the BM during the primary response to T-dependent antigens and the high PFC response in this organ after secondary immunization with the same antigen [6, 25]. Several experiments revealed a coincidence between the occurrence of B and T memory cells and the capacity to respond to challenge with the relevant antigen with antibody formation in the BM [8, 10]. By priming and boosting of mice with heterologous hapten-carrier complexes, we recently showed that B memory cells, but not necessarily T memory cells, must be present before booster immunization for PFC to appear in the BM [32].

The origin of PFC which appear in the BM during secondary immune responses has been studied in parabiotic mice congenic for the Igh-1 locus. From analysis of the allotype of antibodies produced by PFC in the BM of such pairs of parabionts, it appeared that antibody formation in the BM is dependent on the immigration into the BM of B memory cells reactivated by antigen in peripheral lymphoid organs [32]. After arrival in the BM these cells mature into PFC and produce larger amounts of IgM, IgG and IgA antibodies [6, 11]. This sequence of events is depicted in figure 1.

The migration of cells from the spleen into the BM takes place during

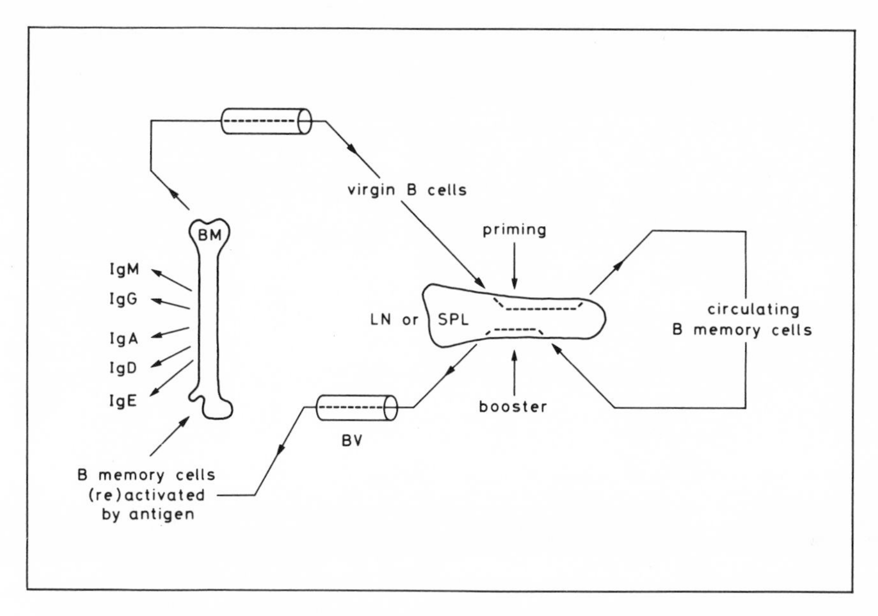

Fig. 1. Schematic representation of the mechanism underlying antibody formation in bone marrow. See text for explanation. BM = Bone marrow; BV = blood vessel; LN = lymph nodes; SPL = spleen.

the first few days of the secondary response, as can be concluded from experiments with mice splenectomized at different intervals after the booster injection. These experiments revealed that splenectomy 4 or more days after the booster injection does not influence the BM response, whereas splenectomy on day 2 can completely prevent BM antibody formation [11]. Consistent with such a migration of activated cells, radioautographic studies in guinea pigs have demonstrated an influx of newly formed mononuclear cells into the BM via the blood stream during the first 3 days after intravascular antigen administration [32].

The above line of evidence shows that antibody formation to T-dependent antigens in BM, in contrast to peripheral lymphoid tissues, is dependent upon immigration of antigen-activated B lineage cells from elsewhere, instead of local induction of antibody formation. Presumably the BM lacks the appropriate microenvironment and/or quantity or quality of cells (T lymphocytes? macrophages?) required for the early steps in the induction of immune responses. In contrast to mouse BM, human BM has features

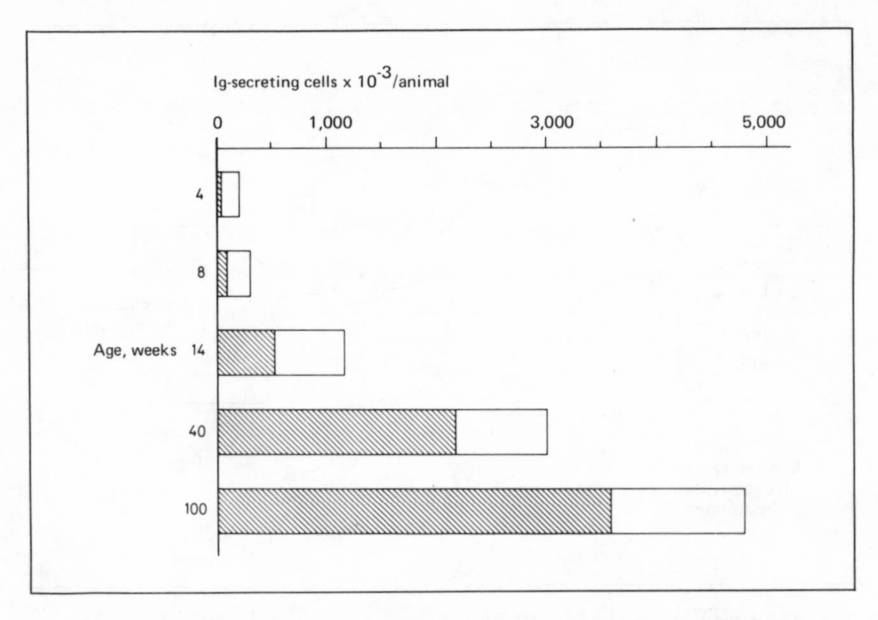

Fig. 2. Age-related increase of the total number of Ig-secreting cells in BALB/c mice. Ig-secreting cells were assayed in spleen, lymph nodes, bone marrow, and Peyer's patches as described elsewhere [14]. Hatched areas represent the contribution of the bone marrow, open areas the contribution of spleen, lymph nodes and Peyer's patches together.

which are characteristic of peripheral lymphoid tissues, e.g. the occurrence of follicles with germinal centers [20, 36]. Therefore, antibody formation in the human BM may not be similarly dependent upon an influx of antigen-activated B cells from the periphery.

Influence of Age upon the Organ Distribution of Ig-Secreting Cells

The total number of Ig-secreting cells per individual is dependent upon age. In normal, specific-pathogen-free BALB/c mice the total number of Ig-secreting cells per animal increased 20-fold between 8 and 100 weeks of age (fig. 2).

The distribution of the Ig-secreting cells among the various lymphoid organs is also highly dependent upon the age of the individual. This is most evident for the BM. In 4-week-old mice only 17% of all Ig-secreting cells are localized in this organ. At that age spleen and lymph nodes are the major sites of Ig production. The absolute and relative contribution of the BM, however, increases enormously with increasing age. At 8, 14, 40, and 100 weeks of age the relative contribution of the BM was found to be 29, 46, 72, and 75%, respectively (fig. 2). A similar age-related pattern has been observed in CBA mice, which are a long-lived strain with no specific pathology of the immune system [23]. Also in congenitally athymic mice, Ig-secreting cells accumulate in the BM during aging, although the phenomenon was clearly retarded as compared with their heterozygous littermates [15, 24]. A similar age-related increase of C-Ig cells has been observed in the human BM [59].

The gradually increasing importance of the BM as site of Ig synthesis throughout the life span probably reflects the gradual adaptation of the individual to its antigenic environment. As an individual becomes older, more antigenic stimuli from the environment will have been experienced and secondary-type responses will prevail. As outlined in the previous sections, such secondary-type responses involve a quantitatively important antibody production in the BM.

Epilog

It is remarkable that the BM has never generally been accepted as a site of antibody formation, in addition to spleen, lymph nodes, gut-associated lymphoid tissue, and bronchus-associated lymphoid tissue. This may be due to the fact that until recently the BM of the most popular experimental animal in immunological research, the mouse, was found to show little or no PFC activity after immunization. Almost all initial studies in mice were limited in one respect, and we now know that it is a crucial one: antibody formation in the BM was studied only during the primary response. However, there is probably another reason as well: it is much more laborious to isolate BM cells than spleen cells. It is unfortunate that such a triviality leads many immunologists to neglect a most important site of antibody formation.

Acknowledgements. I most sincerely thank Mr. *A. van Oudenaren* for his skillful assistance and important contributions throughout our studies on antibody formation in bone marrow. Furthermore, I thank Dr. *G. Koch,* Prof. Dr. *O. Vos,* Dr. *J.J. Haaijman,* and Prof. Dr. *W. Hijmans* for valuable discussions about the subject, Ms. *Cary Meijerink* for secretarial assistance, and the Netherlands Foundation for Medical Research (FUNGO) for financial support.

References

1 Askonas, B.A.; White, R.G.: Br. J. exp. Path. *37:* 61–74 (1956).
2 Askonas, B.A.; Humphrey, J.H.: Biochem. J. *68:* 252–261 (1958).
3 Askonas, B.A.; White, R.G.; Wilkinson, P.C.: Immunochemistry *2:* 329–336 (1965).
4 Asofsky, R.; Thorbecke, G.J.: J. exp. Med. *114:* 471–483 (1961).
5 Baker, P.J.; Stashak, P.W.; Amsbaugh, D.F.; Prescott, B.: Immunology *20:* 469–480 (1971).
6 Benner, R.; Meima, F.; Van der Meulen, G.M.; Van Muiswinkel, W.B.: Immunology *26:* 247–255 (1974).
7 Benner, R.; Meima, F.; Van der Meulen, G.M.; Van Ewijk, W.: Immunology *27:* 747–760 (1974).
8 Benner, R.; Van Oudenaren, A.: Cell. Immunol. *19:* 167–182 (1975).
9 Benner, R.; Van Oudenaren, A.: Immunology *30:* 49–57 (1976).
10 Benner, R.; Van Oudenaren, A.; De Ruiter, H.: Cell. Immunol. *33:* 268–276 (1977).
11 Benner, R.; Van Oudenaren, A.; De Ruiter, H.: Cell. Immunol. *34:* 125–137 (1977).
12 Benner, R.; Haaijman, J.J.: Dev. comp. Immunol. *4:* 591–603 (1980).
13 Benner, R.; Van Oudenaren, A.; Koch, G.: in Lefkovits, Pernis, Immunological methods, vol. 2, pp. 247–261 (Academic Press, New York 1981).
14 Benner, R.; Rijnbeek, A.-M.; Bernabé, R.R.; Martinez-Alonso, C.; Coutinho, A.: Immunobiology (in press).
15 Benner, R.; Van Oudenaren, A.; Haaijman, J.J.; Slingerland-Teunissen, J.; Wostmann, B.S.; Hijmans, W.: Int. Archs Allergy appl. Immun. (in press).
16 Brahim, F.; Osmond, D.G.: Anat. Rec. *168:* 139–159 (1970).
17 Chaperon, E.A.; Selner, J.C.; Claman, H.N.: Immunology *14:* 553–561 (1968).
18 Deutsch, L.: Annls Inst. Pasteur, Paris *9:* 689–727 (1899).
19 Donnelly, N.; Sussdorf, D.H.: Cell. Immunol. *15:* 294–302 (1975).
20 Duhamel, G.: Presse med. *76:* 1947–1950 (1968).
21 Eipert, E.F.; Klempau, A.E.; Lallone, R.L.; Cooper, E.L.: Cell. Immunol. *46:* 275–280 (1979).
22 Gollapudi, V.S.S.; Kind, L.S.: J. Immun. *114:* 906–907 (1975).
23 Haaijman, J.J.; Schuit, H.R.E.; Hijmans, W.: Immunology *32:* 427–434 (1977).
24 Haaijman, J.J.; Slingerland-Teunissen, J.; Benner, R.; Van Oudenaren, A.: Immunology *36:* 271–278 (1979).
25 Hill, S.W.: Immunology *30:* 895–906 (1976).
26 Hijmans, W.; Schuit, H.R.E.; Hulsing-Hesselink, E.: Ann. N.Y. Acad. Sci. *177:* 290–305 (1971).
27 Hijmans, W.; Schuit, H.R.E.: Clin. exp. Immunol. *11:* 483–494 (1972).
28 Jankovic, B.D.; Paunovic, V.R.: Ann. Immunol., Paris *124C:* 133–152 (1973).
29 Jankovic, B.D.; Isakovic, K.; Petrovic, S.: Immunology *25:* 663–674 (1973).
30 Kind, L.S.; Malloy, W.F.: J. Immun. *112:* 1609–1612 (1974).
31 Knothe, R.; Herrlinger, J.D.; Mueller-Ruchholtz, W.: Int. Archs Allergy appl. Immun. *59:* 99–103 (1979).
32 Koch, G.; Osmond, D.G.; Julius, M.H.; Benner, R.: J. Immun. (in press).

33 Landy, M.; Sanderson, R.P.; Jackson, A.L.: J. exp. Med. *122:* 483–504 (1965).
34 Langevoort, H.L.; Asofsky, R.M.; Jacobson, E.B.; De Vries, T.; Thorbecke, G.J.: J. Immun. *90:* 60–71 (1963).
35 Lüdke, H.: Berl. klin. Wschr. *49:* 1034–1035 (1912).
36 Maeda, K.; Hyun, B.H.; Rebuck, J.W.: Am. J. clin. Path. *67:* 41–48 (1977).
37 McMillan, R.; Longmire, R.L.; Yelenosky, R.; Lang, J.E.; Heath, V.; Craddock, C.G.: J. Immun. *109:* 1386–1394 (1972).
38 McMillan, R.; Longmire, R.L.; Yelenosky, R.: J. Immun. *116:* 1592–1595 (1976).
39 McMillan, R.; Yelenosky, R.J.; Longmire, R.L.: in Battisto, Streilein, Immuno aspects of the spleen, pp. 227–237 (Elsevier, Amsterdam 1976).
40 Mellbye, O.J.: Int. Archs Allergy appl. Immun. *40:* 248–255 (1971).
41 Miller, S.C.; Osmond, D.G.: Cell Tiss. Kinet. *8:* 97–110 (1975).
42 Morell, A.; Skvaril, R.; Hijmans, W.; Scherz, R.: J. Immun. *115:* 579–583 (1975).
43 Orlans, E.; Peppard, J.; Reynolds, J.; Hall, J.: J. exp. Med. *147:* 588–592 (1978).
44 Osmond, D.G.: in Trnka, Cahill, Monogr. Allergy, pp. 157–172 (Karger, Basel 1980).
45 Pfeiffer, R.; Marx: Z. Hyg. InfektKrankh. *27:* 272–297 (1898).
46 Press, O.W.; Rosse, C.; Clagett, J.: Cell. Immunol. *33:* 114–124 (1977).
47 Radl, J.; Schuit, H.R.E.; Mestecky, J.; Hijmans, W.: in Mestecky, Lawton, The immunoglobulin A system, pp. 57–65 (Plenum Press, New York 1974).
48 Reiter, H.: Z. ImmunForsch. exp. Ther. *18:* 5–61 (1913).
49 Röpke, C.; Everett, N.B.: Cell Tiss. Kinet. *7:* 137–150 (1974).
50 Skvaril, R.; Morell, A.: in Mestecky, Lawton, The immunoglobulin A system, pp. 433–435 (Plenum Press, New York 1974).
51 Sprent, J.; Miller, J.F.A.P.: J. exp. Med. *139:* 1–12 (1974).
52 Thorbecke, G.J.; Keuning, F.J.: J. Immun. *70:* 129–134 (1953).
53 Thorbecke, G.J.; Keuning, F.J.: J. infect. Dis. *98:* 157–171 (1956).
54 Thorbecke, G.J.; Asofsky, R.M.; Hochwald, G.M.; Siskind, G.W.: J. exp. Med. *116:* 295–310 (1962).
55 Turesson, I.: Acta med. scand. *199:* 293–304 (1976).
56 Van Camp, B.G.K.; Schuit, H.R.E.; Hijmans, W.; Radl, J.: Clin. Immunol. Immunopathol. *9:* 111–119 (1978).
57 Van Furth, R.; Schuit, H.R.E.; Hijmans, W.: Immunology *11:* 19–27 (1966).
58 Vaughan, J.H.; Chihara, T.; Moore, T.L.; Robbins, D.L.; Tanimoto, K.; Johnson, J.S.; McMillan, R.: J. clin. Invest. *58:* 933–941 (1976).
59 Vossen, J.M.: Boll. Ist. sieroter. milan. *53:* suppl. I, pp. 152–157 (1974).

Dr. R. Benner, Department of Cell Biology and Genetics, Erasmus University, Postbus 1738, Rotterdam (The Netherlands)

The Immune System, vol. 1, pp. 372–378 (Karger, Basel 1981)

Current Concepts of Lymphocyte-Endothelium Interactions

A.A. Freitas[1]

Department of Immunology, Faculty of Medical Science, University of Lisbon, Lisbon, Portugal

Multipotential stem cells from the fetal liver or adult bone marrow differentiate within the primary lymphoid organs into functionally different T and B cells. These cells are continuously produced and released in large numbers into the blood stream. Once in the blood, the cells are transported rapidly through the whole circulation, traversing different organs with an order of priority that conforms to the anatomical distribution of blood vessels, their size and their relative position to the direction of flow. This process is repeated continually and in one unit of time the number of opportunities for a lymphocyte to reach the more peripheral capillary networks increases with the speed of circulation.

Eventually most lymphocytes leave the blood stream and migrate into the specialized structures of the secondary peripheral lymphoid organs, i.e. spleen, lymph nodes and gut-associated lymphoid tissue (GALT). The rate of migration of lymphocytes from blood to the lymphoid organs has been shown to be a linear function of the concentration of lymphocytes in the blood perfusing the lymphoid tissue. Those lymphocytes migrating through the spleen travel between the intricate pathways of the white pulp and return directly into the blood (minor recirculation) while those lymphocytes migrating to the nodes traverse the substance of the node, enter the efferent lymphatics and recirculate via the lymph back to the blood. This phenomenon has been demonstrated in several species of adult mammals, in duck and plaice, and in fetal lamb, indicating that lymphocyte recirculation appears in the early embryonic life and evolved phylogenetically before mammals.

[1] *Antonio Freitas* has, on several occasions, worked as a visiting scientist at the Basel Institute for Immunology.

Different lymphocytes recirculate at different speeds according to their state of maturation, their surface make-up and their interactions with other cells in their recirculatory pathways. Since the transit time of lymphocytes in blood or lymph is negligible (seconds) the total recirculation time is defined as the time one lymphocyte takes to cross the spleen plus the time it takes to cross from blood to lymph through the nodes. The so-called 'spleen-seeking' lymphocytes are slow-moving cells which because of their stickiness are retained longer within the spleen than fast-moving cells. Such a name is obviously inappropriate, since such cells have been shown to enter lymph nodes in splenectomized recipients. The speed of migration can also be related to the different pathways of lymphocytes within the secondary lymphoid organs. Thus, T and B lymphocytes have different speeds of migration and divergent migratory pathways within the nodes. The sorting out of T and B cells into separated zones is determined by the special interactions that exist between the surface of T and B cells and the reticulin mesh of T and B areas.

Random vs Oriented Lymphocyte Migration

It has been generally believed that small lymphocytes recirculate at random throughout the different lymph nodes and lymphatics of an intact organism. This opinion has been challenged by the report that in sheep there are two distinct pools of recirculating cells, one present in intestinal lymph which recirculates preferentially through the small intestine, and another present predominantly in efferent lymph from peripheral lymph nodes which recirculates through the lymph nodes. We have recently investigated the existence of these two different sets of recirculating small T cells in the mouse. Our results indicate that there is only one pool of recirculating small T cells and that this pool recirculates randomly through the lymphoid organs. The finding that in rodents only large, activated lymphocytes show a non-random pattern of migration, plus the finding that in the fetal lamb (free of exogenous antigen) nodal and peripheral lymphocytes recirculate randomly between intestinal and peripheral lymphatics suggest that only after antigenic stimulation can lymphocytes acquire the capacity to distinguish one tissue from another and thus show a preferential (oriented) pattern of migration. In support of this idea is the observation that mice housed under germ-free conditions have no T cells in the small intestine. We have suggested that after contact with antigen, T cells become sensitized both to foreign antigen and to local tissue determinants, which are recognized later and are responsible for causing the oriented migration and selective localization of tissue-sensitized cells.

The selective migration of lymphoblasts can be mediated at the level of lymphocyte interaction with organ-specific endothelium. Strain differences in the lymphocyte-high endothelial venules (HEV) adherence were described and tentatively mapped to chromosome 7. Further cellular studies may elucidate the functional importance and the maturational sequence of the lymphocyte structures in control of migration.

Lymphocyte HEV Interactions

Lymphocytes leave the blood stream and enter the lymph nodes through the HEV. The process is highly selective for lymphocytes, i.e. other cells, such as leucocytes or monocytes, are efficiently excluded. The nature of the interaction between the recirculating lymphocyte and the HEV endothelium has been the object of extensive studies, and it is now clear that it requires the active participation of the lymphocyte.

The specialized HEV endothelium is formed by polygonal cells linked together by discontinuous junctional complexes. Ultrastructurally, most high-endothelial cells have abundant cytoplasm and contain a prominent Golgi apparatus, vesicles, mitochondria and free and clustered ribosomes. Non-specific esterase activity is readily detected in their cytoplasm, and they have been shown to readily incorporate sulphate molecules. These peculiar structural and metabolic features of the HEV endothelium are more likely to be a consequence, rather than the cause, of lymphocyte emigration since although initially absent in the lymph nodes of athymic or thymectomized rodents, they develop after transfer of large number of T lymphocytes. Furthermore, HEV have not been found in some animal species, i.e. sheep, where lymphocyte recirculation is well demonstrated.

Two alternative, but not necessarily opposing, hypotheses can be considered to explain the selective lymphocyte-HEV interactions.

The 'One Specific Receptor – Two Signals' Hypothesis. In 1968 *Woodruff and Gesner* found that exposure of lymphocytes to trypsin inhibited their migration into the lymph nodes but not into the spleen. In 1969 the same authors reported that the early localization of neuraminidase (N-ase)-treated cells in the spleen and lymph nodes was less than that of untreated lymphocytes. Hence, it was suggested that the unique interaction between lymphocytes and the HEV involved a selective recognition step between a specific receptor (glycoprotein?) on the lymphocyte surface and its complementary structure on the surface of the high endothelial cell. Thereafter, most experiments on the migration of modified lymphocytes consisted of an in vitro treatment of lymphocytes designed to alter surface glycoproteins in an attempt to characterize the specific lymphocyte receptor for the HEV. Thus, the finding that lymphocytes are less localized in the lymph nodes after exposure to concanavalin A (Con A) was interpreted as further evidence for the existence of a glycoprotein receptor on the lymphocyte surface specific for the HEV. Con A would cover such a receptor and thus inhibit the adhesion of the lymphocyte to the endothelial cell. Precise interpretation of the data obtained by these experiments was, however, not possible.

First, inhibition of lymph node entry might be caused by the rapid removal of lymphocytes from the circulation before the cells have had the opportunity of reaching the nodes. Thus, the effect of N-ase on lymphocyte migration is probably through a rapid and temporary removal of the N-ase-treated lymphocytes from the blood and sequestration in the liver (this happens with asialoglycoproteins or N-ase-treated red cells) rather than a specific failure of the cells to enter the node, since N-ase treatment does not alter lymphocyte entry into the isolated and perfused rat mesenteric node.

Second, there was no way to distinguish effects which might be due to destruction of the putative homing receptor from those arising secondarily through non-specific and conformational changes in the membrane of the lymphocyte. For example, it was demonstrated that Con A exposure does not impair the entry of lymphocytes into the nodes of splenectomized recipients or the adhesion of lymphocytes to HEV of the isolated and perfused lymph node; Con A slows down the rate of lymphocyte recirculation and enhances retention of cells in the lungs and spleen, not by covering lymphocyte surface glycoprotein residues, but by modifying membrane fluidity.

Recently it was demonstrated that recirculating lymphocytes bind selectively in vitro to the HEV of frozen sections of lymph nodes. The meaning of these findings is not yet clear, as the binding occurs only under particular experimental conditions and the precise location of the endothelial binding site is not known. It could be either the cell surface or the cytoplasmic space of the endothelial cells exposed in the node sections. These results have been interpreted as confirming the existence of a lymphocyte-specific receptor for the endothelial cell of the HEV. This hypothesis seemed to be reinforced by the observation that trypsin treatment inhibits directly the in vitro adhesion of lymphocytes to the HEV of frozen sections of lymph nodes. The question remains whether trypsin causes nonspecific damage to the cell surface which must be repaired before normal adhesion can ensue or whether it releases molecules specifically required for the recognition of the high endothelial cells of the lymph node. In support of the first alternative, it has been shown that trypsin treatment can increase cell deformability, induce conformational changes and enhance adenylate cyclase activity in the plasma membrane or inhibit cell uptake of palmitate; alterations in the fatty acid composition of the plasma membrane can produce substantial changes in adhesion. Moreover, other agents whose mechanism of action differs from trypsin affect the lymphocyte→HEV adhesion both in vivo (sodium azide phospholipases A and C cytochalasin B) or in vitro (sodium azide, cytochalasin B). Conclusive experimental evidence for the existence of a specific lymphocyte receptor for the endothelial cells of the HEV is, therefore, still lacking and can only be obtained after the isolation and purification of a lymphocyte surface structure which can compete with lymphocytes for binding to the HEV.

The existence of a specific receptor, although sufficient for understanding the lymphocyte → HEV adhesion, cannot, by itself, explain either the actual mechanism of entry of lymphocytes into the lymphoid organs or the unidirectional migration of lymphocytes from blood to lymph. Thus, the one specific receptor hypothesis requires two signals to explain the whole process of blood to lymph passage of lymphocytes: The adhesion of the lymphocyte to the endothelia would act as a first signal and somehow modify the lymphocyte, making it susceptible to the action of a second signal, which would trigger the lymphocyte into active movement and induce, first, lymphocyte entry into the organs, and, second, the release of the lymphocyte from the adhesion sites and the unidirectional migration of the cell from blood to lymph.

The 'Chemotaxis' Hypothesis. In contrast to the one specific receptor hypothesis in which the adhesion of the lymphocyte to the HEV is the initiating signal of the process of lymphocyte emigration, in the chemotaxis hypothesis the lymphocyte → HEV adhesion, although an essential step in that it provides the traction sites necessary for lymphocyte locomotion, is a consequence of the lymphocyte response to a chemotactic gradient. The chemotaxis hypothesis postulates, then, that the nature of the lymphocyte

→ HEV adhesive interactions is more likely determined by the bulk properties of the circulating cell.

The chemotaxis hypothesis is based on microscopic observation of the entry of lymphocytes into the nodes at the level of the HEV; at that particular moment, the lymphocyte has the shape of a cell undergoing an active process of movement as if it were responding chemotactically. Chemotaxis implies an active and oriented movement of a cell in response to a signal exterior to the cell and requires the existence of intact locomotive properties of the responsive cell. It is known that lymphocyte migration requires the integrity of microfilaments (it can be modified by cytochalasin B), and that it is an energy-dependent process (it can be impaired by sodium azide). Furthermore, isoproterenol, theophylline, prostaglandin E_1 and dibutyril-cAMP, agents which increase intracellular levels of cAMP, modify the traffic of lymphocytes and interfere directly with their entry into the nodes, and there is some evidence, though controversial, that intracellular levels of cyclic nucleotides may control the directional movement of the cell. Finally, phospholipase C, which also inhibits in vitro chemotaxis of neutrophils, also impairs the exit of lymphocytes from the blood. There is, however, no direct evidence for the existence of a chemotactic gradient across the HEV.

The hypothesis of the chemotactic gradient fails to explain the restriction of lymphocyte entry to the HEV. One must postulate then that either the chemotactic gradient exists only at the level of the HEV or that the special anatomic structure of the HEV favours particular blood flow conditions which allow the lymphocyte to receive and respond to the chemotactic signal delivered across the endothelial walls. On the other hand, the HEV may provide the only luminal surface compatible with the lymphocyte surface, a requirement that may be essential for the exit of lymphocytes from the blood in response to a chemotactic gradient. The adhesion sites on the lymphocyte surface must represent a small membrane micro-environment in which the special arrangement of the surface molecules will determine the ideal configuration and electric charge for adhesion to occur. Conformational changes of the plasma membrane, induced either by modification of membrane fluidity and movement or by enzymatic blockade, may modify its dielectric constant, the formation of adhesion sites and, consequently, alter cell adhesion.

The one specific receptor hypothesis and the chemotactic hypothesis I have discussed represent two extreme positions. Other intermediate alternatives can also be considered.

Perspectives

Most research on lymphocyte traffic has been mainly concerned with its pattern and dynamics and has given special emphasis to the selective lymphocyte→endothelial interactions. The final purpose of lymphocyte recirculation remains unclear. It may serve the dissemination of immunological memory, as originally suggested, and it obviously serves as a surveillance mechanism through the wide distribution of lymphocytes in the body and their assembly in the nodes where they can meet antigen transported from the periphery. However, both these aims could probably be achieved without the active passage of lymphocytes to lymph. It may well be that lymph has an important survival value for lymphocytes. Recent findings that lymph does alter the in vivo migration of lymphocytes [personal observations] may provide a new insight to this issue. What we need to find out is how long a lymphocyte needs to recirculate while, as *Niels Jerne* would say, 'waiting for the end'.

Acknowledgements. I thank Prof. *D.M.V. Parrott* and Dr. *B. Rocha* for reviewing this manuscript.

Selected Reading

De Sousa, M.: Cell traffic; in Cuatrecasas, Greaves, Receptors and recognition, vol. 2, p. 105 (Chapman & Hall, London 1976).
Ford, W.L.: Lymphocyte migration and immune responses. Prog. Allergy, vol. 19, p. 1 (Karger, Basel 1975).
Jerne, N.K.: Eur. J. Immunol. *1:* 1 (1971).
Jerne, N.K.: Annls Immunol. *125C:* 373 (1974).
Ciba Foundation Symp.: *71:* 265 (1980).

Dr. Antonio Freitas, Universidade Nova de Lisboa, Faculdade de Ciências Médicas, Departamento de Immunologia, Campo de Santana 130, P–1198 Lisboa (Portugal)

Growth of Cells in Culture

The Immune System, vol. 1, pp. 379–382 (Karger, Basel 1981)

Antibodies, Hormones and Cancer

Gordon H. Sato[1]

Department of Biology, University of California, San Diego, La Jolla, Calif., USA

In 1954, *Niels Jerne* came to Cal Tech. We were receiving news from *Watson* via *Max Delbrück* on the structure of DNA, *Luria* had recently come through reporting on *Benzer's* idea for measuring the size of a gene, *Dulbecco* was initiating his studies on animal virology, and *Delbrück* had just sent off his paper on the *Visconti-Delbrück* theory. It was difficult for us graduate students to accommodate *Jerne's* ideas on the selection theory of antibody production into what we were perceiving to be a golden age of the new biology centered on DNA and phage. We were not alone. We heard that *Pauling's* reaction to the manuscript was disinterest. For most of us, it would be many years before we would come to appreciate the very special nature of that intuitively incisive genius. We were, however, immediately captivated by his warmth, wit and unpretentious intelligence. When *Taj,* as he was known in those days, went for coffee, a coterie of admiring graduate students trooped after him. Those conversations over coffee at Cal Tech's old greasy spoon evoke pleasurable memories to this day. I decided that someday I would work on immunology with *Taj* and the following is a belated attempt to still do so.

The original intention of my laboratory some 12 years ago was to study hormone-dependent growth in tissue culture. Some clever members of my laboratory – *Clark* [4], *Armelin* [2], *Sirbasku,* and *Nishikawa* [8] – gradually shifted our emphasis to an inquiry into the role of serum in culture. During a sabbatical leave at the Basel Institute for Immunology in 1974, I was able to sort out some of these ideas concerning the function of serum in cell

[1] *Gordon Sato* is a long-time friend and associate of *Niels Jerne* and was a member of the Basel Institute for Immunology in 1974–1975.

culture [9]. This line of thought culminated in experiments by *Hayashi and Sato* [6] in which she replaced serum with mixtures of hormones in the cultures GH_3, HeLa, and BHK cells.

Our ideas about cellular requirements have expanded considerably since those initial experiments. At the present time, we believe that the requirements of any cell in culture can be met by an appropriate combination of factors from the following four classes: (1) hormones (growth factors); (2) binding proteins; (3) basement membrane components, and (4) nutrients. The requirements for growth in serum-free medium have now been determined for a large number of cell types, both in long-term and primary culture. A more detailed account of this work can be found in a recent review from our laboratory [3]. Suffice it to say for our present purpose that the consequences of replacing serum with defined factors (mostly hormones) are: (1) it should be possible to establish cell types in culture which had never been established previously; (2) unforeseen complexities of cellular endocrinology should be uncovered by this experimental approach, and (3) the discovery of new hormones should be facilitated.

The use of serum is manifestly a poor method of providing cells with the factors they require for survival, growth, metabolism or expression of differentiated function. Serum is generally toxic to cells and is usually diluted by a factor of ten to obviate its toxicity. This necessity for dilution imposes some severe limitations on the usefulness of serum. For instance, the pituitary is exposed to high levels of various hypothalamic hormones through a specialized vasculature connecting the two organs. In the general circulation, the levels of hypothalamic hormones are very low. It therefore is not possible to reproduce the specialized environment of the pituitary cell with a one to ten dilution of serum and such a dilute serum solution is not likely to support pituicytes in culture. Providing the factors required by a cell type directly rather than through the artifice of serum should, by these arguments, yield superior results. A case in point is the beautiful work of *Ambesi-Impiombato* et al. [1] in which normal rat thyroid follicular cells were established in media in which the bulk of the serum was replaced by defined factors. Included among the requirements for thyroid follicular cells in culture are insulin, transferrin, hydrocortisone, glycyl-histidyl-lysine, somatostatin and TSH. Normal thyroid cells would be expected to require TSH but the other requirements are beyond the ordinary expectations of endocrine physiology.

Implicit in our discussion is the assumption that hormonal responses in vitro are an accurate reflection of in vivo physiology. We would expect that insulin, transferrin, hydrocortisone, glycyl-histidyl-lysine and somatostatin would be involved in the function of the thyroid gland and such possibilities should be tested in appropriately designed animal experiments. All our experience indicates that the endocrine physiology of cells, normal or cancerous, is complex. This complexity is readily revealed in medium in which responses to various factors are not obscured by the presence of serum. In fact, the deletion of serum should not only unmask new reponses to old hormones but responses to novel ones as well. Fibroblast growth factor (FGF) was discovered, for instance, in experiments in which efforts were made to obviate the confusing influence of serum [4].

My own motivations for pursuing these studies are an interest in generating systematic approaches to cancer therapy and an abiding desire to fulfill the ambitions inspired by *Niels Jerne* during my graduate school days.

The recent development of monoclonal antibody techniques [7], I believe, provide a long-sought opportunity to combine these interests. Antibodies to hormone receptors can, in some cases, block the binding and action of hormones, and in other cases, mimic the action of the hormone [5]. A systematic effort to produce monoclonal antibodies to hormone receptors, using a growth assay screen in serum-free, defined media would add a new dimension to hormone mechanism studies.

The very demonstration that an antibody mimics the action of a hormone rules out the necessity for internalization of the actual hormone although it may be necessary to internalize appropriately clustered receptors to bring about biological effects. In cases where the identity of a hormone required for growth is unknown, it may be possible to isolate a growth-stimulatory monoclonal antibody. An anti-idiotypic antibody might be directed against the putative hormone and could be used for its isolation. Hormone studies in serum-free media indicate that cell membranes contain a diversity of hormone receptors. The presence of any one of these receptors may not be specific for the given cell type. However, the combination of receptors for any given cell should be relatively specific. The lessons learned about hormones and their receptors in cell culture may be extended in our thinking to surface antigens of tumor cells. An approach that is suggested by this line of thought is to develop libraries of monoclonal antibodies to tumor cells, none of which, by themselves, may be specific for the tumor cell in question. However, it may be possible to combine individual monoclonal antibody types so as to achieve relative quantitative specificity for site-directed chemotherapeutic agents.

The time is ripe for a combined endocrinological and immunological attack on the problems of cancer therapy, and I look forward to the developments over the next few years with great anticipatory excitement.

25 years ago, *Niels Jerne* was probably not aware that he was my teacher. I am ever grateful for that experience.

References

1 Ambesi-Impiombato, F.S.; Parks, L.A.M.; Coon, H.G.: Proc. natn. Acad. Sci. USA *77:* 3455–3459 (1980).
2 Armelin, H.A.: Proc. natn. Acad. Sci. USA *70:* 2702–2706 (1973).
3 Barnes, D.; Sato, G.: Analyt. Biochem. *102:* 255–270 (1980).
4 Clark, J.L.; Jones, K.L.; Gospodarowicz, D.; Sato, G.H.: Nature new Biol. *236:* 180 (1972).
5 Flier, J.S.; Kahn, C.R.; Roth, J.: New Engl. J. Med. *300:* 413–419 (1979).
6 Hayashi, I.; Sato, G.: Nature, Lond. *259:* 132 (1976).
7 Milstein, C.: Sci. Am. *243:* 66–74 (1980).
8 Nishikawa, K.; Armelin, H.A.; Sato, G.: Proc. natn. Acad. Sci. USA *72:* 483–487 (1975).
9 Sato, G.: in Litwak, Biochemical action of hormones, vol. III, pp. 391–396 (Academic Press, New York 1975).

G.H. Sato, PhD, Biology Department, Q-058, University of California, San Diego, La Jolla, CA 92093 (USA)

The Immune System, vol. 1, pp. 383–390 (Karger, Basel 1981)

Growth Factor Control of Cell Proliferation by Control of Cellular Nutrient Requirements

Wallace L. McKeehan[1], *Kerstin A. McKeehan*

W. Alton Jones Cell Science Center, Lake Placid, N.Y., USA

We are pleased to dedicate this paper to *Niels Jerne*. Although we work with cultured fibroblasts, we hope that he will enjoy the paper and its implications for control of activation and growth of lymphocytes.

Nutrients and the Control of Cell Proliferation Rate

The proliferation of unicellular organisms is controlled by the availability of extracellular nutrients. In higher organisms, the nutrient supply is homeostatically controlled, and except under starvation conditions, the nutrient supply does not limit the growth of the organism. However, the proliferation of different cell types within multi-cellular organisms is restrained, even in the presence of a constant nutrient supply. Cell growth occurs only on the demand of distal signals such as hormone-like growth factors or disruptions in the cell-to-cell or cell-to-matrix relationships. One hypothesis to account for this is that multiplication-limiting permeability barriers, intracellular metabolic compartmentation and specialized metabolic requirements for nutrients for proliferation all developed concurrently with cell specialization. Therefore, the constant external supply of nutrients is suboptimal and inadequate to support cell proliferation. Specific external growth signals penetrate these permeability and metabolic barriers such that the constant external supply of nutrients is optimal for cell proliferation until the external signal is removed.

[1] *Wallace McKeehan* was a member of the Basel Institute for Immunology from 1971 to 1973.

Using isolated cells in culture we have directly tested this hypothesis. We have viewed individual cells within a population in the exponential phase of growth as catalysts which interact with nutrients (substrates) to produce new cells (products) from the substrates [11]. In contrast to simple reactions that are catalyzed by enzymes, where the number of catalysts (enzyme molecules) remains unchanged during the course of the reaction, the multiplication of a population of cells is an autocatalytic process in that the catalysts and the products of the reaction are identical. After each round of the reaction, cell + substrate → 2 cells (products), double the number of cells emerge that are ready for another round of catalysis of the process (cell division). When mammalian cells are isolated from the blood supply and tissue matrix, their proliferation can be controlled directly by manipulation of the concentration of nutrients (substrates) in the environment. When all defined nutrients and other conditions are optimized for cell survival and multiplication, cell multiplication rate (velocity) increases with increasing external levels of single nutrients in a manner that indicates a dissociable, saturable interaction between cells and the nutrients. A plot of multiplication rate versus nutrient concentration yields a rectangular hyperbola that is described by the Henri-Michaelis-Menten equation of enzyme kinetic analysis. From the multiplication response, kinetic parameters equivalent to the K_m and V_{max} of enzyme kinetic analysis can be assigned to each nutrient. The K_m for cell multiplication, which we refer to as the $S_{0.50}$ value for a single nutrient under a defined condition, is the concentration of the nutrient that is required to promote a half-maximal rate of proliferation. The V_{max}, which we refer to as R_{max}, is the maximal rate of proliferation of the cell population that is possible under the conditions.

Serum Growth Factors Specifically Modify the Cell Multiplication Requirement for Ca^{2+}, K^{+}, Mg^{2+}, Pi and 2-Oxocarboxylic Acids

Regulatory effectors of the velocity of enzyme reactions exert their effects by modification of the K_m ($S_{0.50}$) for substrate or the V_{max} (R_{max}) of the reaction. Thus, if some growth effectors exert their influence on cell proliferation by the control of the cellular requirement for nutrients, then the relationship should be detected by a growth factor-induced change in the $S_{0.50}$ value of the relevant nutrient for cell multiplication. Using the non-dialyzable macromolecular fraction of fetal bovine serum as a complete source of multiple growth regulators for human diploid fibroblasts, we

Table I. Comparison of $S_{0.50}$ values for Ca^{2+}, K^+, Mg^{2+}, Pi and 2-oxocarboxylic acids for multiplication of human fibroblasts to plasma levels[a]

Nutrient	[Plasma]	$S_{0.50}$ (10 μg/ml)	$S_{0.50}$ (100 μg/ml)	$S_{0.50}$ (1,000 μg/ml)	$S_{0.90}$ (1,000 μg/ml)	[Plasma] / $S_{0.90}$ (1,000 μg/ml)
Ca^{2+}	1,180 [16]	≫1,180	300	10	90	13
K^+	1,400 [3]	≫1,400	3,800	300	2,700	1.5
Mg^{2+}	530 [16]	80	80	30	270	2
Pi	1,170 [16]	0.81	0.81	0.35	3.2	365
Pyruvate	140 [3]	200	20	0	0	–

[a] $S_{0.50}$ and $S_{0.90}$ values and plasma concentrations are given in μmol/l. $S_{0.50}$ and $S_{0.90}$ is the concentration of nutrient required to promote a 50 and 90% of maximum multiplication rate, respectively. Values in parentheses denote the serum factor concentration at which the $S_{0.50}$ or $S_{0.90}$ was determined. $S_{0.50}$ values at 10 μg/ml were estimated by extrapolation of data in [10]. Plasma values for Ca^{2+}, K^+, Mg^{2+} and Pi are the free ion concentrations. The plasma concentrations were taken from the references in brackets. $S_{0.50}$ and $S_{0.90}$ values were from [10].

found that of over 30 individual nutrients examined, serum factors modify the cellular requirement ($S_{0.50}$) for only a few specific nutrients. These are Ca^{2+}, K^+, Mg^{2+}, phosphate ions and 2-oxocarboxylic acids [10]. In contrast, the $S_{0.50}$ value for the majority of nutrients including the 20 amino acids, glucose, purines, pyrimidines, choline and inositol were unaffected by the level of serum factors in the medium [10]. This is surprising because many of the latter nutrients have been proposed as potential mediators of the action of external growth signals and other homonal effects (e.g. insulin) via an increased activity of plasma membrane transport [2, 6].

The regulatory significance of non-substrate effectors of the velocity of enzyme reactions which exert their influence by a reduction of the K_m of a reaction for substrates is greatest when the K_m of the enzyme for the substrate is near or below the normal physiological levels of the substrate in the microenvironment of the reaction. Unfortunately, the normal levels of nutrients in the microenvironment around tissue cells such as the fibroblasts used in this study cannot be precisely determined with present analytical technology. As a best estimate, table I compares the requirements for Ca^{2+}, K^+, Mg^{2+}, Pi and 2-oxocarboxylic acids for multiplication of human fibroblasts to the reported levels of these nutrients in human plasma or serum. The use of plasma levels as an estimate of the concentration of these nutrients around cells in tissue is likely a maximum figure. Concentration gradients from the blood supply to the cells probably result in lower values in the cellular microenvironment, especially in the case of Ca^{2+} and Mg^{2+} where only the free, unbound concentration of the cations may be important. On the other hand, the microenvironmental concentration of 2-oxocarboxylic acids such as pyruvate is likely determined by diffusion gradients from the blood supply and the efflux of cell-derived 2-oxocarboxylic acids into the local environment. Table I indicates that, when the external concentration of serum growth factors is low, the $S_{0.50}$ values for Ca^{2+}, K^+ and 2-oxocarboxylic acids are at or above the levels of these three nutrients in plasma. The multiplication requirement for Ca^{2+} and K^+ becomes very large (approaches infinity) as the levels of serum factors drop below 100 μg/ml, the lowest level where the $S_{0.50}$ value could be accurately determined by kinetic analysis in vitro [10]. By extrapolation of the inverse linear relationship between the multiplication requirement for pyruvate and serum factors, at 10 μg/ml serum factors, the $S_{0.50}^{pyr}$ is also higher than the reported value of the pyruvate levels in plasma (table I). Thus, if the plasma levels of Ca^{2+}, K^+ and pyruvate are valid maximal estimates of the external levels of the nutrients around human fibroblasts in tissue, cell multiplication in vivo may be limited by the external concentration of Ca^{2+}, K^+ or 2-oxocarboxylic acids when the levels of serum-derived growth signals in the environment are limiting. Increasing levels of serum factors in the medium cause a drop in the $S_{0.50}$ values for Ca^{2+}, K^+ and pyruvate below the normal plasma levels (table I). At 1,000 μg/ml serum factors, plasma levels of Ca^{2+} and K^+ are 13 and 1.5 times, respectively, the requirement for external Ca^{2+} and K^+ to promote a 90% of maximum rate of cell multiplication ($S_{0.90}$). Thus, the range in which serum factors modify the cellular requirement for Ca^{2+}, K^+ and pyruvate in vitro are within the range that could be of regulatory significance under estimated physiological concentrations of the nutrients in the cellular microenvironment in vivo. In contrast to Ca^{2+}, K^+ and 2-oxocarboxylic acids, even when the level of external serum growth signals is low, the cell multiplication requirement for both Mg^{2+} and Pi is below the level of both nutrients in plasma (table I). The Pi requirement is so far below the estimated extracellular levels that it is unlikely that the external Pi could limit cell multiplication, even if the actual level of extracellular Pi around cells in tissue is much lower than that in plasma. However, the potential regulatory significance of Mg^{2+} may deserve further consideration. The free [Mg^{2+}] in the microenvironment around tissue cells could be much smaller than plasma free Mg^{2+} levels and on the order of the $S_{0.50}^{Mg2+}$ (80 μ*M*) for

Table II. Comparison of $S_{0.50}$ values and plasma concentrations for nutrients other than Ca^{2+}, K^+, Mg^{2+}, Pi and 2-oxocarboxylic acids[a]

Nutrient	$S_{0.50}$	[Plasma]	$S_{0.90}$	$\frac{[Plasma]}{S_{0.90}}$
Arg	1.5	87	14	6
Cys	2.4	33	22	1.5
Gln	11	568	99	6
Gly	16	205	144	1.4
His	1.5	74	14	5
Ile	0.4	68	4	17
Leu	0.5	129	5	26
Lys	15	186	135	1.4
Met	0.1	25	0.9	28
Phe	3.0	51	27	2
Ser	0.8	52	7	7
Thr	1.8	117	16	7
Trp	0.006	54	0.05	1,080
Tyr	0.1	57	0.9	63
Val	1.5	246	14	18
Ade	0.01	0.07–1 [5, 15]	0.09	0.8–100
Hypoxanthine	–	25 [13]	–	–
Choline	1.0	440	9	49
dThd	0.01	2 [12]	0.09	22
Glc	27	540	243	2
Inositol	0.3	26	3	9

[a] $S_{0.50}$ and $S_{0.90}$ values and plasma concentrations are in units of µmol/l. $S_{0.50}$ and $S_{0.90}$ are defined in table I. Plasma levels of nutrients except those indicated by a separate reference in brackets were from [3]. $S_{0.50}$ and $S_{0.90}$ were determined at 250 µg per FBSP in the medium [10].

cell multiplication. Recent improvements in non-invasive analytical methods to quantitate the free and bound cation levels in cells and tissue [4] may be applicable to the determination of the actual external free [Mg^{2+}] in the tissue microenvironment and may shed light on this question.

Besides Ca^{2+}, K^+, Mg^{2+}, Pi and 2-oxocarboxylic acids, the requirement for the majority of common nutrients for multiplication is a constitutive property of normal cells and unaffected by the level of serum growth factors in the medium. The constitutive multiplication requirement is below the levels of the nutrients in plasma (table II). Plasma levels of most organic nutrients are equal to or greater than the external requirement ($S_{0.90}$) to promote a 90% of maximum multiplication rate of isolated cells. If plasma levels of the nutrients in table II are an overestimate of up to a factor of 10 of the microenvironmental nutrient concentrations around cells in tissue, then the concentration of Arg, Cys, Gln, His, Lys, Phe, Ser, Thr, Glc or inositol could be potentially limiting to cell multiplication rate ([plasma]/$S_{0.90} \leq 10$). However, this is unlikely since the plasma levels of most organic nutrients probably approximate the actual levels of the nutrients in the microenvironment around cells in tissue relative to the inorganic ions. The effective concentrations of organic nutrients are less likely to be affected by the glycoproteins, proteoglycans and glycosaminoglycans of the extracellular matrix than the free levels of inorganic nutrients especially Ca^{2+} and Mg^{2+}. The [plasma]/$S_{0.90}$ ratio of Ile, Leu, Met, Trp, Tyr, Val and choline is sufficiently high that external concentrations of these nutrients are unlikely to ever limit cell multiplication under any physiological conditions. The high ratio of [plasma] to the $S_{0.50}$ or $S_{0.90}$ values in addition to the fact that the $S_{0.50}$ values for most nutrients are unaffected by serum factors is consistent with the argument that the external concentrations of nutrients other than Ca^{2+}, K^+, Mg^{2+}, Pi and 2-oxocarboxylic acids are probably not limiting to fibroblast multiplication in vivo and that these nutrients are unlikely to be direct mediators of the stimulation of cell multiplication by external growth signals.

Ca^{2+}, K^+ and 2-Oxocarboxylic Acids as Primary Mediators of External Signals for Cell Proliferation

From the foregoing arguments, we propose that of over 30 nutrients commonly found in the external environment, Ca^{2+}, K^+ and 2-oxocarboxylic acids may directly mediate the action of specific external growth signals. Some external growth factors may simply reduce the cellular requirement for the three nutrients by increasing access of them to key intracellular processes. Alternatively growth signals may activate gene products that reduce or bypass the requirement for Ca^{2+}, K^+ and 2-oxocarboxylic acids in multiplication-limiting cellular processes. Mg^{2+} and phosphate ions may also be involved in the mechanism of growth factor action, but less directly than Ca^{2+}, K^+ and 2-oxocarboxylic acids [1, 8, 14].

Ca^{2+} and K^+ have been extensively implicated as mediators of cellular responses to most external stimuli. Cell proliferation is no exception and the role of both ions in cell multiplication processes has been reviewed in a recent symposium volume [7]. The role of Ca^{2+} is intimately associated with other intracellular signals that regulate cell division including cyclic nucleotides, calmodulin (CDR: calcium-dependent regulatory protein), prostaglandins and protein kinases [17]. The role of K^+ (and related Na^+) in cell proliferation has been studied more intensely than anything other than Ca^{2+} and has also been discussed in recent reviews [7].

From a rigorous study of the role and interplay of serum growth factors and nutrients in the multiplication rate of human fibroblasts, 2-oxocarboxylic acids emerge as a third key nutrient (in addition to Ca^{2+} and K^+) that we propose to be directly related to the mechanism by which serum growth factors control cell multiplication. Because they are such common intermediary metabolites, which cells can synthesize, the potential importance of 2-oxocarboxylic acids as quantitative regulators of cellular metabolism and, therefore, cell proliferation have been overlooked. Elsewhere we showed that the cellular requirement for multiplication is not specific for any single 2-oxocarboxylic acid, but is for any one of a class of 2-oxocarboxylic acids of five carbons or less [9]. Glyoxylic, pyruvic, 2-oxoglutaric and oxalacetic acids are most active. The counterpart amino acids or reduced 2-hydroxycarboxylic acids will not substitute for the four most active 2-oxocarboxylic acids. In addition, several other products of the cellular metabolism of the active 2-oxocarboxylic acids were inactive. These include acetate (or acetyl CoA), citrate, isocitrate, succinate (or succinyl CoA) and fumarate [9]. Thus, it is neither a deficiency of the reduced or aminated products of 2-oxocarboxylic acids nor other metabolic products of 2-oxocarboxylic acids that normally limit cell proliferation, but perhaps a common consequence of either their reduction or transamination. The common result of reduction of 2-oxocarboxylic acids that participate in pyridine-linked oxidoreductase reactions is a shift in the $NAD^+/NADH$ ratio. The common consequence of the transamination of glyoxylic, pyruvic, 2-oxoglutaric and oxalacetic acids is a different 2-oxocarboxylic acid. In this respect, 2-oxocarboxylic acids may play an indirect and catalytic role in metabolism and, in turn, cell multiplication, in much the same way as Ca^{2+} and K^+. Neither Ca^{2+}, K^+ nor 2-oxocarboxylic acids actually play direct nutritive roles as 'building-blocks' or energy-producing substrates for increasing cell mass. Instead, the three nutrients act as regulatory effectors of key processes whose normal products or consequences are rate-limiting to cell proliferation. This makes them well-suited as mediators of higher order effectors of cell proliferation such as hormones and polypeptide growth factors as well as local growth signals, i.e. disruption of the cell-to-cell or the cell-to-matrix relationships.

References

1 Barsh G.S.; Cunningham, D.D.: J. supramol. Struct. *7:* 61–77 (1977).
2 Bhargava, P.M.: J. theor. Biol. *68:* 101–137 (1977).
3 Dittmer, D.S. (ed.): Blood and other body fluids (Federation of Am. Soc. for Experimental Biology, Washington 1961).
4 Gupta, R.J.; Yushok, W.D.: Proc. natn. Acad. Sci. USA *77:* 2487–2491 (1980).
5 Hamet, M.: Annls Biol. clin. *33:* 131–138 (1975).
6 Holley, R.M.: Proc. natn. Acad. Sci. USA *69:* 2840–2841 (1972).
7 Leffert, H.L. (ed.): Growth regulation by ion fluxes. Ann. N.Y. Acad. Sci. *339:* (1980).
8 McKeehan, W.L.; Ham, R.G.: Nature, Lond. *275:* 756–758 (1978).
9 McKeehan, W.L.; McKeehan, K.A.: J. cell. Physiol. *101:* 9–16 (1979).
10 McKeehan, W.L.; McKeehan, K.A.: Proc. natn. Acad. Sci. USA *77:* 3417–3421 (1980).
11 McKeehan, W.L.; McKeehan, K.A.: J. supramol. Struct. *14* (in press).
12 Mitchell, R.S.; Balk, S.D.; Frank, O.; Baker, H.; Cristine, M.J.: Cancer Res. *35:* 2613–2615 (1975).
13 Murray, A.W.: A. Rev. Biochem. *40:* 811–826 (1971).
14 Rubin, H.: J. cell. Physiol. *89:* 613–626 (1977).
15 de Verdier, C.H.; Ericson, A.; Niklasson, F.; Westman, M.: Scand. J. clin. Invest. *37:* 567–575 (1977).
16 Walser, M.: J. clin. Invest. *40:* 723–730 (1961).
17 Whitfield, J.F.; Boynton, A.L.; MacManus, J.P.; Sikorsky, M.; Tsang, B.K.: Mol. cell. Biochem. *27:* 155–179 (1979).

Dr. Wallace L. McKeehan, W. Alton Jones Cell Science Center, Old Barn Road, Lake Placid, NY 12946 (USA)

The Immune System, vol. 1, pp. 391–393 (Karger, Basel 1981)

Long-Term Growth of Murine Spleen B Lymphocytes

B.J. Weimann[1]

Pharma Research, Hoffmann-La Roche, Basel, Switzerland

In vitro growth of thymus-derived lymphocytes from normal human bone marrow and peripheral blood [6] as well as from murine origin [3] was achieved using conditioned media harvested from mitogen-stimulated spleen cells. In these and in numerous reports that followed the original observations, it was shown that the T cells were completely dependent on the growth factor present in the conditioned medium and that they maintained their morphologic and functional characteristics. I, therefore, examined whether conditioned media from mitogen-stimulated spleen cells also could support the growth of normal murine B lymphocytes. Small resting cells from spleens of C57BL nu/nu mice were used as a source of B cells. Approximately half of these cells express immunoglobulins on their surface membrane and can be stimulated by mitogens such as lipopolysaccharide (LPS), *E. coli* lipoprotein, or dextran sulfate to give rise to plaque-forming cells secreting IgM, IgG or IgA.

Single cell suspensions prepared from athymic mice, C57BL nu/nu, were applied to 2–4% continuous bovine serum albumin gradients and cells were separated according to their velocity sedimentation in a 1 *g* gravitational field [5]. Small, resting lymphocytes were then either exposed to mitogens in order to determine the number of plaque-forming cells [2, 4], examined for surface markers (sIg or Thy-1.2 antigen) by immunofluores-

[1] *Bernd Weimann* was a member of the Basel Institute for Immunology from 1974 to 1978.

cence, or propagated in RPMI-1640 containing 10% fetal calf serum (FCS) and 20% conditioned medium from concanavalin A (Con A)-stimulated mouse spleen cells. No attempts were undertaken to separate the growth promoting activity from Con A in the medium. Conditioned media were prepared by stimulation of 5×10^6 mouse spleen cells/ml with 5 µg Con A/ml in RPMI-1640 containing 10% FCS for 24 or 48 h.

In the presence of 20% Con A-conditioned medium, cells could be grown in suspension continuously in RPMI-1640 containing 10% FCS for periods up to 6 months. Medium was replaced every 4th–5th day. Cultures were split when they reached densities of about 8×10^5 cells/ml. However, splitting the cultures at lower densities than around 2×10^5 cells/ml resulted in poor growth rates. Their survival was strictly dependent on the activity in the Con A-conditioned medium. The cell density maximally obtained was $8\text{–}10 \times 10^5$ cells/ml. The cells could be frozen in RPMI-1640 medium supplemented with 20% FCS and 10% dimethyl-sulfoxide and stored in liquid nitrogen. They could be thawed and successfully cultured again for periods of months. About 2–4 weeks after initiation of the cultures, staining for surface immunoglobulin and Thy-1.2 antigen was performed. Around 96% of the cells were found to be sIg-positive and 2–3% were Thy-1.2-positive. After 14 weeks, 80–85% of the cells were sIg^+ and 10–15% were Thy-1.2-positive. When these cells (3×10^5/ml) were exposed to *E. coli* lipoprotein (5 µg/ml) only a small proportion was found to secrete IgM (around 70 plaque-forming cells on day 5). No cells could be induced to Ig secretion by LPS.

These continuously growing cells are not tumorigenic when injected into congenic C57BL nu/nu mice. Neither particles of C-type RNA tumor viruses nor reverse transcriptase activity was found in supernatants of cultured cells stimulated by mitogens. The karyotype remained diploid (40 chromosomes).

It appears from these experiments that B lymphocytes can also be continuously grown in suspension cultures. The nature of the growth-supporting activity and the cells from which it derived are unknown. Evidence that the cells belong to the B lymphocyte lineage is the fact that they display immunoglobulins on their surface. The inability of polyclonal activators like LPS and *E. coli* lipoprotein to induce immunoglobulin synthesis in these cultured cells may have several explanations. It could be that the cells lose their mitogen receptors during the cultivation period. It is also possible that the in vitro growth conditions favour the proliferation of B cell subsets which lack mitogen receptors also in vivo. The loss of mitogen receptor activity could also be a phenomenon of aging interpreted as an accumulation of errors and may be comparable to findings in mice [1], where the proportion of LPS- and lipoprotein-reactive cells in the B cell population decreased with age.

References

1 Andersson, J.; Coutinho, A.; Melchers, F.: J. exp. Med. *145:* 1511 (1975).
2 Andersson, J.; Lafleur, L.; Melchers, F.: Eur. J. Immunol. *4:* 170 (1974).
3 Gillis, S.; Smith, K.A.: Nature, Lond. *268:* 154 (1977).
4 Gronowicz, E.; Coutinho, A.; Melchers, F.: Eur. J. Immunol. *6:* 588 (1976).
5 Miller, R.G.; Phillips, R.A.: J. cell. Physiol. *73:* 191 (1969).
6 Morgan, D.E.; Ruscetti, F.W.; Gallo, R.: Science *193:* 1007 (1976).

Dr. Bernd Weimann, Pharma-Forschung, F. Hoffmann-La Roche & Co. A.G.,
CH–4005 Basel (Switzerland)

Cancer

The Immune System, vol. 1, pp. 394–398 (Karger, Basel 1981)

In Search of a New GOD

E.S. Lennox[1]

MRC Laboratory of Molecular Biology, Cambridge, England

Dear Niels,

Writings for occasions like this should be personal so you will understand if this is a self-centred story in which I tell you why, after many years of thinking about the sources of antibody diversity, I became fascinated with the diversity of tumour antigens and have been trying for the last 6 years to discover its origin(s).

I will tell you what I think the questions are, what I guess the answers are and why I care.

One of the reasons for caring about the antigens on tumour cell surfaces is that transformed cell lines are so much used as models for normal cells with differentiated functions and it is this array of antigens that we use as an index of relationship of the tumour cell to its normal counterpart. It is thus important to know how faithful is the correspondence of one with the other. This leads to the question whether the antigens on tumour cells are just the ensemble of those on the normal counterparts or are composed of other, additional elements. It is in this context that the transplantation antigens of chemically induced rodent sarcomas seem so extraordinary for they are so diverse that it is difficult to imagine them as normal differentiation antigens. It is the nature and origin of this diversity that has drawn the attention of many people including me.

A summary of the properties of the tumour-specific transplantation antigens (TSTA) of chemically induced tumours is the following [3]: (1)

[1] *E.S. Lennox* was a member of the Basel Institute for Immunology in 1976.

They are very diverse. In fact it is claimed that repeats are never found. I think this is probably an exaggeration for these antigens are detected and measured only in transplantation assays and these are notoriously difficult to put into the quantitative terms needed for thinking about numbers of antigenic molecules and their degree of similarity. The making of specific antisera so necessary for the needed quantitation has been almost entirely unsuccessful. (2) While most observations are on tumours induced in vivo, and hence in a potentially heterogeneous normal cell population, carcinogen treatment in vitro of cloned normal fibroblasts also yields tumours with diverse non-cross-reacting antigens. Although there are not many such in vitro examples there is at least no support for the view that the antigenic diversity already exists in a heterogeneous population before transformation. (3) These antigens are stable. Passage in animals or in tissue culture does not change them. (4) They are not restricted to sarcomas but similarly diverse antigens are seemingly expressed on epithelial and lymphoid mouse tumours.

There have been many attempts to explain these diverse TSTAs in terms of 'inappropriate' expression of H-2, that is of H-2 alloantigens not expressed on normal cells of the mouse in which the tumour is induced. In fact if you were to hear the discussions of this possibility that go on you would feel a flush of déja vu for they turn on the question whether the varieties of H-2 antigens are encoded in germ line genes or are somatically generated. With germ line coding there is room for regulatory events to allow 'inappropriate' H-2 expression and the TSTAs might be explained in those terms. There is, in fact, little to support this correspondence and a lot against it, yet in the gathering of the evidence, 'inappropriate' H-2 expression has been used as a net to collect a wide variety of diverse phenomena with causes as unrelated as, on the one hand, varying expression of antigenic determinants of H-2 molecules depending on its neighbours and on the other hand, H-2 typing sera with unexpected specificities or even tumours from mice with unexpected genes, probably from contaminated stocks. In following this literature I have been struck how naming a phenomenon draws in under the same name a lot of unrelated observations that would probably be forgotten on their own.

Without the varieties of H-2 molecules to draw on as the source of diversity for TSTA where else should we look? Is there another generator of diversity yet to be found?

Before I tell you what I think is the answer to these questions let me fill in the background a little more so you will understand what blinkers were narrowing my vision (and that of many others, I suspect) [3]. Hence this diversion on RNA tumour virus proteins and glycoproteins on mouse cells, normal and tumour. As you know, the expression of these is common, more so in tissue from some mouse strains than others. And while it may be especially marked in tumours induced by these viruses, it is not restricted to them. In fact it has long been known that many chemically induced sarcomas – that also express TSTA – do express these viral components. In this situation they are generally regarded as a bloody nuisance for they are serodominant and antibody response to them defeats almost all attempts to make antisera with specificities corresponding to those of TSTA. Moreover, they have been dismissed as possible carriers of TSTA specificity on the grounds that not only are they not varied enough but also there are tumours clearly not reacting with sera recognizing the virus components but which nonetheless have specific transplantation antigens. In spite of this, two sets of observations have shifted my perspective from the 'bloody nuisance' view of RNA tumour virus components on tumour cells to the view that one of them is the carrier of TSTA diversity.

One is the observation that gp70, the envelope glycoprotein of these viruses, is much more varied than previously thought [1, 2]. In fact each mouse genome has genes coding for several different viruses and probably many copies of each. In addition, rather frequent recombination among these genes yields viruses with new host ranges and recombinant gp70 molecules and indeed different tissues of the same mouse express different recombinants.

The other comes from our own experiments that led directly to a gp70 as a good candidate for the TSTA of one of the tumours we studied. I mentioned before that most attempts to make antisera with specificities corresponding to TSTA had failed, in my laboratory as well as in many others. In fact all we ever made was mostly anti-viral and cross-reacting. By hard work and good luck we finally did manage in a syngeneic mouse immunization to make an antiserum containing some antibodies specific for the immunizing tumour. From this mouse we derived a number of monoclonal antibody producing lines and several of these antibodies did indeed seem specific for the immunizing tumour. In extensive testing, this specificity held up well for we could not find the recognized antigen on any normal tissue and found it only very rarely in a large collection of tumours. To pursue the possibility that this restricted epitope might be on the molecule that carries the transplantation specificity, we did do transplantation experiments using the few tumours that do express this cross-reacting antigen. While these have been limited in number, the correlation has been very good, that is, if the tumour expresses the antigen recognized by the 'specific' monoclonal antibody it cross-reacts with the original immunizing tumour. As for the chemistry of the molecule carrying this epitope, I shall cut a long story short and tell you that the monoclonal antibody recognizes a gp70 like molecule on the tumour cell surfaces and in some purified viruses.

Well, that is the main part of the story. Thus, from thinking that the RNA tumour virus components on tumour cell surfaces are a barrier that kept us from uncovering the underlying cause of diversity in tumour-specific transplantation antigens, I now think that on the contrary one of them (gp 70) is the molecule carrying that diversity and that recombination events in viral genomes are the underlying genetic event for generating that diversity. You will understand that I am less than certain about this, of course, for I am guilty of generalizing from one example. Not only this, but even for this one case I have not tied tight the threads linking gp70 and TSTA. In any event I find the idea attractive for it draws, as a generator of diversity of TSTA, on a ready source likely having at its disposal recombination, insertions, deletions and all the tricks we are learning about.

None of this, of course, answers the question whether all this variety serves any function. Since the variety occurs in normal tissue, it is argued by some people that it does. It is striking for example that spontaneous thymic lymphomas in AKR mice are very heterogeneous in their antigens and seemingly each tumour reflects a recombination event in the viral genomes. [4]. Is this an effect or a cause of transformation?

One final question: Do human tumours have a similar generator of diversity? As you know, attempts to identify a similar system of RNA tumour viruses in man have been inconclusive so it is likely this source of diversity will not cloud analysis of human tumour antigens. In fact, detailed analysis so far by monoclonal antibodies have revealed only what seem to be normal assemblies of differentiation antigens although it is still too early to be sure.

Well, that is the story as it stands. It will take more time to see whether things are as I have told you. If they are, it will be another example of how events before our eyes can escape our attention until we have a point of view for understanding them. Of course, this is one of the things you have been telling us all these years and have yourself helped us so much in our groping to understand immunology by giving us these points of view.

Cordially yours,

Ed

References

1 Elder, J.D.; Gautsch, J.W.; Jensen, F.C.; Lerner, R.A.; Hartley, J.W.; Rowe, W.P.; Proc. natn. Acad. Sci. USA *74:* 4676–4680 (1977).

2 Hartley, J.W.; Walford, N.K.; Old, L.J.; Rowe, W.P.: Proc. natn. Acad. Sci. USA *74:* 789–792 (1977).

3 Lennox, E.S.: in Fougereau,Dausset, Immunology 80. Progress in immunology, vol. IV, pp. 659–667 (Academic Press, New York 1981).

4 Zielinski, C.C.; Waksal, S.D.; Templis, L.D.; Khiroya, R.H.; Schwartz, R.S.: Nature, Lond. *288:* 489–491 (1980).

Dr. E.S. Lennox, MRC Laboratory of Molecular Biology, Hills Road, Cambridge CB2 2QH (England)

The Immune System, vol. 1, pp. 399–405 (Karger, Basel 1981)

The Immune System Is a Prerequisite of Cancer: the Hypothesis of Neoplasitropic Determinants

J. V. Spärck[1]

Immunological Laboratory, Statens Seruminstitut, Copenhagen, Denmark

Just how the immune system is related to the occurrence and growth of malignant tumours is still problematic. The general opinion has been that the animal organism is equipped with an immune defence against neoplasia. The 'immune surveillance' hypothesis [2] implies that it is a *main* function of the immune system to survey and destroy neoplastic cells which are continuously produced by somatic mutation or by a transformation mediated by oncogens. It is assumed that such neoplastic cells usually have specific antigens which are recognized as 'foreign' by the immune mechanism. Thus, in the case of reduced immune responsiveness, such neoplastic cells should have the possibility of avoiding the immune surveillance and of multiplying and forming tumours.

However, very many findings are accumulating which are in contrast to what would be expected if immune surveillance of neoplasia existed. There are many data to show, on the contrary, that a direct positive relation does exist between immune function and tumor growth, i.e. that neoplastic development is dependent on the immune response. Such data make it necessary to consider a different basic concept of cancer, and I have proposed a new hypothesis of neoplastic growth as a reaction phenomenon, a product of the immune system [9, 10]. I suggest that there is an occasional occurrence in the animal organism of variant tissue constituents, which display ultra-weak antigenic determinants that stimulate the host reaction to an inflammatory proliferation but which cannot be rejected because of insufficient antigenicity. Therefore, the reaction continues, resulting in a

[1] *J. V. Spärck* was a colleague of *Niels Jerne* at the State Serum Institute in Copenhagen from 1952 to 1956.

growing tumor. This particular class of ultra-weak antigens should be named *neoplasitropic determinants* because of its capacity to stimulate the lymphoreticular tissue to a neoplastic growth.

The proposal of the hypothesis that neoplasia is caused by a particular type of immune response was based on the result of experiments showing that lymphocyte depletion does not stimulate but, on the contrary, inhibits tumour growth. When spontaneous mammary carcinomas were transplanted to irradiated or cortisone-treated recipient mice [9, 10], it was demonstrated that tumour growth was retarded in immunosuppressed recipients, when spontaneous, *syngeneic* carcinomas were transplanted, while in the allogeneic combination the opposite was the case: tumour growth was increased in immunosuppressed recipients, that is, in a situation resembling the natural tumour situation – the syngeneic transplantation – there is a positive relation between the immune function of the host and tumour growth (an example of these experiments is shown in fig. 1). Obviously, the tumour formation is dependent on, and is favoured by, an active immune response of the host.

Indeed, a tumour-specific immunogenicity has been established in a number of experimental and transplantable neoplasias, and it has been shown that even the organism in which a tumour has arisen may in some experimental systems react specifically against its own tumour [4]. However, most cases in which such immunogenic potential of tumours has been demonstrated have been concerned with chemically or virally induced tumours which may prove to be laboratory artifacts not representative of spontaneous tumours [10]. Usually it is not possible to show immunogenicity of spontaneous tumours and, since they very seldom regress, it is difficult to maintain that an immunological defence against neoplasia exists in the animal organism. It is, in any case, obvious that the immune surveillance against the incipient neoplasia does not function.

From the experiments on the growth of murine mammary carcinomas, as already mentioned – the first studies of immunosuppression in an immunogenetically well-defined tumour transplantation system – I suggested that the weak immune stimulus, which the ultra-weak antigenic determinants of the tumour represent to the autochtonous or syngeneic host, releases a mild, proliferative immune response supporting or contributing to tumour growth. The strong immune response, on the other hand, which is released by the histo-incompatible, allogeneic graft results in regression of the tumour [9, 10]. This function of the lymphoreticular response, either promoting or inhibiting tumour growth, according to the strength of the

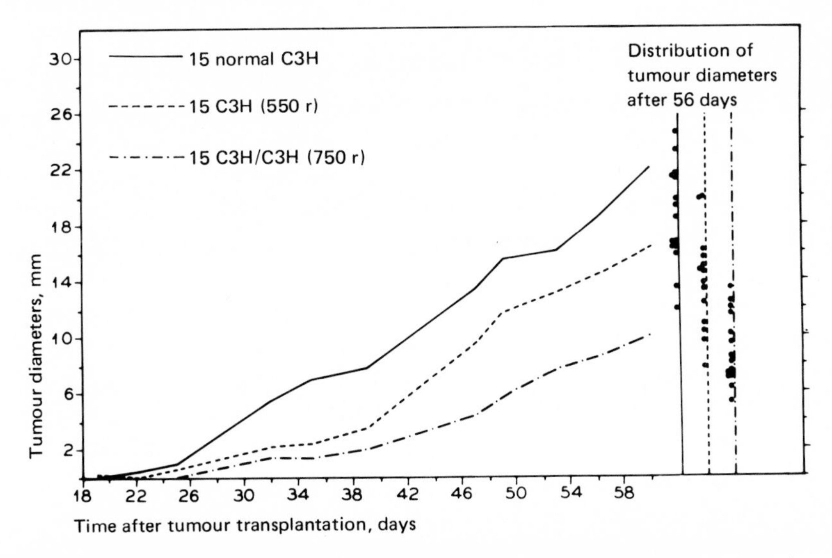

Fig. 1. Effect of whole-body X-irradiation on the tumor growth following syngeneic transplantation. C3H mice were irradiated 56 days before the inoculation subcutaneously of 10^6 cells from a primary, spontaneous C3H mammary carcinoma.

response, seems to apply in general to growth. The lymphoreticular system has not exclusively defensive functions but has also growth stimulating and regulating functions in relation to both normal and malignant tissues.

Prehn [6] later obtained similar results in experiments on tumour growth in immunosuppressed mice, namely reduced growth of murine carcinomas in irradiated syngeneic hosts. He was also able to demonstrate that in vivo sensitized spleen cells would enhance tumour growth in syngeneic hosts [7]. The recipient mice were thymectomized and then given 450 R whole-body irradiation. When spleen cells from mice with the same tumour were added to the tumour inoculum, a significant and reproducible increase of tumour growth was obtained in the recipients mentioned. *Prehn* found, however, that the spleen cell/tumour cell ratio was all-important: low ratios resulted in tumour stimulation, while high ratios of spleen cells admixed with tumour cells caused tumour inhibition. His interpretation also was that a weak immune reaction promotes tumour growth, while a strong immune response inhibits the tumour.

A similar biphasic effect of spleen cells from tumour-bearing mice on the growth of tumour target cells in vitro was found by *Larsen and Spärck* [5]. An inhibitory effect was seen regularly when high concentrations of these spleen cells were added, while the effect was stimulatory when low concentrations were added to the tumour target cells.

The above-mentioned finding of reduced tumour growth in immunodepressed syngeneic recipients and the conclusion that tumour development is dependent on the immune system is supported by many observations on the influence of experimental and spontaneous immunodepression on malignant growth. Indeed, there are conflicting reports as to the effect of thymectomy on tumour growth, and one explanation may be the varied immunogenicity of the tumours studied, although very often the effect is not in accordance with the predictions of the 'immune surveillance' hypothesis. *Heppner* et al. [3] for instance, found that neonatal thymectomy of Balb/c mice resulted in reduction and a marked delay in the development of spontaneous mammary carcinomas. *Umiel and Trainin* [12] found that removal of the thymus from adult mice before syngeneic tumour transplantation led to a delay in tumour growth and to a reduction in the number of tumour takes. *Balner and Dersjant* [1] report similar effects of thymectomy on chemical carcinogenesis.

It is most important, in this connection, that the nude mouse, which is an athymic mouse mutant with a deleted cellular immune system, has no increased spontaneous tumour frequency. *Rygaard and Poulsen* [8] report, from the observation of several thousand nude mice, a complete absence of spontaneous tumours. It is also very interesting that *Stutman* [11] was able to induce tumours by means of Moloney sarcoma virus in these immunologically crippled nude mice, but the tumours developed much later than in normal mice and did not regress as they do in normal hosts. This course of events is comprehensible if the formation of the tumour is a result of the host reaction, which is repressed in the nude mouse.

The previously discussed hypothesis that a weak antigenic stimulus releases a type of immune response that favours the development of a tumour, whereas a strong stimulus leads to an inhibitory response, was supported by the results of my studies on the local lymphatic response to transplantation antigens. Groups of mice of inbred strains received, in their left hind footpad, a suspension of spleen cells from F_1 hybrid donors and simultaneously, as control, the same number of syngeneic spleen cells in their right hind footpad. At different times after injection groups of mice were sacrificed, the draining (popliteal) lymph nodes removed and mean weights of lymph nodes determined.

The weight increase response was found to be related to histoincompatibility, i.e. the increase was consistently greater and more prolonged in the

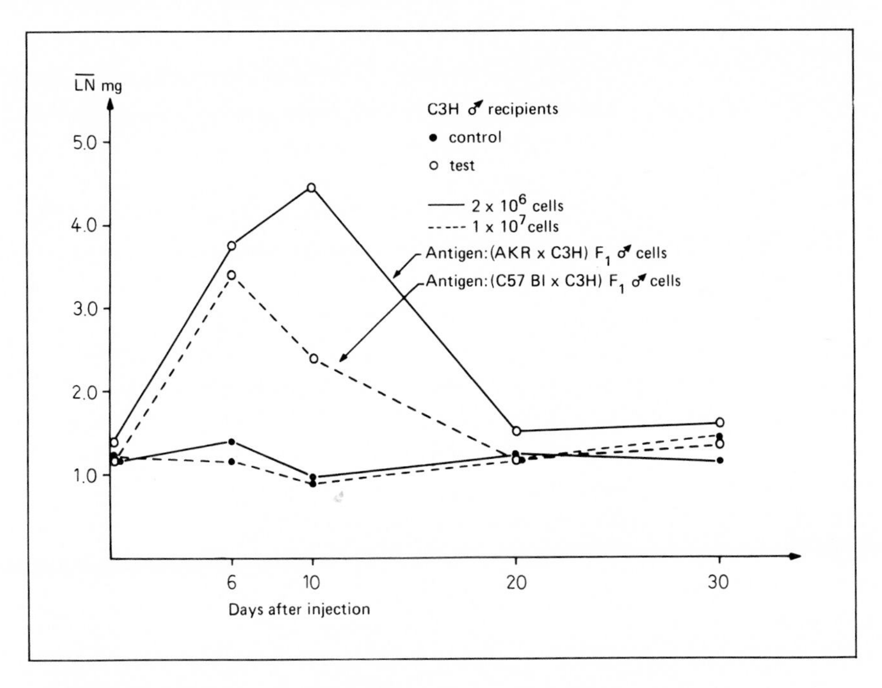

Fig. 2. The local lymphatic host response in C3H mice as weight increase of the draining popliteal lymph node following injection of hybrid cells into the hind footpad. Cells representing a weak, non-H-2, antigenic difference (AKR × C3H) and cells representing a strong, H-2, antigenic difference (C57Bl × C3H) are compared.

case of weak (non-H-2) antigenic differences between donor and recipient than in the case of strong (H-2) antigenic differences (fig. 2). This relationship of host lymphatic response to antigenic stimulation is in close agreement with the concept that the distinction between a weak and a strong antigen is that the recipient has fewer lymphoid cells with specific reactivity to the weak antigen than to the strong antigen. The introduction of the foreign graft results in an expansion of the reactive lymphoid cell population. But obviously it must take longer in the case of a weak antigenic stimulus to mature a sufficient number of reactive cells to destroy the graft than in the case of strong antigenic stimulus. Hence the difference in size and duration of the lymph node hyperplasia in the two situations. It seems justified to extrapolate to a situation of still weaker stimulation: the occurrence of an ultra-weak antigen which must give rise to a still more continuous, unceasing production of lymphoid cells, resulting in a tumorous development of the node.

In suggesting that tumours are formed as a reaction of the immune system I had carried out experiments to examine the extent to which the tissue of a transplantation tumour is of host origin and to what extent of donor origin [9, 10]. The experimental system consisted in transplanting spontaneous mammary tumours from C3H donors to recipients that were either F_1 hybrids of C3H and DBA mice or radiation chimaeras, i.e. C3H mice reconstituted following whole-body irradiation with DBA bone marrow. In this system, cell surface (H-2) antigens could be used as markers by exposing the tumour cells to specific antibodies produced by cross-immunizing the two mouse strains. The test was the cytotoxic antibody test using the dye exclusion criterion. By exposing cell suspensions prepared from tumours of the hybrid or chimaeric recipients to both anti-DBA and anti-C3H antibodies, it was found that almost all the cells recovered were of host type (by disrupting the tumour tissue into a cell suspension, about 40% of the cells were damaged, but the remaining 50–60% viable cells all reacted like host cells).

There is no convincing evidence of the existence in neoplasia of an autonomous cancer cell clearly distinguishable from normal cells. There is also no convincing evidence available to show that the infiltrating lymphocytes around tumours have a defensive function or that the close spatial relationship between tumour cells and infiltrating lymphoreticular cells has a toxic effect likely to retard tumour growth in the spontaneous tumour situation. On the basis of the data I have described, I therefore propose the hypothesis that tumours are formed by a particular type of inflammatory proliferation of the host organism. A particular mode of reaction of the immune system is released by ultra-weak antigenic determinants causing the reacting lymphoreticular cells to become transformed into tumour tissue.

It is a fundamental characteristic of the immune system of the animal organism that it is capable of a self-non-self discrimination. Foreign matter is rejected, self material is catabolized by enzymatic mechanism. Considering the immense variability of macromolecular patterns it is only reasonable to anticipate the occasional occurrence of ultra-weak antigens, only slightly changed tissue constituents which, on the one hand, stimulate the lymphoreticular system to a reaction – an inflammatory proliferation – but which, on the other hand, lack sufficient antigenicity to stimulate rejection and thereby a termination of the inflammatory process. Thus, for immunological reasons, the reaction becomes continuous and the inevitable consequence is the transformation into a tumour. It is suggested that this type of ultra-weak antigen be named *neoplasitropic determinant.*

The contributions of *N.K. Jerne* to immunological theory, including the suggestion of a need for autoreactivity both to histocompatibility antigens [3a] and to idiotypes [3b], have brought about a general acceptance of autoreactivity as part of the normal regulation mechanism. The hypothesis proposed in the present paper links the formation of cancer with the phenomenon of autoreactivity by suggesting the development of the malignant tumour to be a case of distorted immune autoreactivity.

References

1 Balner, H.; Dersjant, H.: J. natn. Cancer Inst. *23:* 513 (1966).
2 Burnet, F.M.: Immunological surveillance (Pergamon Press, Oxford 1970).
3 Heppner, G.H.; Wood, P.C.; Weiss, D.W.: Israel J. med. Scis *4:* 1195–1203 (1968).
3a Jerne, N.K.: Eur. J. Immunol. *1:* 1–9 (1971).
3b Jerne, N.K.: Annls Immunol. Inst. Pasteur *125C:* 373–389 (1974).
4 Klein, G.; Sjögren, H.O.; Klein, E.; Hellström, K.E.: Cancer Res. *20:* 1561–1572 (1960).
5 Larsen, F.S.; Spärck, J.V.: Acta pathol. microbiol. scand., C, Immunol. *86:* 131–139 (1978).
6 Prehn, R.T.: J. natn. Cancer Inst. *43:* 1215–1220 (1969).
7 Prehn, R.T.: Science, N.Y. *176:* 170–171 (1972).
8 Rygaard, J.; Poulsen, C.O.: Transplant. Rev. *28:* 3–15 (1976).
9 Spärck, J.V.: Immunity and host response in the growth of transplanted tumours (Munksgaard, Copenhagen 1962).
10 Spärck, J.V.: Acta path. microbiol. scand. *77:* 1–23 (1969).
11 Stutman, O.: Nature, Lond. *253:* 142–144 (1975).
12 Umiel, T.; Trainin, N.: Transplantation *18:* 244–250 (1974).

Dr. J.V. Spärck, Immunological Laboratory, Statens Seruminstitut,
Amager Boulevard 80, DK–2300 Copenhagen S (Denmark)

The Immune System, vol. 1, pp. 406–412 (Karger, Basel 1981)

Purification of Tumor Antigens Recognized by Activated Macrophages

Béla J. Takács[1], *Theophil Staehelin*[1], *Jörg Schmidt*[1]

Pharma Research Department, Hoffmann-La Roche, Basel, Switzerland

The cell surface plays a crucial role in a number of well-defined cellular functions, including the immune response. The cytoplasmic membrane is also involved in a variety of physiological properties which directly relate to neoplastic transformation such as the progressive growth of malignant cells in vivo, their escape from host immunological surveillance, and their metastasis to secondary sites. This organelle has attracted a great deal of interest because it is believed to provide the most useful avenue for combating malignant disease. Not surprisingly, the literature teems with reports on differences between 'normal' and malignant cells [for reviews, see 6, 23]. In fact, the problem no longer consists of finding differences but of discerning which of these differences are responsible for malignancy. The concerted effort of many of the world's scientists has not produced an unequivocal answer to this question. There is, however, a cell, the macrophage, which in a certain state of activation can distinguish between tumor and normal cells. Furthermore, it can do this independently of strain or even species origin. Therefore, we have attempted to use this property of macrophages in an attempt to purify the cell surface structure which they recognize on tumor targets. We employed glutaraldehyde-fixed or formalin-treated macrophages and detergent solubilized target cell membrane extracts to purify the target cell structure. In this contribution to a Festschrift in honor of *Niels Jerne* we report for the first time our findings in the hope that he will be able to fit macrophages into his network theory.

[1] *Béla J. Takács* was a member of the Basel Institute for Immunology from 1974 to 1977, *Theophil Staehelin* was a member of the Institute from 1970 to 1977, and *Jörg Schmidt* was a student at the Institute from 1974 to 1977.

Results

Cultured tumor cell lines, which are inhibited or killed by activated macrophages, provide a continuous and highly uniform source of antigen. Since the molecules that are recognized by cytolytic macrophages are most likely to be intrinsic membrane glycoproteins, separation of the cytoplasmic membrane was chosen as an initial purification step. However, since membrane proteins are associated with the lipid bilayer and with each other by strong forces, they must first be extracted from the membrane matrix in a soluble form before they can be purified. This solubilization step is best achieved by the use of agents that do not grossly alter the proteins' tertiary structure. The relatively mild, neutral or nonionic detergents fulfill this requirement. Nonidet P40 or Triton X-100 are relatively mild, yet highly effective solubilizing agents for membrane proteins [14]. At the concentration used, 0.5%, they are thought to react mainly with the hydrophobic, lipophilic portions of proteins [8], without interfering with the hydrophilic portions. Furthermore, at this concentration, they do not solubilize nuclear membranes, an important consideration when extraction of intact cells is undertaken. Since these detergents do not interfere with antigen antibody interactions [4], we hoped that they would also not destroy the structural integrity of the cell surface tumor antigens that activated macrophages recognize on the intact cell.

Intrinsic membrane proteins require the continuous presence of lipid or neutral detergent for solubility. Their removal results in protein aggregation which might be irreversible [14]. Therefore, prior to the incubation of solubilized target cell antigens with activated macrophages, it is necessary to stabilize the macrophage membrane. For this purpose we used glutaraldehyde fixation or formalin treatment under controlled conditions. The use of these agents for immobilization of either antigens or antibodies for the purification of the respective antibody or antigen is well documented [21]; the binding of cytotoxic lymphocytes to formalin-fixed target cells has also been reported [16].

Initially we used either mouse or guinea pig peritoneal macrophages that were activated by the injection of living *Mycobacterium bovis,* strain BCG, according to published procedures [9]. Later we obtained several mouse macrophage cell lines from Prof. *S. Fazekas de St.Groth* of the Basel Institute for Immunology. These cell lines are apparently immortalized in an activated state as judged by their ability to kill tumor cells. However, it was found that on prolonged culturing, these macrophage cell lines changed morphologically and functionally. The cells became round, and they adhered to plastic only weakly or they actually floated. Although these cells were still phagocytic, as assessed by the ingestion of fixed yeast cells, their cytotoxicity against tumor cell targets was diminished and their capacity to bind cell surface antigens from detergent solubilized tumor cell extracts was lost.

The concentration of the fixative used proved to be critical, not only in preventing the dissolution of the macrophage membrane by detergents, but also in influencing their ability to adsorb components from detergent lysed tumor cell extracts. About 0.1–0.2% glutaraldehyde or 0.5–1% formaldehyde were found to be equally effective. However, at higher concentration, 5%, the formalinized macrophages completely lose the ability to bind target cell structures.

For some of the experiments, macrophages were grown on the surface of microcarrier beads (Pharmacia), fixed on the surface of these beads, and employed for adsorption in a column. In these experiments detergent extracts of biosynthetically labelled tumor cell targets were passed through the column and recirculated 2–3 times. Unspecific material was then removed by washing the column with lysis buffer followed by lysis buffer containing high salt (0.5 and 1.0 *M* NaCl). Each washing step was continued until the level of radioactivity in the effluent reached background levels. Specifically retarded material was then eluted by low pH and SDS-containing buffers. Fractions with peak radioactive counts were pooled and analyzed by SDS-PAGE as described previously [18]. Autoradiography of the dried gel revealed that macrophages, fixed either in suspension or on the surface of microcarrier beads, adsorbed two polypeptides from detergent extracts of all tumor cell lines tested. The major of these polypeptides, with an apparent molecular size of 100,000 dalton (fig. 1), is present on the surface of tumor cell targets because it can also be labelled by lactoperoxidase catalyzed iodination (fig. 2). The smaller polypeptide, with an apparent molecular size of 60,000 dalton, is either not a cell surface component or not available to iodination, since it could only be labelled biosynthetically. Both of these polypeptides could be adsorbed to immobilized lentil lectin columns and subsequently eluted with the competitive sugar, Me-α-*D*-mannopyranoside, indicating that they are glycosylated. After low pH elution from macrophages, these components could still be adsorbed to immobilized lentil lectin columns. However, after sugar elution from lentil lectin they could no longer be adsorbed to macrophages, suggesting the involvement of carbohydrate residues in the recognition and binding step.

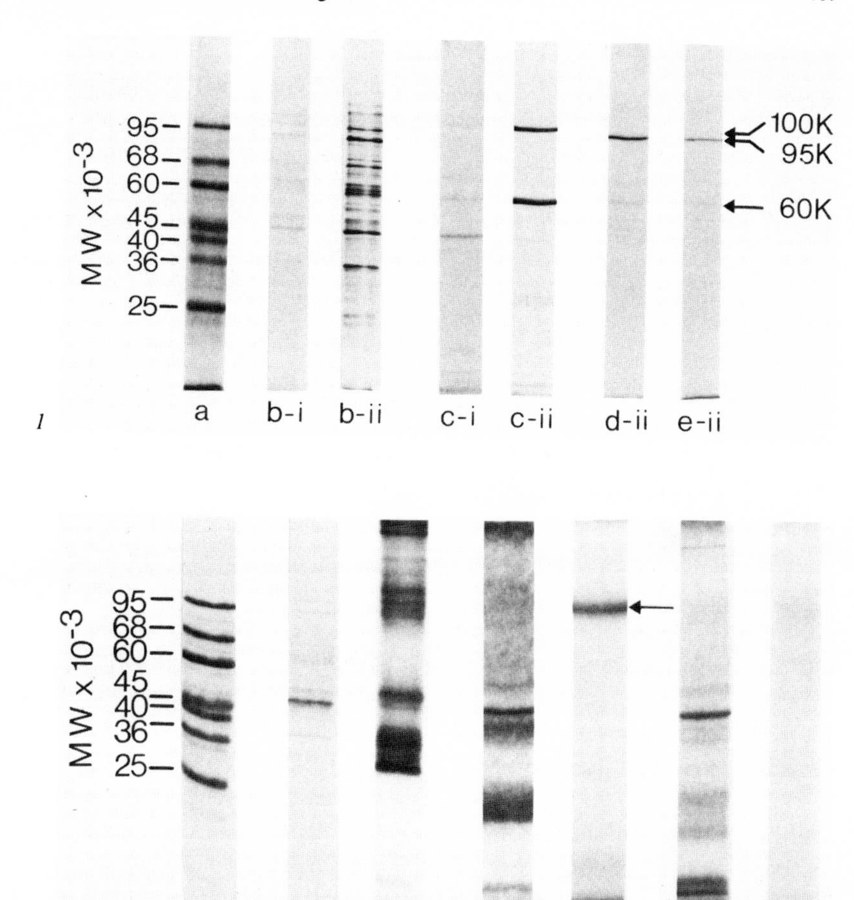

Fig. 1. SDS-PAGE analysis of ^{35}S-methionine-labelled tumor antigens eluted from activated macrophages. *a* Molecular weight standards; *b* NP-40-soluble HeLa extract; *c* NP-40-soluble HeLa extract obtained from glutaraldehyde-fixed macrophages by low pH elution; *d* material retained on BCG-activated, glutaraldehyde-fixed macrophages from HeLa supernatants; *e* material retained on thioglycollate-activated, glutaraldehyde-fixed macrophages from HeLa supernatants. *i* and *ii* represent staining patterns and autoradiographic patterns, respectively, of the same gel track.

Fig. 2. SDS-PAGE analysis of NP-40-soluble extracts from surface-iodinated HeLa and normal peripheral blood lymphocytes (NPBL). *a* Molecular weight standards; *b* NP-40-soluble HeLa extract; *c* NP-40-soluble HeLa extract eluted from glutaraldehyde-fixed macrophages by low pH; *d* NP-40-soluble NPBL extract obtained as indicated in *c*. *i* and *ii* represent staining patterns, and autoradiographic patterns, respectively, of the same gel track.

Incubation of spent culture medium from tumor cell lines with activated macrophages and subsequent elution and analysis of bound material also revealed two glycosylated polypeptides with apparent molecular sizes of 95,000 and 60,000 dalton (fig. 1d, e). We do not yet know whether the 95,000-dalton polypeptide originates from the 100,000-dalton polypeptide as a result of normal turnover of this cell surface protein. However, the smaller molecular size of the shedded component is consistent with previous reports on the enzymatic cleavage and release of certain cell surface molecules [13].

Discussion

During the past decade, considerable information has accumulated concerning alterations in antigens on the surface of transformed cells [for reviews see 6, 23]. There is overwhelming evidence that malignancy is associated with proteins or glycoproteins having molecular sizes of around 100,000 dalton. Using immunoprecipitation with an antitumor serum, *Bauer* et al. [1] identified a tumor-specific surface antigen (TSSA) on the surface of avian tumor virus transformed cells as a 100,000-dalton glycoprotein. The appearance of a similar antigen is also associated with feline leukemia and sarcoma virus infection [5, 15]. The feline oncornavirus-associated cell membrane antigen (FOCMA) is thought to be synthesized as an 85,000- to 100,000-dalton phosphorylated precursor. A glycoprotein with an apparent molecular size of 100,000 dalton was also found on the surface of chick cells transformed by Rous sarcoma virus [12]. A protein with a similar molecular size is associated with tumor rejection of Balb/c 3T3 cells in mice [20]. A cell surface protein of 92,000 dalton size has also been identified on Moloney leukemia virus-transformed cells [19]. Recently *Bramwell and Harris* [3] have undertaken an approach which selects directly for malignancy. They produced non-malignant hybrids between malignant and normal cells, then they subjected these to back-selection for malignancy by injection into syngeneic animals. They were able to identify a malignancy linked acidic glycoprotein with an isoelectric point of about 4.0 and a molecular size of 100,000 dalton. *Omary* et al. [10], using monoclonal antibodies, have recently characterized a similar or identical glycoprotein. *Lerner* et al. [7] have also extracted a cell surface component (MW range 100,000 dalton) from the human breast tumor cell line, BOT-2, which reacted with sera from mammary cancer patients. *Sutherland* et al. [17] were recently able to purify an acute lymphoblastic leukemia (ALL) associated antigen from established cell lines. The antigen, which consists of a single glycosylated polypeptide (MW 100,000 dalton), was found to react specifically with anti-ALL antibodies. *Billing* et al. [2] were also able to demonstrate, using immunoprecipitation and SDS-PAGE, that the ALL antigen consists of a 98,000 dalton polypeptide. *Ritz* et al. [11] recently reported the generation of a monoclonal antibody which is specific for a common ALL antigen and which precipitates a single glycoprotein (MW 95,000) from the cell membranes of ALL patients. *Woodbury* et al. [22] were also able to produce a monoclonal antibody which binds to melanoma cells but not to autologous fibroblasts. The antibody recognized a protein (MW 97,000) which was present on melanoma cell lines, on biopsy material from melanomas, and on the surface of certain other tumors but not on samples obtained from normal donors.

From the foregoing it is evident that proteins with an apparent molecular size of about 100,000 dalton are linked to malignancy. It was, therefore, gratifying to find that one of the components which we purified, using activated macrophages, has the same apparent molecular size.

The serologic detection of tumor antigens has played a central role in cancer research for many years. Since a number of heteroantisera and monoclonal antibodies to tumor antigens, and now also activated macrophages, apparently recognize a similar glycosylated polypeptide (MW 100,000), it is intriguing to speculate about the existence of a structure common to all tumor targets. Activated macrophages might recognize determinants common to all tumor cells irrespective of strain or species origin whereas heteroantisera and monoclonal antibodies might react with variable parts of the same molecule. The existence of such structures on the surface of neoplastic cells is an intriguing but unproved hypothesis. Purification and comparative characterization of these tumor-associated antigens is urgently needed. Such studies should provide the basis for a better understanding of the mechanism of tumor rejection and of the involvement of tumor-associated antigens in transformation, and they may be of major importance for successful immunotherapy of neoplastic disease.

References

1 Bauer, H.; Kurth, R.; Rohrschneider, L.; Pauli, G.; Friis, R.R.; Gelderblom, H.: Cold Spring Harb. Symp. quant. Biol. *39:* 1181–1185 (1974).

2 Billing, R.; Minowada, J.; Cline, M.; Clark, B.; Lee, K.: J. natn. Cancer Inst. *61:* 423–429 (1978).

3 Bramwell, M.E.; Harris, H.: Proc. R. Soc. Lond. B *201:* 87–106 (1978).
4 Crumpton, M.J.; Parkhouse, R.M.E.: FEBS Lett. *22:* 210–212 (1972).
5 Hardy, W.D.; Zuckerman, E.E.; MacEwen, E.G.; Hayes, A.A.; Essex, M.: Nature, Lond. *270:* 249–251 (1977).
6 Haynes, R.O.: Biochim. biophys. Acta Rev. Cancer *458:* 73–107 (1976).
7 Lerner, M.P.; Hill Anglin, J.; Nordquist, R.E.: J. natn. Cancer Inst. *60:* 39–44 (1978).
8 Lerner, R.A.; McConahey, P.J.; Dixon, F.J.: Science *173:* 60–62 (1971).
9 Meltzer, M.S.; Tucker, R.W.; Sanford, K.K.; Leonard, E.J.: J. natn. Cancer Inst. *54:* 1177–1184 (1975).
10 Omary, M.B.; Trowbridge, I.S.; Minowada, J.: Nature, Lond. *286:* 888–891 (1980).
11 Ritz, J.; Pesando, J.M.; Notis-McConarty, J.; Lazurus, H.; Schlossman, S.F.: Nature, Lond. *283:* 583–585 (1980).
12 Rohrschneider, L.R.; Kurth, R.; Bauer, H.: Virology *66:* 481–491 (1975).
13 Sanderson, A.R.: Nature, Lond. *220:* 192–195 (1968).
14 Steck, T.L.; Fox, C.F.: Membrane molecular biology, pp. 27–75 (Sinauer Associates, Stanford 1973).
15 Stephenson, J.R.; Essex, M.; Hino, S.; Hardy, W.D.; Aronson, S.A.: Proc. natn. Acad. Sci. USA *74:* 1219–1223 (1977).
16 Stulting, R.D.; Todd, R.F., III; Amos, D.B.: Cell. Immunol. *20:* 54–63 (1975).
17 Sutherland, R.; Smart, J.; Niaudet, P.; Greaves, M.: Leukemia Res. *2:* 115–126 (1978).
18 Takács, B.: in Lefkovits, Pernis, Immunological methods, pp. 81–105 (Academic Press, New York 1979).
19 Troy, F.A.; Fenyö, E.M.; Klein, G.: Proc. natn. Acad. Sci. USA *74:* 5270–5274 (1977).
20 Van Nest, G.A.; Grimes, W.J.: Biochemistry, N.Y. *16:* 2902–2908 (1977).
21 Weetall, H.H.: in Hair, The chemistry of biosurfaces, vol. 2, pp. 597–631 (Dekker, New York 1972).
22 Woodbury, R.G.; Brown, J.P.; Yeh, M.-Y.; Hellstroem, J.; Hellstroem, K.E.: Proc. natn. Acad. Sci. USA *77:* 2183–2187 (1980).
23 Yamada, K.M.; Ponyssegur, J.: Biochimie *60:* 1221–1233 (1978).

Dr. B.J. Takács, FP/II, Bau 69/202, F. Hoffmann-La Roche & Co. A.G., Grenzacherstrasse, CH–4005 Basel (Switzerland)

Parasitology and Infectious Disease

The Immune System, vol. 1, pp. 413–422 (Karger, Basel 1981)

Immunology in the 1880s: Two Early Theories

Jean Lindenmann [1]

Institute for Immunology and Virology, University of Zürich, Zürich, Switzerland

If a date were to be put on the beginnings of immunology as a concern both theoretical and experimental, 1880 would be a defensible choice. My excursion into the thoughts prevailing 100 years ago will be bracketed by two remarks made by *Louis Pasteur* in his first publication on chicken cholera. I trust that both will please *Niels Jerne,* for whose entertainment the following pages have been written.

Pasteur's first remark deals with the surprising fact that certain diseases occur only once: *'Quelle étrange circonstance! C'est à peine si l'imagination trouve à hasarder de ce fait une explication hypothétique ayant une base expérimentale quelconque'* [17, p. 240].

Immunity following smallpox, variolation and vaccination was familiar to observers during the entire 19th century. Since a widely held theory considered the variolous matter some sort of poisonous refuse that had to be violently evacuated through the pustules, the ensuing immunity posed no particular problems. It would take a long time for the poison, once eliminated, to accumulate again in sufficient quantity. The fact that the eruption of variolation seemed to protect better, although at higher risk, than the much milder disease induced by vaccination, lent support to these views. Both inoculators and vaccinators sought to improve the cleansing capacity of their procedure by performing deep or multiple incisions, thus allowing the disease to run a more severe, hence more beneficial course.

The theoretical problems surrounding immunity became acute with the advent of experimental immunization procedures. That a 'virulent disease'

[1] *Jean Lindenmann* was a member of the Board of Consultants of the Basel Institute for Immunology from 1970 to 1973.

whose cause was bacterial lent itself to the elaboration of a 'vaccine' (a term clearly borrowed from the smallpox field) was at first quite surprising. Thus *Pasteur* wrote: '*... il existe des états variables de virulence dans le choléra des poules: étrange résultat assurément, quand on songe que le virus de cette affection est un organisme microscopique qu'on peut manier à l'état de pureté parfaite ...*' [13, p. 674]. Something of this surprise still transpires 2 years later, in *Pasteur's* lecture before the International Congress of Hygiene in Geneva: '*Un virus, alors même qu'il est constitué par un microbe, peut ... être atténué dans sa virulence ...*' [18, p. 127]. This 'alors même' is quite revealing. What was accepted, although unexplained behavior from the part of a 'virus', was surprising in a bacterium. Surprising, but at the same time stimulating, for now immunity could be thought of in terms of the newly acquired expertise in bacteriology.

One observation which seemed to shed light on immunity had to do with the natural resistance exhibited by certain animal species or breeds towards infectious agents that were highly lethal to others. *Pasteur* stumbled upon a first striking analogy: A yeast broth which was a perfectly good medium for many bacteria, including anthrax, did not support growth of the chicken cholera organism *(Pasteurella multocida)*. '*Chose étrange, ce milieu de culture est tout à fait impropre à la vie du microbe du choléra des poules ... N'est-ce pas l'image de ce qu'on observe quand un organisme microscopique se montre inoffensif pour une espèce animale à laquelle on l'inocule?*' [17, p. 242].

He then went on to his much vaunted and anecdotally embroidered observations on chickens vaccinated with an attenuated chicken cholera bacterium. The first inoculation had caused marked necrosis at the point of injection, but later inoculations remained without effect: '*... le muscle qui a été très malade est devenu, même après guérison et réparation, en quelque sorte inpuissant à cultiver le microbe, comme si ce dernier, par une culture antérieure, avait supprimé dans le muscle quelque principe que la vie n'y amène pas et dont l'absence empêche le développement du petit organisme. Nul doute que cette explication ... ne devienne générale, applicable à toutes les maladies virulentes*' [17, p. 247–248]. By this procedure '*les animaux sont amenés aux conditions de ceux des espèces qui ne contractent jamais le choléra des poules*' [15, p. 954].

Back to his test tubes, *Pasteur* allowed chicken cholera bacteria to grow in an appropriate broth for 4 days, then filtered the broth and attempted to reinoculate it with the same microorganism: The broth remained limpid, it had become immune, so to speak. Hence his conclusion: '*Comment ne pas être porté à croire que par la culture dans la poule du virus atténué on place le corps de celle-ci dans l'état de ce liquide filtré qui ne peut plus cultiver le microbe?'* [15, p. 957].

Of course *Pasteur's* view was broad enough to embrace another possibility, the elaboration of some noxious material by the bacteria during their growth. He even did an experiment to settle the matter: he evaporated (in the cold, under vacuum) his exhausted broth, replenished it with fresh broth, and inoculated it: the bacteria grew abundantly. With this he had proved to his satisfaction that the phenomenon he had observed in vitro was indeed caused by exhaustion of some essential nutrient, and not by formation of a growth inhibitor. By analogy he thought that immunity in vivo could be similarly explained.

It was impossible to do in vivo an exact parallel to *Pasteur's* in vitro experiment: to inoculate a chicken, evaporate it and inject it wholesale into a fresh bird. Strangely, the now so familiar idea of passive transfer with blood, serum or some tissue extract had to wait for several more years before it was attempted. But *Pasteur* nevertheless did an experiment of extraordinary foresight which demonstrated the existence of bacterial toxins 8 years ahead of *Roux:* upon injection of his filtered and concentrated broth chickens would show symptoms of the disease, particularly a pronounced somnolence, but would rapidly recover without ensuing immunity.

Pasteur stated his first immunological hypothesis in the following words: *'J'ai envisagé l'organisme comme un milieu de culture qui, par une première atteinte du mal, perdrait, sous l'influence de la culture du parasite, des principes que la vie n'y ramènerait pas, ou n'y ramènerait qu'après un certain temps. Bonne ou mauvaise, cette explication satisfait l'esprit présentement, parce qu'elle rend compte des premiers faits acquis. Tant qu'on lui trouvera cette vertu, il sera sage de chercher des vérifications expérimentales aux déductions qu'elle suggère'* [14, p. 315].

One of the features of *Pasteur's* analogy which fitted well with accepted facts was the specificity of both immunity in vivo and nutrient exhaustion in vitro: the broth which failed to support a second growth of the chicken cholera organism still allowed development of the anthrax germ. However, *Pasteur* was keenly aware that such negative evidence was of little value for supporting a new hypothesis: '... *les résultats négatifs ne prouvent rien* ...' [20, p. 90]. Upon closer examination he convinced himself that, in the broth exhausted by the chicken cholera bacterium, the anthrax germ was distinctly inhibited. If his theory were correct, then the positive prediction could be made that a chicken vaccinated against chicken cholera would resist challenge with anthrax. This is exactly what he thought he found, and immediately jotted down in a short, admittedly tentative note [14]. *Pasteur's* hens were naturally resistant to anthrax, and in order to do his experiment he had to break their resistance, probably by immersing the wretched birds in cold water; certainly not an elegant experiment, and it was never heard of again. Nor do we find traces of the high hopes for a sort of universal vaccine capable of preventing and curing all maladies which *Pasteur's* fertile mind had entertained. *'Si ce résultat se confirme, et principalement s'il se généralise pour d'autres maladies virulentes, on pourra en espérer les conséquences thérapeutiques les plus importantes, en ce qui concerne même la pathologie des maladies virulentes propres à l'espèce humaine'* [14, p. 315]. An interesting glimpse into *Pasteur's* thinking; he not only generalized immediately to human maladies, but contemplated therapeutic applications of measures that were conceptually prophylactic.

The first objection to the universality of *Pasteur's* theory of immunity came from *Chauveau.* He again argued with the postulated similarity between natural resistance and acquired immunity. He had observed that Algerian sheep were highly resistant to the inoculation of anthrax. However, this resistance could be overcome when the challenge dose was increased. If resistance was due, as *Pasteur* claimed, to the lack of some essential nutrient, would it not follow that the more bacteria you inoculated, the less they would be able to grow? *'Les faits que je viens de faire connaître démontrent, en effet, que la bactéridie charbonneuse se comporte, dans l'organisme des moutons algériens, non pas comme s'il était privé de principes nécessaires à la vie bactéridienne, mais bien plutôt comme si c'était un milieu rendu impropre à cette dernière par la présence de substances nuisibles. En très petit nombre, les bactéridies sont arrêtées dans leur développement par l'influence inhibitoire de ces substances. Très nombreuses, au contraire, elles peuvent surmonter bien plus facilement cet obstacle à leur prolifération'* [3, p. 1530].

In his next communication, *Chauveau* extended his observations to acquired immunity. The Algerian sheep, although they usually survived a first injection of anthrax, nevertheless became visibly sick. A later injection of similar strength produced fewer symptoms. What was more, lambs born of mothers whose natural resistance had been fortified by challenge also showed increased levels of immunity. Since it was known that anthrax bacilli never passed into the fetus, *Chauveau* concluded that direct contact between microorganism and host was not necessary for immunity to be produced. As to the mechanism involved, he suggested that: *'les inocula-*

tions préventives agissent sur les humeurs proprement dites, rendues stériles et stérilisantes, soit par soustraction de substances nécessaires à la prolifération bactéridienne, soit plutôt par addition de matières nuisibles à cette prolifération' [2, p. 151].

It is gratifying to see that, faced with similar problems, the human mind always seems to reach identical conclusions. Compare the following statements scattered over 100 years and dealing with vastly different aspects of resistance to infectious agents: '... *la vie du microbe, au lieu d'enlever ou de détruire certaines matières (...) en ajoute, au contraire, qui seraient pour ce microbe un obstacle à un développement ultérieur'* [15, p. 957]; '... *the virus first upon the scene uses up some essential foodstuff in the cell. An alternative ... hypothesis ... would be, of course, the generation ... of some ... inhibitory substance'* [1,p. 219]. *'There may be a factor ... which inhibits virus replication .... Alternatively, there may be a deficiency ... of a nutrient ... essential for the replication of the virus'* [12, p. 144].

Although *Chauveau* had been careful to pay tribute to the ingeniosity of *Pasteur's* hypothesis, which he had called *'cette séduisante théorie, basée sur une des plus intéressantes séries de ces expériences nettes et décisives dont M. Pasteur est coutumier'* [3, p. 1530], *Pasteur* must have been irked by the little word 'plutôt' in *Chauveau's* conclusions quoted above. *'Pas n'est besoin, comme le pense M. Chauveau, d'invoquer l'existence de matières nuisibles à la vie de la bactéridie. ... le fait de la non-récidive de l'affection charbonneuse ... s'explique ... par la stérilité qu'amènent plus ou moins à leur suite dans le même milieu une ou plusieurs cultures successives d'un organisme microscopique. ... Je n'abandonnerai pas facilement cette théorie ... ; elle repose sur des observations qui lui sont pour ainsi dire adéquates, et elle satisfait l'esprit dans une question qui défiait jusqu'à l'hypothèse'* [19, p. 537].

To this, *Chauveau* replied that he had no intention of opposing another theory to *Pasteur's,* since this would have been premature, but that he had simply drawn attention to a fact which was not easily reconciled with the exhaustion hypothesis. He thought that: *'M. Pasteur n'a pas bien compris ma pensée et mes intentions'* [4, p. 651].

Chauveau's thoughts and intentions were more clearly explained in a paper published 8 years later, but based largely on older work. In this he said: '... *un microbe infectieux, s'étant une première fois multiplié dans son milieu naturel de culture, et l'ayant ainsi rendu rebelle à une culture ultérieure, exerce cette action, non pas en épuisant le terrain, ... mais bien en y laissant des substances nuisibles ...'* [5, p. 67]. As evidence in favor of his interpretation, he reemphasized his experiments on lambs born of immune mothers: these were immune, yet had not been in direct contact with anthrax bacteria. The proof that this must have been due to passage, from the mother to the fetus, of soluble substances noxious to bacterial development, left nothing to be desired [5, p. 68]. From the entire context it is absolutely clear that he never for a moment considered the possibility that these noxious substances were generated by the macroorganism – they were the product of the microorganism, ptomaines that stopped bacterial multiplication just as alcohol stops fermentation.

In a curious paper from the same year, 1888, but also largely based on much older data, *Roux and Chamberland* pointed out that, if something could pass from the mother to the fetus, the reverse was equally true, so that it was still entirely conceivable that, by diffusion, an essential constituent of the lamb's economy was depleted by bacterial growth, even if this was limited to the mother: *'On peut donc tirer, de l'expérience sur les agneaux, un argument en faveur de la théorie de l'épuisement aussi bien qu'en faveur de la théorie de la substance ajoutée ...'* [22, p. 422].

But at that time the exhaustion hypothesis had already been abandoned by its chief supporter, *Pasteur.* The explanation he now championed was, in fact, *Chauveau's,* the one he had himself contemplated and dismissed in his chicken cholera work. In 1886 an English Committee of Inquiry had investigated *Pasteur's* rabies treatment, and this is how *Horsley,* the Committee's secretary, summarized the theory: *'M. Pasteur believes that the virus of rabies is a living microorganism, and that, like some others, it produces in the tissues it invades an excretory substance, by which, when present in sufficient quantity, its own development and increase are checked, as are those of the yeast ferment by the alcohol produced in the vinous fermentation'* [10, p. 236].

The 'antidote' present in infected spinal cords was supposed to withstand desiccation somewhat better than the infectious principle; hence the series of inoculations, beginning with cords desiccated for 14 days and progressing to cords kept for only 1 day, would ensure that increasing doses of infectious virus would always be met by even larger doses of the antidote.

The new theory led to a flurry of experiments. The most remarkable is possibly the report of *Roux and Chamberland* on immunity to gas gangrene caused by the 'vibrion septique' *(Clostridium septicum).* The idea was to introduce into the body of guinea pigs sufficient amounts of the culture filtrate to induce immunity. It must be kept in mind that *Roux and Chamberland* did not intend to 'immunize' in the sense we use now; rather, they were thinking of something like chemoprophylaxis: *'Ne serait-il pas possible d'accumuler ces produits'* (namely products formed during bacterial growth and acting as antiseptics) *'dans le corps d'un cobaye et de rendre le milieu animal stérile comme le milieu inerte, en un mot de conférer ainsi au cobaye l'immunité pour la septicémie?'* [21, p. 565].

The doses injected were huge indeed: guinea pigs weighing 400 g received as much as 120 ml of heated culture fluid in 3 portions intraperitoneally over a few days. Incidentally, this is much more than the 35 ml of diphtheria filtrate that *Roux and Yersin* were to inject a year later, a quantity *de Kruif* thought was verging on the insane [7, p. 175]. The guinea pigs became immune – they even remained immune for at least 30 days. To the present day observer it would seem that only the most tenuous of veils was still hiding the 'truth' from *Roux's* keen eye – but the step from a consideration of the parasite alone to that of an interaction between parasite and host was immense.

Another example of direct filiation from the auto-inhibitor hypothesis was the realization that, on media in which certain microorganisms had grown, certain others were severely inhibited. This allows me to add a touch of local color to my collage. On May 21, 1887, *Garrè* read a paper, enlivened by demonstrations, before the Swiss Central Medical Society in Basle, under the heading 'Über Antagonisten unter den Bacterien'. In it he proposed to study immunity in vitro, and he expected deeper insights: *'wenn es gelingt, das gewöhnliche Nährsubstrat der Bacterien ... durch Mikroorganismen immun für eine weitere Implantation zu gestalten'* [9, p. 386]. And he ended on an optimistic note: *'Gewinnt nicht von diesem Standpunkte aus unsere Vorstellung vom Mechanismus der Immunität grössere Klarheit?'* [4, p. 392]. Indeed it gained, and it is still gaining on the very soil where these words were spoken.

The two early theories of immunity I have just sketched, which we may call the 'exhaustion theory' and the 'auto-inhibitor theory' for short, had some practical consequences. Thus, as long as *Pasteur* believed in the exhaustion theory, he would not accept the idea that a non-living germ could act as a vaccine. How should a vaccine use up some essential nutrient if it were not alive? Therefore, *Pasteur* was highly sceptical of the results reported by *Toussaint* [23, p. 303], who had induced immunity to anthrax with defibrinated anthrax blood heated for 10 min at 55 °C, which he thought sufficient to inactivate all bacteria. *Pasteur* and his collaborators soon showed that the procedure was not as reproducible as *Toussaint* had hoped: sometimes enough bacteria survived to kill the animal, and when animals did become immune, this must be due to growth of heat-'attenuated' germs [16, p. 457].

Chauveau, for theoretically less stringent reasons, was also not convinced of *Toussaint's* claims. His argument was that the doses used, 3 ml twice for sheep, were much too small to introduce enough of the supposed auto-inhibitor. Nevertheless, he seems obscurely to have felt that *Toussaint* was groping towards a revelation that escaped both him and *Pasteur,* for he wrote in 1888: *'L'idée de Toussaint n'en était pas moins légitime au fond; mais elle était en avance sur les faits et sur l'époque'* [5, p. 73]. *Roux and Chamberland,* who had helped *Pasteur* in demoting *Toussaint's* claims, belatedly reported that heat-killed anthrax bacteria did confer immunity [22]. This is perhaps to be seen as an oblique admission that their first judgement in this matter had been too harsh. There is little doubt that *Toussaint,* who became incapacited by disease at an early age, was a star of first magnitude within the dazzling galaxy of early French microbiologists [6].

The 'auto-inhibitor theory', which was endorsed by *Pasteur* in his rabies work, also had immediate consequences. The immunotherapy of patients bitten by rabid animals was a race against time. The original schedule used by *Pasteur* required 14 days to be completed. But in the auto-inhibitor hypothesis time did not enter as a factor. It seemed therefore possible to accelerate injections in order to reach high enough levels of the inhibitor before symptoms set in. This led to the so-called 'intensive treatment', in which several injections of increasing virulence were done on the same day, a practice later abandoned after cases of vaccine-induced rabies could no longer be ignored. According to our present views the merit of *Pasteur's* first scheme was to allow sufficient time for the host to build up immunity; in the light of the auto-inhibitor hypothesis, this time was merely lost.

Here we have two examples of the dangers of preconceived ideas. Dogmatic rigidity which refuses to acknowledge irreconcilable facts and risky extrapolations into uncharted terrain. But consider, on the other hand, the almost inexhaustible fertility of both theories: pursued with vigor, the exhaustion hypothesis had to lead straight into the study of bacterial growth factors, of bacterial metabolism and eventually to the entire field of bacterial physiology. The auto-inhibitor hypothesis, as we have seen, brought to light the existence of toxins and of bacterial antagonism. These in turn initiated the quantum jumps of antitoxins (and thus signalled the demise of the auto-inhibitor theory) and of antibiotics. *Duclaux,* the second director of the Pasteur Institute, wrote after *Pasteur's* death: *'Ce qui fait le mérite d'une théorie nouvelle, ce n'est pas d'être vraie: il n'y a pas de théories vraies; c'est d'être féconde'* [8, p. 643].

This is exactly what so utterly escaped *Löffler,* who at *Koch's* instigation subjected *Pasteur's, Chauveau's* and *Toussaint's* findings to malevolent scrutiny [11]. To the French, this whole work, more than 50 large format pages of solid text, an extraordinary mixture of review, tentative or irrelevant experiments, self-righteous criticism and autopsy reports, must have smacked of bad faith. *Löffler* indeed found more flies than ointment in *Pasteur's* every paragraph. Yet the paper gives abundant evidence that he had read the French contributions most thoroughly, and had understood every word. He quoted some passages verbatim, and he correctly interpreted certain ambiguous statements of *Pasteur* as representing idealized experiments of the mind rather than actual protocols. But there is not the shadow of a hypothesis, fertile or otherwise, of his own, and not even a criticism of *Pasteur's* and *Chauveau's* ideas. The only sentence which shows that *Löffler* was aware of the French theories, but dismissed them as irrelevant, comes at the very end: *'Weder die Gegengifttheorie noch auch die Erschöpfungstheorie haben einen Fortschritt für die Immunitäts- und speciell für die Impf-Frage gebracht'* [11, p. 186].

Much has been written about the antagonisms of the French and German schools of microbiology in the late 19th century. At the time with which I am concerned here, the Germans probably had a slight technical lead. They were better equipped for the identification of infectious agents, in which they succeeded brilliantly. As for immunology, it is difficult to deny the French our admiration. Both groups must have felt envious of each other's achievements. They were itching, without admitting as much, to beat their adversary at his own game. In both cases this led to disaster: to young *Thuillier's* untimely death during the hapless French cholera expedition, and to *Koch's* tuberculin fiasco.

Science is, who ever doubted it, a regulatory network. Novel elements cannot be seen as positive additions without consideration of negative feedbacks. Thus, *Toussaint's* premature foray into killed vaccines elicited counter-measures by *Pasteur's* disciples, which blocked progress in this direction for several years. Sociological and technical factors entered into play: had *Toussaint* belonged to a more prestigious school, had his health been better, or perhaps simply his thermometer more precise, the resulting pull on the scientific network would have been different, and the principle of killed vaccines might have become accepted a couple of years earlier. But the network structure of scientific exchanges ensures that stresses are rapidly distributed. Immunology today would not be much different had *Toussaint* never existed, nor even *Pasteur* or *Roux* – but science as an individual adventure of the mind would be much the poorer for it.

Before closing this historical parenthesis, I wish to draw attention to two key words no information retrieval system will ever think worth indexing, and yet without which there is no good science. They are the words *'étrange'* in my first, second and fourth quotation, and *'imagination'* in my first and in my last, with which I now close my 'hommage à *Niels Jerne*': *'Les combinaisons de la nature sont à la fois plus simples et plus variées que celles de notre imagination'* [17, p. 243].

References

1 Andrewes, C.H.: Br. J. exp. Path. *23:* 214–220 (1942).
2 Chauveau, A.: C.r. hebd. Séanc. Acad. Sci., Paris *91:* 148–151 (1880).
3 Chauveau, A.: C.r. hebd. Séanc. Acad. Sci., Paris *90:* 1526–1530 (1880).
4 Chauveau, A.: C.r. hebd. Séanc. Acad. Sci., Paris *91:* 648–651 (1880).
5 Chauveau, A.: Annls Inst. Pasteur, Paris *2:* 66–74 (1888).
6 Decourt, P.: Archs int. Claude Bernard *5:* 165–184 (1974).
7 De Kruif, P.: Microbe hunters (Harbrace Paperback Library, Harcourt, Brace & World Inc., New York 1926).
8 Duclaux, E.: Revue scient., Paris *56:* 641–645 (1895).
9 Garrè, C.: Correspondenzbl. schweiz. Ärzte *17:* 385–392 (1887).
10 Horsley, V.: Nature, Lond. *36:* 232–235 (1887).
11 Löffler, F.: Mitt. Kaiserl. Gesundheitsamtes Band *1:* 134–187 (1881).
12 Murray, M.J.; Murray, A.B.; Murray, M.B.; Murray, C.J.: Lancet *ii:* 143–144 (1980).
13 Pasteur, L.: C.r. hebd. Séanc. Acad. Sci., Paris *91:* 673–680 (1880).
14 Pasteur, L.: C.r. hebd. Séanc. Acad. Sci., Paris *91:* 315 (1880).
15 Pasteur, L.: C.r. hebd. Séanc. Acad. Sci., Paris *90:* 952–958, 1030–1033 (1880).
16 Pasteur, L.: C.r. hebd. Séanc. Acad. Sci., Paris *91:* 455–459 (1880).
17 Pasteur, L.: C.r. hebd. Séanc. Acad. Sci., Paris *90:* 239–248 (1880).
18 Pasteur, L.: 4e Congr. Int. Hyg. Démogr. *1:* 127–145 (1883).
19 Pasteur, L.; Chamberland, C.: C.r. hebd. Séanc. Acad. Sci., Paris *91:* 531–538 (1880).
20 Pasteur, L.; Chamberland, D.; Roux, E.: C.r. hebd. Séanc. Acad. Sci., Paris *91:* 86–94 (1880).
21 Roux, E.; Chamberland, C.: Annls Inst. Pasteur, Paris *1:* 561–572 (1887).
22 Roux, E.; Chamberland, C.: Annls Inst. Pasteur, Paris *2:* 405–425 (1888).
23 Toussaint, H.: C.r. hebd. Séanc. Acad. Sci., *91:* 301–304 (1880).

Prof. J. Lindenmann, Institute for Immunology and Virology, University of Zürich, POB, CH-8028 Zürich (Switzerland)

The Immune System, vol. 1, pp. 423–429 (Karger, Basel 1981)

Inexperienced Immunologists versus Very Experienced Parasites

Short Odds for New Biological Insights, Longer (but Attractive) Odds for Vaccines, and No Contest for the Sprinter

Graham F. Mitchell[1]

Laboratory of Immunoparasitology, The Walter and Eliza Hall Institute of Medical Research, Melbourne, Australia

The period 1972–1973 in Basel is a particular source of happy memories for myself and family; at the still embryonic Basel Institute for Immunology we made, and have maintained, several close friendships. At the time, a common joke around the laboratory was that if you mentioned the word 'macrophage' or 'accessory cell', the director would slap a SFr. 5.00 fine on you! The real action in immunology centred around cells and molecules with immunological (antigen-binding) specificity. I can only imagine the magnitude of the fine if 'parasite' was mentioned at that time. *Niels* will forgive the exaggeration – his intellect is fully capable of embracing the diversity of immunology and related disciplines just as well as it can the diversity of antibody molecules. The sheer brilliance of the Introductions to each annual Basel Institute report is evidence enough, these Introductions being obligatory reading for immunologists and 'paraimmunologists'.

On returning to Melbourne and the Hall Institute in late 1973, after post-docs in Basel, Mill Hill and Stanford, and with the unlimited assistance, encouragement and enthusiasm of *Gus Nossal* and *Jaq Miller,* I

[1] *Graham Mitchell* was a member of the Basel Institute for Immunology in 1972 and 1973.

decided to embark on a long-term programme with an applied emphasis. The aim was to utilize the power of emerging technologies in the 'New Immunology' to dissect two long-standing immunobiological problems where only fragmentary information was available – namely the immunological intricacies of host-parasite relationships and those of feto-maternal relationships. The former flourished, the latter not so, and the laboratory is now solely engaged in immunoparasitology. As each year goes by, we accumulate experiences with various host-parasite systems, some of which I will briefly describe.

(1) A candidate mechanism has been identified which largely accounts for mouse strain variation in susceptibility to the larval cestode, *Taenia taeniaeformis.* This parasite exists as a cyst in the liver after penetration through the intestines. A 'race against time' exists for the invading parasite and resistent strains of mice differ from susceptible strains in that they are able to produce adequate amounts of T cell-dependent, complement-fixing, host-protective IgG antibodies prior to the full expression of protective mechanisms by the establishing larvae [16]. Even the most susceptible mouse strain can be vaccinated against infection using larval antigen preparations [19].

(2) A candidate mechanism has been identified which accounts in part for the high susceptibility of one strain of mouse, BALB/c, to the intramacrophage protozoan parasite, *Leishmania tropica.* Expression of H-2 antigens on infected macrophages of this strain is reduced such that recognition of parasite antigens on the cell surface may be inefficient [6]. In a genetically resistant strain of mouse, CBA/H, Lyl^{+} T cells, presumably of the macrophage-activating type, are responsible for resolution of infection when injected into nudes [13].

(3) A wide spectrum of susceptibility exists amongst inbred mouse strains to the intestinal parasites, *Giardia muris* (a protozoa) and *Nematospiroides dubius* (a nematode). C3H/He develop a chronic infection with *G. muris* quite unlike many other strains of mice in which infection appears to resolve, this resolution of infection being T cell-dependent [20, 21]. In the *N. dubius* system, mice either develop good resistance to reinfection (and reject the existing worm burden) or accumulate large numbers of parasites in the intestines which may lead to death [18]. After exposure to a single oral dose of *N. dubius* infective larvae, intestinal worms persist for many months, whereas after a single dose of *Nippostrongylus brasiliensis* infective larvae, intestinal worms are rejected within 2 or 3 weeks. The excretory/secretory products of in vitro incubated homologous worms sensitize for

accelerated rejection in the case of *N. brasiliensis* but less readily in the case of *N. dubius* [5, 8].

(4) For all *natural* mouse-parasite systems we have examined, susceptibility/resistance appear to be under polygenic control.

(5) Various nematode larvae are rich in phosphorylcholine (PC) and anti PC antibody responses in *Ascaris suum*-infected mice are largely of IgM isotype. Some evidence has been obtained that one reason for restricted antibody responses may be that PC antigens are relatively toxic for B cells of high antigen-binding capacity. Thus, when PC was conjugated to DNP-carrier, IgG anti-DNP antibody production in a secondary adoptive transfer system was reduced provided the amount of PC was high and the conjugate used was freshly prepared [14].

(6) Using biosynthetically labelled blood stage proteins of *Plasmodium berghei* and *Plasmodium falciparum,* respectively, sera from mice and humans with differing susceptibilities to disease differentially immunoprecipitate a *limited* number of molecules in the homologous system [1, 9]. The identified molecules may well be 'host-protective antigens' in these malaria infections.

(7) A prototype immunodiagnostic reagent has been developed which has apparent absolute specificity and high sensitivity and is based on the inhibition of binding (by sera from infecteds versus non infecteds) of a labelled hybridoma to the crudest of parasite antigen mixtures. Attempts to substitute the parasite antigens with a large pool of anti idiotypic antibodies in the competitive radioimmunoassay were unsuccessful [12].

(8) Using clonally derived parasites, we have not been able to ascribe the very poor expression of resistance to reinfection in mice against the trematode, *Fasciola hepatica,* to antigenic variability within the parasite population [*Chapman, Rajasekariah,* unpublished].

(9) Studies on the extraordinary T cell-dependent and polyclonal IgG_1 hypergammaglobulinaemia in some murine parasitic infections (amounting to more than 30 times the normal serum level of IgG_1) have indicated that the response probably reflects chronic, high-dose, strong antigenic stimulation [2, 3]. We have yet to determine whether such high levels of this reportedly bland Ig isotype have any consequences for the parasite or for the host and how much is antiparasite antibody. In the case of the larval cestode, *Mesocestoides corti,* the parasites in the peritoneal cavity are covered with IgG_1 and at least some of this is antiparasite antibody [15].

(10) A large number of responses in mice infected with the larval cestode, *Mesocestoides corti* are T cell-dependent – e.g. eosinophilia, IgG_1

hypergammaglobulinaemia, liver fibrosis and restrained parasite proliferation [10, 17]. The influence of various T cell subpopulations on these and other manifestations of infection in nude mice has yet to be determined.

(11) Sialic acid changes on the surface of erythrocytes in lethal mouse malarias may exacerbate the anaemia associated with such infections [7]. Moreover, in lethal murine babesia infections, infected nude mice may survive longer than infected intact mice and antierythrocyte autoantibody production is reduced in the nudes [4, 11].

The experimental immunologist familiar with modern concepts and technologies has much to offer the field of immunoparasitology. However, such a person need not assume that the field is currently in a sorry state with nothing really known and that the skills of the immunologist are all that is needed to wrap things up! That which has gone before is not trivial and there is a wealth of background information in many systems to build on. It can be salutary for an immunologist to be asked by parasitologist how to ensure that a particular type of immune response is induced in an animal by a particular purified parasite antigen preparation. We have not come all that far from 'stick it with an adjuvant and hope for the best'. Not too many modern immunologists have had much experience at developing vaccines, devising desensitization strategies or producing and testing immunodiagnostic reagents of high sensitivity and specificity. Thus, if such a person is really serious about making a contribution to the control of the 'great neglected diseases' then that person had better link very closely with people who are familiar with the parasites, the intricacies of parasitism and the needs of the endemic countries for new tools. On evolutionary grounds, and bearing in mind that most medically and economically important parasites cause chronic infections, any single *natural* host-parasite relationship will be complex; the immunologist rushing about firing from the hip with solutions will meet his or her match with the simplest of parasites. It does not take too long to realize that immune responses in a genetically diverse host population against a genetically diverse parasite population (with great structural and life cycle complexity and the potential for growth and differentiation) differ from that of laboratory mice in response to 10^8 sheep erythrocytes or 10 μg GAT or flagellin. Nevertheless, the immunologist can take comfort in the fact that his speciality is one which uses technologies with the potential to revolutionize the analysis of host-parasite relationships. The immunologist cannot obtain really useful information on host immune responses without isolated and characterized parasite antigens; much depends, these days, on the ability of parasitologists to culture parasites or parasite cell lines for antigen production and on the biochemically oriented investigator to manufacture parasite antigens by biological (recombinant DNA) or chemical synthetic methods.

There are only minor differences in the nature of basic research and applied (not to be confused with directed) research in a discipline such as immunoparasitology. Applied research is more difficult and in many cases one can expect fewer publications per year to emerge! There is great 'read-out satisfaction' in many aspects of immunoparasitology. It is usually pretty obvious when a model vaccine works (the parasite is either present or not, the host is either diseased or not) or when a prototype immunodiagnostic reagent shows absolute specificity for a particular infection. In rare moments of reflection on early life as a cellular immunologist, I remember that with many experimental results, one always had serious doubts about their general biological significance. The 'soft edges' of the systems were all too obvious. Was the observation so highly dependent upon the choice of a whole series of special experimental conditions, a highly manipulated experimental animal, and a restricted read-out, that the generalizing interpretation one would like to make is simply unjustified? This dilemma never diminished the fun (and, who knows, maybe the ultimate validity) of building elaborate models of the cellular, and even molecular, events of induction of antibody production. (How preoccupied we seemed to be with induction – who works on catabolism of antibodies or decay of sensitized cell populations?) In experimental immunoparasitology, I suppose the comparable dilemma is deciding on how close to those existing in nature is the parasite isolate currently in use in the laboratory (even in a mouse system). Or, after determining 'what can happen', demonstrating 'what does happen' in vivo in a real-life host-parasite relationship.

Quite apart from the global importance of parasitic diseases and their impact on human populations, there is considerable biological fascination about parasites and the state of parasitism. One feature of most important parasitic infections is chronicity: how does the parasite evade aggressive host-protective immunities but not so effectively that the well-being of the majority of hosts is jeopardized? As in any infectious disease situation, there is no doubt that of the various host-versus-parasite (HvP) reactions, antiparasite immune responses have been a key ingredient in the evolutionary development of balanced host-parasite relationships. One challenge facing the immunoparasitologist is to dissect the various relationships established between genetically diverse parasite populations and their genetically diverse natural host populations and to identify those components which can be exploited so as to increase host resistance (to infection) or host resilience (to disease). It is a reasonable prediction that insights into the basic immunobiology of parasitic protozoan, metazoan (helminth) and ectoparasitic arthropod infections will lead to new tools for disease control.

There are numerous immunoparasitological studies which are of general interest to the cellular or molecular experimental immunologist. Examples are: (1) the biological effects (effector functions) of antiparasite antibodies of various isotypes (e.g. IgE, IgA and the IgGs) and mechanisms underlying their induction and regulation; (2) induction and modulation of granulomatous inflammation in schistosomiasis; (3) influence of activated T cells of various subpopulations (and their products) in chronic inflammation, macrophage activation, immunosuppression and hypergammaglobulinaemia; (4) functions of eosinophils and mast cells; (5) organization and expression of genes coding for variant surface glycoproteins of African trypanosomes and any variant-specific antigens which may be found in human plasmodia; (6) identification of surface molecular changes on parasitized cells; (7) mechanisms of evasion of extant host-protective immunities utilized by parasites resident in tissues, macrophages or intraluminally in the intestines; (8) development of 'functional' hybridoma-derived antibodies and their use in identification and isolation of 'relevant' antigens from complex mixtures of parasite molecules and detection of expression of cloned recombinant parasite DNA, and (9) expression of resistance to infection – i.e. concomitant immunity, sterilizing immunity, parasite-modulating immunity and non-permissiveness – and dissection of the very obvious genetically based variations in the manifestations of resistance.

Thus, immunoparasitology is a rich area for collaborative application of the skills of the immunologist, molecular biologist, immunochemist, geneticist, epidemiologist and parasitologist and is now a growth area in biomedical research. The innate fascination of parasites, the importance of parasitic diseases (there are no parasite vaccines available for use in humans and the field is currently dominated by the quest for vaccines), the challenge of dealing with the complexities of host-parasite relationships, and several new sources of research funding will ensure that this field flourishes. It is hoped that outputs will include new tools for the detection and control of parasitic infection and disease (a difficult task) as well as new information to add to the pool of biological knowledge (an easier task). Immunologists have an enviable reputation when it comes to *biological concepts* and *analytical techniques* and the wise immunoparasitologist will keep well acquainted with new development on both these fronts. The writings of *Niels Jerne* and the research output of the Institute of which he was foundation director will always be a source of inspiration to investigators applying the 'New Immunology' in the field in which immunology has a long-standing proven track record – namely the control of infectious disease.

References

1 Brown, G.V.; Anders, R.F.; Stace, J.D.; Alpers, M.P.; Mitchell, G.F.: Parasite Immunol. (in press).
2 Chapman, C.B.; Knopf, P.M.; Hicks, J.D.; Mitchell, G.F.: Aust. J. exp. Biol. med. Sci. *57:* 369–387 (1979).
3 Chapman, C.B.; Knopf, P.M.; Anders, R.F.; Mitchell, G.F.: Aust. J. exp. Biol. med. Sci. *57:* 389–400 (1979).
4 Cox, K.O.; Howard, R.J.; Mitchell, G.F.: Cell. Immunol. *32:* 223–227 (1977).
5 Day, K.P.; Howard, R.J.; Prowse, S.J.; Chapman, C.B.; Mitchell, G.F.: Parasite Immunol. *1:* 217–239 (1979).
6 Handman, E.; Ceredig, R.; Mitchell, G.F.: Aust. J. exp. Biol. med. Sci. *57:* 9–29 (1979).
7 Howard, R.J.; Day, K.P.: Expl Parasit. (in press, 1980).
8 Hurley, J.C.; Day, K.P.; Mitchell, G.F.: Aust. J. exp. Biol. med. Sci. *58:* 231–240 (1980).
9 Knopf, P.M.; Brown, G.V.; Howard, R.J.; Mitchell, G.F.: Aust. J. exp. Biol. med. Sci. *57:* 603–615 (1979).
10 Mitchell, G.F.: Adv. Immunol. *28:* 451–511 (1979).
11 Mitchell, G.F.: Int. Archs Allergy appl. Immun. *53:* 385–388 (1977).
12 Mitchell, G.F.; Cruise, K.N.; Chapman, C.B.; Anders, R.S.; Howard, M.C.: Aust. J. exp. Biol. med. Sci. *57:* 287–302 (1979).
13 Mitchell, G.F.; Curtis, M.; Handman, E.; McKenzie, I.F.C.: Aust. J. exp. med. Sci. *58:* 521–532 (1980).
14 Mitchell, G.F.; Lewers, H.M.: Int. Archs Allergy appl. Immunol. *52:* 235–240 (1976).
15 Mitchell, G.F.; Marchalonis, J.J.; Smith, P.M.; Nicholas, W.L.; Warner, N.L.: Aust. J. J. exp. Biol. med. Sci. *55:* 187–211 (1977).
16 Mitchell, G.F.; Rajasekarian, G.R.; Rickard, M.D.: Immunology *39:* 481–489 (1980).
17 Pollacco, S.; Nicholas, W.L.; Mitchell, G.F.; Stewart, A.C.: Int. J. Parasitol. *8:* 457–462 (1978).
18 Prowse, S.J.; Mitchell, G.F.: Aust. J. exp. Biol. med. Sci. *58:* 603–605 (1980).
19 Rajasekariah, G.R.; Mitchell, G.F.; Rickard, M.D.: Int. J. Parasitol. *10:* 155–160 (1980).
20 Roberts-Thompson I.C.; Mitchell, G.F.: Gastroenterology *75:* 42–46 (1978).
21 Robert-Thompson, I.C.; Mitchell, G.F.: Infect. Immunity *24:* 971–973 (1979).

Dr. Graham Mitchell, Laboratory of Immunoparasitology, The Walter and Eliza Hall Institute of Medical Research, P.O. Royal Melbourne Hospital, Parkville 3050, Vic. (Australia)

The Immune System, vol. 1, pp. 430–436 (Karger, Basel 1981)

Are Parasites Malevolent? – New Problems for Immunology

L. Hudson[1]

Department of Immunology, St. George's Hospital Medical School, London, England

Prologue

Although immunology has made an obvious and far-reaching contribution to the present quality of life in its control of many infectious diseases of viral and bacterial origin, its application, in immunoprophylaxis, remains an art and has yet to evolve into a true science. Many of the most spectacular successes in disease control were achieved before the molecular and cellular processes of protection were fully understood: they derived from observation and empiricism.

Against this lack of recent progress in applied immunology may be set the prodigious advances made in the last two decades in basic immunology, due in no small part to the recipient of this Festschrift.

In this review I wish to argue that we may now have sufficient knowledge of basic immune mechanisms to add on a second order of complexity, the interaction of these mechanisms of immunity with parasitic organisms. At present the balance between man and his parasites is decidedly in favour of the parasite; it is to be hoped that this balance might be altered by immunological intervention.

[1] *Leslie Hudson* was a member of the Basel Institute for Immunology in 1974 and 1975.

Introduction

The most injurious and economically important diseases of parasitic origin are caused by the protozoan genera *Trypanosoma, Plasmodium, Leishmania, Theileria, Anaplasma* and *Babesia* which occur almost exclusively in tropical countries. Each parasite has a complex life cycle, often involving several morphologically and biochemically distinct stages which may be either intra- or extracellular. The most important helminth diseases are probably those caused by *Schistosoma* in man and *Fasciola* (and other liver flukes) in domestic ruminants. Although schistosome infection is limited to human populations in tropical rural areas, liver fluke infection may occur in animals throughout the world and is responsible for serious economic losses, especially in young animals.

Disease Control

Past efforts aimed at the prevention or cure of parasitic diseases have been based almost exclusively on vector control or the use of drugs. In the campaign for the eradication of malaria, for example, mosquito control measures and the use of insecticide sprays resulted in the eradication or control of the disease in many formerly endemic areas. Even so, malaria still presents a major public health problem in many countries either because eradication schemes have not been implemented or, following the relaxation of control measures for political or economic reasons, there has been a rapid resurgence of acute infection associated with lessened herd immunity.

For only some of the diseases in question, notably malaria and African trypanosomiasis, are safe and effective chemotherapeutic or chemoprophylactic agents available. In many cases, for example, Chagas' disease *(Trypanosoma cruzi)* the results of treatment with chemotherapeutic agents have been disappointing; the available compounds reduce parasitaemia but have to be used at a dose near the drug's toxic concentration.

Although parasites are undoubtedly complex and have evolved highly effective mechanisms for survival (see below) the slow progress towards the control of tropical infections is not due to scientific and medical factors alone. Parasitic infections rarely cause acute symptoms in humans or animals and often go unrecognised against a general background of poor health due to economic deprivation. This, and a long-held belief that parasites provoked only a weak immune response, gave little reason to suppose that these infections would yield to immunoprophylaxis. It is now known, however, that parasites are excellent immunogens and elicit a very powerful immune response. Even so, the development of immunity is completely obscured by the ability of the parasites to survive in an immunologically hostile environment.

The mere fact that these diseases are endemic in economically underdeveloped countries would be sufficient to explain the relative paucity of scientific research of all types: pharmacological, immunological, entomological, etc. For many of the diseases it is only possible to make a rough estimate of the size of the infected population because of lack of epidemiological data.

Immunological Studies

The elaboration of a general model for effector and control functions within the immune response, plus the recent advances in cellular and molecular biology – for example, monoclonal antibody and gene cloning techniques – offer new possibilities for the study of immunoparasitology. These studies are certain to be important in three major areas of disease control: diagnosis, prophylaxis and pathology. In addition, recent experimental findings in malaria [2, 14] indicate that human monoclonal antibodies might even be used therapeutically.

Diagnosis

The isolation and demonstration of the infecting organism remains the only unequivocal means of differential diagnosis of infectious diseases. In parasitic diseases, particularly during the chronic stage, this ideal can be obtained only rarely. Instead great reliance must be placed on the serological detection of anti-parasite antibodies in host serum. This, however, has several major disadvantages: (a) it is impossible to differentiate between a past or present infection unless serial samples are examined; (b) cross-reactive antibodies or antigens are often detected thus giving either a falsely positive result, or implicating the wrong micro-organism; (c) it is often necessary to use tests at less than their maximum sensitivity, thus increasing the danger of missing patients early in the infection or during the chronic stage. It is unfortunate that with conventional serological techniques any increase in test sensitivity brings a concomitant decrease in serological specificity.

The introduction of homogenous antibodies from monoclonal hybridomas will remove all these problems and should revolutionise immunodiagnosis. In addition to the other, well recognised advantages associated with the use of defined antibody reagents [reviewed in 9], there is the exciting prospect of immunodiagnostic assays with sufficient sensitivity to allow the detection of parasite antigen in the patient's plasma. Thus, in some diseases, it might be possible to estimate the parasitological status of the patient both at initial presentation and during therapy. Real advances have been made in this area within the last 2 years, as evidenced by the production of monoclonal antibodies specific for a wide variety of parasites, including *Plasmodium* [2, 14], *Trypanosoma* [7] and *Schistosoma* [13].

Immunoprophylaxis

As mentioned previously, parasites can provoke an immune response of such intensity that the host may completely resist a secondary challenge – even so the primary infection persists and often kills the host. This paradoxical situation can be explained by the parasite's ability to avoid or compromise the host's immune response. So versatile have parasites been during evolution that there are almost as many mechanisms for thwarting the immune response as there are parasite species [reviewed in 3]. Antigenic variation was one of the first of these mechanisms to be described, originally in trypanosomes [4], but later also in *Plasmodium, Babesia,* and *Schistosoma.* In African trypanosomes the variant antigens are a special class of cell surface glycoprotein that covers virtually the whole surface of the parasite. When a host is infected, a population of trypanosomes arises, each member of which expresses the same variant antigen. Under immune pressure the parasites are eliminated, only to be replaced by a second and subsequent population each displaying antigenically dissimilar coat proteins. Recent evidence from in vitro culture [1] suggests that some members of the parasite population may vary the coat protein phenotype by switching through a predetermined series of genes.

An equally effective mechanism for survival is shown by schistosomes; the invading stages of this parasite disguise themselves by adsorbing host antigens onto their surface before the immune response is mobilised. This mechanism neatly explains the ability of adult worms to survive in an immunological milieu capable of resisting a second wave of invading schistosomes.

Table I. T and B lymphocyte dependence of immunity against *Trypanosoma cruzi* infection

Reference	Parasite strain	Mouse strain	Effect of reconstituting syngeneic mice with		
			immune serum	immune T lymphocyte	immune B lymphocyte
10	Y	CBA × C57Bl	protective	no effect	protective
12	Brazil	C3H	ND	protective	protective
8	Tulahuen	C57Bl	no effect	protective	no effect

ND = Not determined.

In many cases the diversity of the immune response itself confers survival benefits on parasites. For example, it has been shown that the formation of 'blocking antibodies' or immune complexes may interfere with the induction or expression of cell-mediated immunity, in a manner analogous to that demonstrated experimentally in tumour systems [6]. Thus, for each parasite system it will be necessary to identify those immune pathways that mediate parasite killing. Selective stimulation of these pathways will be crucial if the degree of immunity gained by vaccination is to out-perform that produced by natural infection. It is sobering to note, however, that when one reviews the literature for evidence of protective immunity yet another mechanism for parasite survival emerges. This is illustrated by data published within the last 6 months (summarised in table I) on cell transfer studies using lymphocytes from animals infected with *T. cruzi*.

Three laboratories measured the same parameters in apparently similar ways but obtained totally contradictory results. Here is the true malevolence of parasites – their mechanism of immuno-confusion.

Candidate Vaccines

There is encouraging experimental evidence that the immune response is intrinsically efficient in some systems and can ameliorate disease or protect against infection. In each case, however, protection is equated with survival of an otherwise lethal challenge; unlike viral or bacterial immunity it does not imply sterile immunity.

For a variety of reasons [reviewed in 3], vaccines for veterinary use are far in advance of those intended for human application. An irradiated lung-worm vaccine for use in large farm animals has been in commercial production for over 20 years, similarly bovine babesiosis and anaplasmosis is routinely controlled by immunisation with attenuated organisms [5].

Although attenuated vaccines are acceptable for veterinary use because of the relatively short life expectancy of farm animals, it is doubtful whether they could be used for human vaccination because of the significant risk of parasites either escaping the attenuation process or reverting back to full virulence. Indeed, in some cases, for example *T. cruzi* and Chagas' disease, it is doubtful whether killed, whole–organism vaccines will be acceptable because of the known cross-reactivity between parsite and host antigens [11]. There is a real possibility that immunisation might prevent infection but still produce the pathology of Chagas' disease.

Definition of Protective Antigens

The definition and isolation of protective antigens is desirable both to avoid the potential side-effects of vacination, as above, and to achieve a selective and effective stimulation of the immune system in the absence of antigenic competition. The lack of past progress in this area is easily explained; parasites are antigenically very complex and often difficult to obtain in sufficient quantities for biochemical investigation. Monoclonal antibodies might allow us to bypass the considerable problems encountered when attempting to isolate protective antigens from small amounts of complex mixtures. In malaria, monoclonal antibodies have been shown to have specific protective capabilities [2, 14] and so have provided both a clue to the mechanism of protection and a means of isolating the relevant antigen by affinity chromatography. The next step, that of cloning the relevant parasite genes and monitoring the bacteria for antigen synthesis using monoclonal antibody, is no doubt underway.

Immunopathology

Several of the human parasites under discussion are not directly pathogenic; the lesions associated with the disease are often caused by the host's immune response against infection. These lesions are caused by a variety of mechanisms [reviewed in 3]; for example, the deposition of immune complexes (malaria), tissue fibrosis (in response to schistosome eggs) or apparent auto-immunity precipitated by the adsorption of parasite antigens onto host cells (Chagas' disease).

This final complication of parasite infection serves to emphasise the need to strive for complete sterilising immunity. There is the real possibility that, in some diseases, vaccination may reduce the level of acute parasitaemia but not affect the long-term immunopathology.

References

1 Doyle, J.J., et al.: Parasitology *80:* 359–370 (1980)
2 Freeman, R.R., et al.: Nature, Lond. *284:* 366–368 (1980).
3 Hudson, L.; Snary, D.: in Marchalonis, Warr, Antibody as a tool: appliations of immunochemistry (Wiley & Sons, Chichester 1981).
4 Mansfield, J.M.: Cell. Immunol. *39:* 204–210 (1978).
5 Miller, L.H.; Pino, J.A.; McKelvey, J.J.: in Adv. Exp. Med. Biol., vol. 93 (Plenum Press, New York 1977).
6 Nelson, D.S.: Transplant. Rev. *19:* 226–254 (1974).
7 Pearson, T.W., et al.: J. immunol. Methods *34:* 141–154 (1980).
8 Reed, S.G.: Infect. Immunity *28:* 404–410 (1980)
9 Rowe, D.S.: Immunol. Today *1:* 30–33 (1980).
10 Scott, M.T.: Parasite Immunol. (in press, 1981).
11 Teixeira, A.R.L.: Bull. Wld Hlth Org. *57:* 697–710 (1979).
12 Trischmann, T.M.; Bloom, B.R.: Expl Parasit. *40:* 225–232 (1980).
13 Verwaerde, C., et al.: in Hybridoma technology with special reference to parasitic diseases, p. 139 (WHO, Genève 1980).
14 Yoshida, N., et al.: Science *207:* 71–73 (1979).

Dr. L. Hudson, Department of Immunology, St. George's Hospital Medical School, Cranmer Terrace, London SW17 0RE (England)

Author Index